God Gives Each of Us
One Lifetime

God Gives Each of Us
One Lifetime

*Living out God's Plan
in Alaska and Siberia*

Written By:
David & Kay (Temple) Henry

XULON PRESS

Xulon Press
555 Winderley Pl, Suite 225
Maitland, FL 32751
407.339.4217
www.xulonpress.com

© 2024 by David & Kay (Temple) Henry

All rights reserved solely by the author. The author guarantees all contents are original and do not infringe upon the legal rights of any other person or work. No part of this book may be reproduced in any form without the permission of the author. For more information, contact David Henry, PO Box 71028, Fairbanks, AK 99707.

Due to the changing nature of the Internet, if there are any web addresses, links, or URLs included in this manuscript, these may have been altered and may no longer be accessible. The views and opinions shared in this book belong solely to the author and do not necessarily reflect those of the publisher. The publisher therefore disclaims responsibility for the views or opinions expressed within the work.

Unless otherwise indicated, Scripture quotations taken from *The Holy Bible, New King James Version*® (NKJV). Copyright © 1982 by Thomas Nelson, Inc. Used by permission. All rights reserved.

Unless otherwise indicated, Scripture quotations taken from the King James Version (KJV) – *public domain.*

The cover photo shows the authors, David and Kay Henry, by their log home in Fairbanks, Alaska. Every year, the tall purple and blue delphiniums adorn their residence.

Paperback ISBN-13: 979-8-86850-119-7
Ebook ISBN-13: 979-8-86850-120-3

Acknowledgments

First, we want to acknowledge and praise the Lord Jesus Christ for His love and work in our lives. In spite of our shortcoming, He has given us a story to tell. God, in His infinite wisdom, gives each of us "one life—one lifetime" on this earth with a major choice: to live for Christ or ourselves.

The Lord used the following Bible verses to encourage us to write *One Lifetime:*

> "O God, You have taught me from my youth;
> And to this day I declare Your wondrous works.
> Now also when I am old and grayheaded,
> O God, do not forsake me,
> Until I declare Your strength to this generation,
> Your power to everyone who is to come."
> Psalm 71:17-18

We also want to give special thanks to those who encouraged us to write our life stories, especially our family members, friends, Indians in Alaska, and the Sakha (Yakut) people in Siberia. It has taken more than four years to complete.

We want to thank our oldest daughter, Debbie Rathbun, for her help in fine-tuning what we wrote by proofreading, fact-checking, gathering information, recalling events, critiquing, and offering excellent suggestions.

Special thanks to Jill Hayes, Sarah Milan, and Patsy Underwood for using their editorial skills to improve this book.

Dedication

We dedicate this book to our four children and their spouses: Debbie and John Rathbun, Dan and Linda Henry, Elizabeth and Andy Cogan, and Steve and Cheryl Henry.

To our 12 grandchildren and their spouses: Melissa and James Hendrickson, Cheryl and David Mount, Andrea and Ryan Circle, Milo Cogan, Kara and Aaron Fields, Tim and Bambi Henry, Suzanna and Andrew DeCarlo, Jonathan Henry, Wesley and Lara Henry, Curtis Henry, Danielle and Isaac Reynolds, and Joshua Rathbun.

To our 23 great-grandchildren: Nyomi Hendrickson, Azryl Hendrickson, Caleb Mount, Kristen Mount, Abagail Mount, Angelica Head, Elijah Head, Audrina Fields, Alana Fields, Emerson Fields, Ethan Fields, Sarah Henry, Clara Henry, George Henry, Vincent DeCarlo, Evelyn DeCarlo, Noelle DeCarlo, Cyrus DeCarlo, Adelaide Henry, Solomon Henry, Abigail Reynolds, Lydia Reynolds, Timothy Reynolds, and future descendants.

To our relatives scattered across the US and in other countries.

To our Native friends, especially the Koyukon Athabascans in Alaska.

To our Sakha/Yakut and Russian friends mostly living in the Republic of Sakha/Yakutia in Siberia.

And many other friends, from long ago to the present time, with whom we have crossed paths.

We were helped, encouraged, and influenced by so many of you, especially in growing stronger and deeper in our relationship with Jesus Christ.

Among our greatest joys would be to see and be with all of you again in heaven.

Table of Contents

Growing Up
Kay's Michigan Years (1933~1951) 1
Best Decision Ever (1941) 12
David's Early Life (1936~1953) 27
My Heaven or Hell Choice (1949) 46

Higher Education
Bryan College (1951~1957) 55
Beyond College (1957~1958) 71

North to Alaska
~ The Koyukon People
Arctic Training Camp: Nenana (1958~1959) 76
Yukon River: Kokrines (1959~1960) 85
Yukon River: Koyukuk (1960~1964) 104
Floods Change Everything (1963) 124
In the Thumb of Michigan: Vassar (1964~1965) 135
On the Arctic Circle: Allakaket (1965~1969) 139
Life in the Big City: Fairbanks (1969~1972) 171
Bilingual Training (1972~1975) 193
Translating the Gospel of Mark (1976~1979) 214
Building Our Log Home (1980~1983) 247
Around the World (1984) 278
Going Mobile (1985~1987) 286
Return to Koyukon Country (1988~1990) 303
Yukon River: Galena (1990~1991) 312
Back to the Big City: Fairbanks (1991~1993) 319

Sent to Siberia
~ The Sakha People
Khabarovsk (1993) 331
First Sakha Christian Book (1994) 346
Total Commitment (1994~1995) 358
Surviving Siberia (1996) 365
Young Again (1996~1997) 386
Out of Gas at 30 Below! (1998) 408
Sakha Church Growth (1999~2001) 413
Living by Faith (2002~2004) 434
Two on Crutches (2005) 463
Golden Wedding Anniversary (2006) 471
Let's Sing Sakha (2007~2008) 476
Visitors from Alaska (2009) 486
Our Time Has Come (2010) 490

Home in Alaska
Fairbanks (2010~2012) 493
Fairbanks Native Bible Church (2012~2016) 501
God's Frozen Chosen (2016) 517
Finishing Well (2017~2020) 523
Kay's Heavenly Home (2021) 539

Appendices
A. How to Know God 544
B. Letters from Grandma Kay 546
C. "Jesus Loves Me" 549
D. Family Sharing 550

Growing Up

Kay's Michigan Years (1933–1951)

1933
Half Day Today

"Half day today—everyone go home! I have become an aunt!" The first- through eighth-grade students excitedly raced out of the one-room school in Mayville, Michigan. Aunt Leola (Wall) Erb sped over the forty miles to be my first visitor at the Women's Hospital in Flint, Michigan.

I, Marilyn Kay Temple, the eldest of six children, was born at 11:55 p.m. on October 12th, 1933, and weighed almost seven pounds. Robert David Temple and Vera Mae (Wall) Temple, my parents, were married in Mayville in December 1932.

My birth also made my dad's youngest brother, John Paul Temple, become "Uncle John" at the tender age of seven. He didn't want to be an uncle, so he stated, "I'm going to eat horse chestnuts and die!" But he never did. He was always a favorite uncle.

We lived in a small house at 1622 Indiana Avenue in Flint. My father worked in the production department at the Chevrolet plant. He was hired for 42 cents an hour and received 18 dollars a week. My mother resigned from her teaching job and became a stay-at-home mom, taking care of me. We moved to 2302 Torrance Street, the house I remember as a young child.

My first birthday cake

1

1934
Some Firsts for Me

My first words spoken in July 1934 were: "Daddy," "Bye," and "Mamma." My first five baby steps were going to Auntie Leola on October 2nd, 1934. I was baptized as a baby at the Methodist Episcopal Church in Mayville by Rev. Manahan on Children's Day 1934.

As a one-year-old, I almost choked to death. The oatmeal my mother fed me was too thick. I started to turn blue, and I couldn't breathe. Mom shouted to Dad, "Hold her up by her feet and slap her back!" He was afraid he would hurt me, but he did it anyway. The oatmeal flew out of my mouth! He saved my life. God gave me many more years to live.

Mom held me next to my grandmother, Edith (Maiers) Wall, and my great-grandmother, Mary (Lauber) Maiers Kauffman.

Four Generations

I posed for a four-generation photo on my mother's side of our family. Lots of people had fancy birdhouses when I was growing up. We enjoyed watching the birds and hearing them sing in our yard.

1935
Shirley Ann Temple

My sister, Shirley Ann Temple, was born in January 1935. You guessed it—she was named after an internationally famous child star of the 1930s, best known for sentimental musicals. There were fifteen and a half months between us, and our mother often dressed us alike in cute dresses.

When I was four, the clerk in a store asked my mother, "Are those girls twins?" I piped up and said, "No, there is a month between us."

Coming out of the Great Depression

My father had a steady job, but his pay was barely enough for our family. Rent was cheap enough, as most people were limited financially. Mother would buy an orange, give us kids the juice, and then eat the rest of it. Times were tough!

Glasses

When I was only two, I started wearing glasses. My eyes crossed, with one called a "weak" eye. They did not work together as they should. I liked wearing my glasses because I could see much better.

Kay in front of a 1930 Buick

Daddy Turns to Jesus

In November 1935, Mom invited Dad to attend special evangelistic meetings. He volunteered to stay at home with us little girls. Vera pleaded, "Oh, please come. This may be the last time we can go together." We had recently been exposed to whooping cough and expected to come down with the disease. Sure enough, one of us whooped, but not before the evangelist came down from the front and asked Daddy, "Don't you want your two daughters to grow up in a Christian home?" That night, he went forward and accepted the Lord.

Dad was raised in church and always attended. He went to the altar when he was 10 in an evangelistic meeting with his grandfather, Heman Lewis, in Silverwood, Michigan, at the United Methodist Church. Daddy said, "I either got away from the Lord or I was never really saved." I'm thankful he made sure he knew the Lord.

The following day, he rode with his carpool buddies to the factory in Flint. One of them asked, "Bob, where's your pipe?" He replied, "I got saved last night; I don't need it anymore."

At work, others laughed at him for getting "religious." But their superintendent, who was not a believer, stopped them from making fun of Dad. Several weeks later, that superintendent also turned his life over to the Lord.

1936
A Little Stubborn

My mother had to punish me for my stubbornness in not saying, "Excuse me." Consequently, she wrote this poem about me on November 20th, 1936:

> "Little girl of mine, you're so dear to me.
> As I gaze upon your tiny brow,
> Tears well in my eyes; I can scarcely see.
> As I ponder wonderingly at you now.
>
> So tiny, and yet so headstrong, too.
> Such determination in one so wee,
> It hurts when I think of punishing you,
> Dear little girl, when you're only three.
>
> 'Twould be the easiest, but not the best.
> To let you always do as you like,
> But I must guard your future happiness,
> Even when you're such a little Tyke.
>
> If I forgot my duty of guiding your feet
> On the right path, day by day,
> Someday we'd both be sorry, sweet,
> Because you couldn't face life that way."

My Christian Heritage

Our history has had a significant influence on our present lives. We can't get away from our background or family tree. We need to learn from our past and build upon it.

Most of our relatives lived in Michigan and Ontario, Canada, just over the border. We visited our relatives frequently and heard many of their stories.

I knew my great-grandfather, Robert Hope Temple when I was young. I heard about his father-in-law, my great-great-grandfather,

Heman Janes Lewis, a fiery Free Methodist preacher. He was a circuit-riding preacher in the thumb region of Michigan. (The state of Michigan is shaped like a big mitten.) He rode on horseback and preached in Clifford and elsewhere, while also having a church in Applegate. Sometimes, he would climb up on the railing in front of the church while twirling his white handkerchief in the air as he preached. I'm sure he got people's attention, and I think many turned to the Lord.

Their consumption tent is decorated with American flags.
My great-great-grandfather Heman Lewis is seated on the left; his second wife, Martha (McNeil) Lewis, is standing in the center; and great-grandmother Eva Logy Temple is on the right. I don't recognize the two young boys.

Bold Faith

One of Heman's daughters was Captain Eva Logy (Lewis) Temple, my great-grandmother. She was appointed as the first commanding officer for the Salvation Army in Montreal, Quebec, Canada when she was only 19 years old. Grandma Eva and four others marched in the streets singing hymns while playing tambourines, accompanied by a drummer, in the summer of 1884. They concluded their marches with open-air services, telling about Jesus in the public market square. Their group was pelted with rotten eggs, flour, soot, and rocks. They were accused of "disturbing the peace" and even jailed for a brief time. Their bold witness enabled them to hold Christian meetings in Weber Hall in December of that year. Their activities were written up in a serial story in the 1921 issues of *The War Cry*, the publication of the Salvation

Army, titled "A Valiant Soldier of the Cross." Her bold witness for our Lord encouraged me to take a stronger stand in following Jesus Christ.

People of her generation died of consumption, and the sick were quarantined in strict isolation in tents during the summer. Around World War I, consumption became more commonly known as tuberculosis, or TB.

My great-grandmother, Mary Kauffman, was very special to me. I would ride the bus by myself from Vassar to her little home near the railroad tracks in Mayville. Starting in the third grade, I stayed with her for several weeks. She raised strawberries and had a large raspberry patch, so I would pick berries for her every other day to trade at the local grocery store.

Afternoons, I sat with her on the swing out on her stoop. That's an old word for a small porch. I listened to stories about her life while we ate lunch. Her old oak dining room table is in our home in Fairbanks. We use it, but it creaks.

As she neared the end of her life, she lived in a nursing home close to Mayville High School. While there, she would write Bible verses on little slips of paper and toss them out the window so people walking on the sidewalk could pick them up and read them.

1937

Great-grandfather Robert Hope Temple on a blanket with Carol Fox, Joanne Temple, John Lewis Temple, Kay, and Shirley

Church

We attended Kearsley Park Evangelical Church in Flint. For the Christmas program, I recited the following poem: (At that time, the word "gay" meant "joyful.")

"I am a little girl,
Happy and gay;
I wish you all
A Merry Christmas Day."

1938
Playing School

Shirley and I enjoyed playing school with our mother who was a former schoolteacher. I even taught my sister to read before she went to school. The three of us did everything together.

Christmas Memories

In 1938, Shirley and I participated in the Christmas program at the Methodist Episcopal Church in Mayville when I was five years old.

My part was:

"Bright angels sent from heaven
Before the break of dawn,
Sang Peace, Good Will, and Glory,
For Jesus Christ was born."

The Christmas program at my paternal grandparents' church in Silverwood was wonderful. I was young and shy. Mom told me, "If you say a piece in front of the church, you'll get a gift from under the tree." I fondly remember receiving a children's poem book from under the tree with my name on it. We especially liked the poems "The Raggedy Man" and "Little Orphan Annie." They had words with rhythm and repetition.

1939
Tragedy Strikes

In Flint, I had a great playmate. Her name was Lorena Siver. Even though she was only three years old while I was six, we enjoyed playing together. Her parents, Leon and Helen Siver, and my parents were good friends, too.

Unfortunately, Lorena had problems with allergies. Her mother took her to the doctor for a Brazil nut allergy shot test, and her reaction to the shot was mild.

The following morning, before she went back to the doctor for her second shot test, I played with her. The doctor and nurse didn't read their notes carefully and gave her the same shot as they had given her the previous day. Then they waited twenty minutes to see if she had an allergic reaction. But her response was too strong!

Lorena died in her mother's arms in the doctor's office. We were all shocked! I told my mother, "But I just played with Lorena this morning, and she was fine!" That was an immense tragedy to get over. Even later in life, as an adult, I would get flashbacks and memories of that sad day. Of course, it was much worse for Lorena's parents.

First Grade

I started first grade on September 5th, 1939. My teacher was Mrs. Gladys Emery. I enjoyed reading before I went to school, so I ended up in the best reading group for the year.

My full name is Marilyn Kay Temple. My relatives called me, Marilyn Kay. My sister Shirley had trouble saying my name and called me "Kay." I knew she would learn someday, so it didn't bother me. In school, one of my teachers asked, "Are you, Marilyn Kay or Kay? What should we call you?" My classmates voted for Kay. So I'm called Kay, except on official documents.

Christmas

When I was six, my sister Shirley got a drum for Christmas. While Mother was away on an errand, we decided to get Mother's knitting needles—first to play on the drum, next to poke holes all over the drum. The Christmas drum was ruined, and Mother wasn't happy, either.

1940
My Second Sister

My second sister, Nancy Lou Temple, was born in October, shortly after my seventh birthday. Now there were three Temple girls.

More Family for Christmas

We usually went to Grandpa Pete and Grandma Bertha (Snover) Temple's home in Silverwood, Michigan, for our celebration with everybody on the Sunday before Christmas. The family dinner was turkey, dressing, squash, and side dishes. They had a piano, and sometimes Grandpa or my aunts, Marion (Temple) Fox and Emma (Temple) DeGrow, would play while we sang Christmas carols. Grandpa read the Christmas story about Jesus' birth.

It seemed that each year, a new baby was added to one of our families. Finally, there were 20 of us grandchildren, plus all the adults. Earlier in the year, we would draw names to buy a gift for someone—the adults for the adults and the children for the children. Grandpa and Grandma Temple would buy gifts for all of the grandchildren.

One Christmas, several beautiful dolls were set up on the piano. I counted them and was concerned because there weren't enough for all of the girls Nancy's age. I wondered who would not get one. Yes, it was Nancy, my little sister. She received a wood-burning set instead. I felt sorry for her not receiving one of those beautiful dolls. In later years, since I've had my grandchildren, I've realized that Grandpa and Grandma could only find four dolls and expected to find another one, but couldn't. If they'd given one to Nancy, then one of her cousins would not get one. She was a year older than the others in that group, but I know Nancy was disappointed.

One gift I especially remember was a small cedar chest from my Aunt Alma Temple's family. It seemed she always knew what gift someone would like. Another favorite was a black poodle dog made of porcelain from Grandpa and Grandma Temple. Mom didn't like it, but I did.

My sister Shirley and I loved to read and always appreciated receiving books as gifts. She liked the *Honey Bunch* books, while I favored the *Pollyanna* books and Louisa Mae Alcott's books. I especially remember *Under the Lilacs*.

'Twas the Night Before Christmas

We would drive to Grandpa Sanford and Grandma Edith (Maiers) Wall's house in Mayville and spend the night each Christmas Eve. They had two bedrooms upstairs and their own room downstairs. The firemen would come to the house dressed as Santa Claus and give us gifts and candy. Later in the evening, just before bed, Grandpa Wall would read the Christmas story from Luke chapter two, and we'd all sing Christmas carols with either Mom or her sister, Auntie Leola, accompanying us on the piano.

As the families grew, Shirley and I would sleep in the living room near the Christmas tree. Sometimes, Cousin David Erb or my sister Nancy would sleep in two easy chairs pushed together to make a bed. There was an old metal baby bed upstairs, where Mom had slept when she was six and had diphtheria. The youngest child would usually sleep there.

On Christmas Eve, I opened the back door to get something off the porch when I spied a brand new two-wheel bike. I quietly came back in and didn't tell anyone, but evidently, I was excited. During the night, Shirley and I, who were sleeping end to end on the couch, woke up. We saw the bike standing in the corner of the living room, behind the Christmas tree. We took turns sitting on the bike, dreaming of riding it down the sidewalk. We noticed David's cash register, which we had to try out. When we played it, it made a ringing sound.

Soon Auntie Leola and Uncle Ed, who were sleeping in the room directly above, heard us. They woke up our parents, who shouted, "Get back to bed! Go to sleep." It was 3:00 a.m. In the morning, we rode that new bike! Until then, we had a three-wheeler, but having a two-wheeler was a dream come true for Shirley and me.

Dolls and More Fun

My favorite toys were: Betsy Wetsy dolls, who wet their diapers when I gave them a bottle of water; Kewpie dolls, whose hair came up to a curl on top of their heads with a ribbon; tiny dolls wrapped in a blanket; and Horsman dolls with beautiful faces. My last doll was a preteen doll when I was 12.

I played with paper dolls by the hour. They were usually a little girl or boy. Shirley and I once received Peter Rabbit paper dolls when we purchased a new pair of shoes.

My sister and I made dollhouses out of wooden orange crates, with a wooden divider in the middle. We would stand two of them upright on one end, which gave us four rooms. To resemble a roof, some boards were added, and the rooms were painted like houses. We created little rugs, blankets, pillows, and wastebaskets. Going to the dime store to buy wooden furniture with drawers, one or two pieces at a time, was a special treat.

We played with tiny dishes and pretended to drink tea. Grandma Edith had a set of china doll dishes that we could play with on special occasions.

My favorite games were Old Maid, Parcheesi, Chinese Checkers, Rook, Tiddly Winks, Go to the Head of the Class, and other table games. It was always fun when Cousin Carol Fox came over. Shirley, Carol, and I went upstairs and played Monopoly late into the night. I looked forward to playing with Cousin David's electric train when we'd visit them.

My favorite outdoor games included roller skating on the sidewalk, ice skating, riding bikes, walking on stilts, jumping rope, jacks, marbles, hopscotch, swinging, climbing trees, and hanging by my knees from a tree limb.

Piano Lessons

My parents gave me piano lessons, which I enjoyed, but sometimes I preferred playing outside more than practicing. The lessons, however, benefited me throughout my lifetime.

Best Decision Ever (1941)

1941
Jesus Died for Me

Each night, Mother read Bible stories to Shirley and me. When I was eight years old and in the third grade, my mother read the Easter story and stated, "Jesus died for our sins, and that includes all of them. Would you like to ask Jesus to forgive your sins and come into your life?" I felt awful because Jesus suffered for the things I had done; I knew I was a sinner.

I eagerly replied, "Yes! I want Jesus Christ to forgive my sins." Shirley also wanted to turn to Jesus. So, right there we believed as we knelt and prayed with our mother.

I knew God had forgiven me of my sins and changed my life. Turning to Jesus was the most important decision I ever made. It affected me for the rest of my life. The next day, when Dad was driving us to the Easter service, we told him, "Last night while you were working, we accepted Jesus Christ as our Savior."

The Seeds That Grew

My mother planted a few vegetables in our backyard. I helped her and wanted to plant my own seeds to see them grow. I kept thinking about those seeds. Here's the story I wrote, inspired by my experience:

"Kay and Her Watermelon Seeds"

Early one summer, I watched my mother and father plant their garden. They dug up the dirt and prepared the soil. Then they made straight rows and planted seeds.

I liked to play in the sand. One day, I decided to plant a garden. So, I dug up the sand and made some rows.

Then I thought, "What can I plant in my garden?"

I went inside and saw little packets of seeds on the shelf. When I saw the watermelon seeds, I took them and planted them in my sandbox.

Two weeks later, Mother asked me, "Kay, did you see my watermelon seeds?" I replied, "No, I didn't see them."

When my mother came into the house, she looked at me and said, "Come with me." She took me by the hand, and we went outdoors to the sandbox. Pointing to the little green sprouts, she asked, "What are those?"

My mother did not see me take the seeds and plant them. She didn't know what I had done. But someone had seen me take them from the shelf and plant them. God saw me. He also heard when my mother asked me about the seeds. God knew that soon my mother would also find out. God made the sun warm the ground and sent the rain to water the seeds. They would start to grow. God saw what I did. I also knew I had done something wrong. I had lied to my mother. God saw me when I took cookies or candy from the kitchen. He also saw me when I fought with my sister. He knows everything. We can't hide anything from him. He sees us in the dark just as well as in the light.

My mother disciplined me because I had stolen the seeds and lied to her. My sin had found me. The Bible says, "Be sure your sin will find you out" (Numbers 32:23). I still like eating watermelon, but sometimes I think about when I stole the seeds and planted them in my sandbox.

**I enjoyed watermelon with my sisters,
Nancy and Shirley when I was nine years old**

1942
Handmade Clothes

During World War II, it wasn't easy to buy ready-made clothes. The pastor's wife, Mrs. Westfall, of the Kearsley Park Evangelical Church in Flint, helped Mom sew dresses for her girls. They sewed pretty teal coats for Shirley and me from two blankets. There was even enough material to make a matching coat for our little sister, Nancy. What fun for all three of us to be dressed alike!

We always enjoyed visiting the Westfalls' home. Their children were grown, so they had lots of children's books. While Mom and Mrs. Westfall sewed, we read their books about the Bobbsey Twins, Honey Bunch, and Bunny Brown and His Sister Sue.

Goodbye, Great-Grandpa

My great-grandfather, Robert Hope Temple, passed away in London Township, Ontario, Canada, on May 24th, 1942. I was eight years old.

Our spring picture with cousins; back row: Kay, Shirley, and Carol Fox; front row: Janet DeGrow, Nancy, and William Fox. We liked all the buttons that gave us a military look, especially during the war.

A Special Birthday
When I turned nine and came home from school, my mother informed me, "Today, you have a new cousin who was born on your birthday!" I was thrilled to have Mary Jean Fox share my birthdate.

Sergeant LeGree
The "Singing Cop," otherwise known as Sergeant LeGree, visited schools all over Flint. He taught us about safety on our streets, with such classic jingles as "The Boys of the Safety Patrol" and "When You Cross the Street." He was also a Christian and had special meetings at our church after school. Sometimes he took us home in his police car!

1943
A Brother Is Born
In February 1943, my first brother, Thomas Robert Temple, was born in Saginaw, Michigan. While Mom was pregnant with him, I overheard Dad tell her, "If it's another girl, I won't give out any of the candy bars I bought for the men at work." At that time, they could not tell if the baby was a boy or girl before birth. What a joyful day it was when Tom was born!

When my parents dedicated Tom to the Lord in front of the church, the preacher asked Dad, "When was he born?" Dad couldn't come up with the date right away. As a nine-year-old girl, I was surprised he didn't remember which day Tom was born since he wanted a boy so badly! I was pleased that I could tell them the date of Tom's birth.

A New School
On June 1st, 1943, we moved four miles to 1327 Welch Boulevard in Flint. I enrolled in Civic Park School and made new friends.

Mother's Day
For Mother's Day, we went to church with Grandpa and Grandma Temple in Silverwood. Grandpa Pete was a pillar of the church. Grandma Bertha would give us a fresh flower from her garden to wear. She must have started them on her porch to have them ready by Mother's Day. Then we'd go to church in Mayville with Grandpa Sanford and Grandma Edith Wall. Both churches were Methodists. They had the same pastor,

who went to Silverwood first and then to Mayville. This meant we could go to church on Mother's Day with both sets of grandparents.

Those City Buses

When I was ten, we would catch the bus for choir practice or special meetings for children. One day during fall vacation, Mother drove us to the nearest bus stop, and we scampered to the bus. Several adults were ahead of us, and the driver didn't see me. When I took the first step in, the driver began to close the door. I jumped off, but my foot caught in the door as it was shutting. I saw the big bus wheels and screamed. Thankfully, someone noticed and shouted at the driver to stop before I was helplessly dragged along the street.

Heading to the Hills

My mother's cousin, Dorothy Robinson, served as a missionary with Child Evangelism Fellowship in Pilgrim, Kentucky. She taught Bible lessons in the surrounding area. Our church in Michigan and relatives donated a car for her. Shirley and I rode with Dad in the 1930 Ford Model A that we delivered to her, while my mom rode with her parents in their car.

Dorothy had to boil her drinking water, even from the well. Then we had to wait for it to cool before getting a drink. The roads were poor, and to get to schools where she taught Bible lessons, she had to ford or drive through shallow creeks since there were no bridges. The water would come up to the running boards, but we made it.

Easter in Silverwood

I always liked going to Mayville or Silverwood to visit our grandparents. One Easter at Grandpa and Grandma Temple's home stands out to me. Shirley and I dressed up in our new matching navy blue coats and hats. Cousin Carol was also wearing her navy blue coat and hat, matching ours. My parents bought ours at the Smith-Bridgman's Department Store in Flint, while Carol got hers at the same company store in Pontiac, Michigan.

Dad always made sure we had nice clothes. Mom used to change Shirley's and my dresses each afternoon. That was before "wash and wear" clothes, so she had to wash and iron many dresses each week just for us. She loved her little girls and enjoyed taking us to visit in the afternoons.

Christmas 1943

Mother kept a detailed scrapbook with my birthday and Christmas cards, with a running history of what I did, where I went, and a list of my presents. My mother's generation would keep these mementos and record them before our modern days of Facebook, TV, or too many other things in life trying to get our attention. I was the firstborn, so my book was thicker and contained more details than later books for my sisters and brothers. Mother wrote the following for my Christmas in 1943:

Marilyn Kay's 11th Christmas

"Marilyn Kay, along with the rest of her family, went to Auntie Leola Erb's house on Christmas Eve. The family members presented a program. Grandpa Wall and Daddy read the Scriptures along with the Christmas stories. Uncle Ed Erb prayed, Auntie Leola Erb played the piano, and David Erb sang. Marilyn Kay played 'Silent Night, Holy Night.'

She received a jumper, blouse, and pencils with her name on them from Grandpa and Grandma Wall, Uncle Ed and Auntie Leola Erb.

When she awoke Christmas morning, under the Christmas tree was a large doll, three books of *The Three Baers* (adventures of the Baer family triplets with Bible lessons by Bertha B. Moore), a pair of red gloves, and the Wide World board game from Diane (Erb) Pedersen.

She went to Aunt Emma and Uncle Al DeGrow's Christmas afternoon. She received a silk slip from Grandpa and Grandma Temple, as well as one from Carol Fox.

She attended the Girl Scout Party at Civic Park School on Monday, before Christmas, where she got *Andersen's Fairy Tales*."

1945
A One-Room School

In the summer of 1945, we moved to Juniata, Michigan, a town of less than one hundred in a farming community. Even though World War II was over with Germany and winding down with Japan, housing was scarce. We rented the parsonage of the Juniata Baptist Church since the pastor was single and didn't need as much room as our growing family with four children. We wanted to move to Vassar, but no housing was available.

We were delighted because we could attend a one-room school as Mom had, and Auntie Leola Erb had taught there! So we happily walked to the Juniata Country School each morning. Shirley was in the sixth grade, Nancy was in kindergarten, and I was in the seventh grade. Our desks would seat two children. The youngest students sat in the front, while the older ones sat in the back of the room. Students would go to the front of the classroom to recite their lessons. Nancy learned to read by listening to the first graders read.

Nancy was always very friendly with everyone, and she got head lice. Mom was diligent and washed her hair with some strong stuff. She also combed her hair with a fine-toothed comb to get rid of the lice eggs.

A Simple House

Life was simple. We had a hand pump in the kitchen for water, as well as an outhouse out back that took care of our basic needs. The garage had a deck at the back, where we liked to sit in the sunshine and read. We also planted flowers in small boxes back there.

Upstairs was an attic with stairs to it from the kitchen. We would go up and jump on the extra bed when Mom wasn't home. Sometimes she would go into Vassar, six miles away, to shop and do our laundry at Auntie Leola's.

My parents both graduated from Mayville High School, only six miles to the east. Grandpa Pete Temple was the postmaster in Silverwood, another six miles farther to the east, and owned a general merchandise store there. So, we saw a lot of our family. We knew many of the people in the area since my parents grew up in the thumb of Michigan.

A Large Bag of Groceries

After Juniata, we stayed with Grandpa and Grandma Wall in Mayville for a short time. We were trying to get resettled after Dad was laid off from the Chevrolet plant in Flint. Dad was helping Ralph Maiers build his house in Mayville. Shirley and I briefly went to school there. We went to a drawing at Uncle George's trucking business, Maiers Freight. They drew Dad's name, and we received a large bag of groceries. We were all thrilled. The Lord provided for us.

1946
The Move to Vassar, Michigan

We moved to Vassar during the summer of 1946. Dad rented a three-bedroom house from an elderly lady living in a rest home. Her home had been vacant, but we soon made it livable. The kitchen was cold! Mom wore boots to keep her feet warm. The cellar under the kitchen likely contributed to the cold floor. I remember the men coming over to wallpaper several rooms. Dad also had a bathroom installed in the massive master bedroom. Up until then, we had been using the outhouse. Mom especially appreciated it, as she was pregnant again.

Grandpa and Grandma Wall moved to Vassar when we did. Previously, he worked for Maiers' Freight in Mayville as a truck driver. Grandpa now worked for Uncle Ed at Erb's Food Shop. Then he got a job driving a bus and doing janitorial work for Vassar Public School. We saw him often while attending school.

A New Car: Kay

Fast-forward: My grandfather, Sanford Wall, wanted to buy a car, so he visited all the car dealerships in town and purchased a small Ford. He signed up for all the giveaways. A few weeks later, the local Chrysler-Dodge dealer called him with the exciting news, "You've won a new car!"

He won a 1960 Dodge Dart Seneca four-door sedan. This vehicle was better than his Ford. He was delighted and said, "The first thing I ever won in my life!"

Vassar Electric Company

My parents bought Vassar Electric Company in February of 1946, located at 112 North Main Street. Dad had saved money while working for Chevrolet in Flint to purchase the store. He expanded the business to do all types of residential and commercial electrical work in the area, plus selling supplies and winding electric motors in the shop. Mother often finalized the bookwork. I liked math, so Dad or someone would tell me the wholesale price for the inventory items. Then I would take the spiral notebook to school and add up the items during study hall.

My cousin Bill Robinson was in our youth group and worked at Dad's store. His mother, Amy (Wall) Robinson, was widowed and moved back to Vassar to be near her family. She was a real prayer warrior! Uncle John Temple, Dad's youngest brother, also worked at the store.

Since my parents sold the business, the store has changed hands several times. Many years later, my husband and I were back in Vassar when there was an open house at the same address. A high school classmate of mine, Dale Newton, had opened Newton's Plumbing and Heating there.

Becoming a Gideon
Shortly after moving to Vassar, Jubal Lester Kauffman, Mom's uncle, urged my dad to join the Gideons and help distribute Bibles in Michigan. The Gideons became an important part of his life as he served the Lord. This naturally involved Mom and many of our family activities, such as attending conventions in other states.

Young Builders
Behind our house was a large field with a small creek we could easily jump over, except when the water was too high in the spring or after lots of rain. A big apple tree soon became our tree fort. We saw some old boards, which we thought were abandoned, and took them to use for the floor of our treehouse. Then we realized they weren't even on our land but on the neighbors'. We had to go and apologize for taking them and see what we needed to pay for them. He was very kind and responded, "You don't have to pay for them, since they were just lying around."

We took our younger brother, Tommy, to play with us. He fell on a rusty nail and screamed when it pierced his hand. I quickly picked him up and started to run toward our house. He was crying so hard. Then I stopped and pulled his hand away from the board with the nails in it. Mom took him to see Dr. Swanson for a tetanus shot. After that, we were more careful with leftover boards with nails.

Enjoying the Candy
One evening, Shirley and I had the unusual fun of riding in the back of Dad's company truck to Silverwood to deliver a new refrigerator to Grandpa and Grandma Temple. Mom, Nancy, and Tommy were in the cab with Dad. It was the day after Halloween. We each had a big bag of candy from trick-or-treating and ate all the candy on that trip!

Train Tracks

We looked forward to having a freight train pass by. We always waved at the engineers, and they'd wave back. The tracks were just on the other side of our backyard fence. We heard many stories about my great-grandfather, Robert Hope Temple, who was a train engineer for 45 years in Canada with the Grand Trunk Railway and then the Canadian National Railway.

A big snowstorm in Vassar canceled school because the roads had too much snow. Dad, Shirley, and I, with our sleds, walked up the embankment to the train tracks.

It was special to walk on the railroad tracks with Dad because we were prohibited from doing so any other time. Then we went sledding.

Another Brother

Our youngest brother, James David Temple, was born in December 1946. He had a full head of black, bushy hair, making him unique in our family. Two boys named Temple were born at the same hospital. A visitor came to the hospital and said, "I want to see the Temple baby, but not the one with the awful hair!"

1947
A Calf Named "Marilyn"

I looked forward to spending time with Aunt Emma (Temple) DeGrow, her husband Alton, and my cousins Janet and Berneice. Emma was easier to pronounce than her official name, Emaranda. They had a lovely farm between Mayville and Silverwood. I would help harvest potatoes in the fall, but I got an aching back from bending over so much. Aunt Emma bought colored marshmallows for a treat. That was the first time I had ever had colored ones.

They also raised milking cows. "We named one of our calves 'Marilyn' in honor of you," they announced. I was surprised and happy!

Family Reunions

The Temple-Lewis family reunions in Canada were fun! We went to an amusement park in London, Ontario. There was a train, horse rides, and a big pavilion where we had our reunions. Adults and children all played games according to their age category, and we laughed together. Those were exceptional times. One uncle and aunt never had any children, but they bought tickets for the train and the Canadian version of Cracker Jack for everyone.

Home on the Hill

Soon we purchased an older home at 925 West Huron Avenue on the main highway, M-15, which runs through Vassar. This home was on top of a steep hill.

Years later, my parents moved to 904 Prospect Street in Vassar. We knew our old home was on prime real estate, but my parents couldn't hold on any longer and needed the money to pay for their new home.

Finding a Good Church

We attended First Baptist Church in Vassar. Pastor Peter Ypma taught the Bible and emphasized Jesus Christ. He encouraged us to live the Christian life, and he used people in the Bible as examples. This move was an excellent choice for a church because it helped me grow stronger in my faith.

When I was 13, our church asked me to teach preschoolers during the summer Vacation Bible School. This enabled me to earn a free week at summer camp, and it also gave me confidence in teaching children.

After one turns to Jesus Christ, it is imperative to grow spiritually in your life as you study your Bible, memorize Scripture, and interact with other growing believers serving the Lord.

My Sunday school teacher was Irene Esckelson, a single lady who loved the Lord and the Bible. Her parents lived near Vassar, and we enjoyed fun events on their farm. Irene was preparing to become a missionary with the Navajo Indians in Arizona, and her life challenged me to think about Indians and serving the Lord.

God Wanted Me to Be a Missionary

I realized God wanted me to be a missionary, so I started preparing myself. I studied what God's Word said and applied what I learned to my life. I read biographies of missionaries. My parents often invited visiting missionaries over for dinner or gave them a place to stay while in Vassar. We asked about things that piqued our interest. I learned more about missions at home. I wanted to know how to tell people about God.

My mother had wanted to be a missionary when she was young, but her parents were against it. I'm thankful my parents encouraged and helped me become a missionary.

1948
The Second TV in Town

We had the second TV in town, while a business competitor had the first one. The teens from church would come to our house to watch Dr. Percy Crawford's Christian programs. They were specially designed to help us teens do well spiritually. We enjoyed watching them together. It didn't seem to matter that there was "snow" in the pictures—we were watching TV! High-definition TV came along much later.

Getting to Church Early

Once, Shirley and I babysat our three younger siblings. We had fun together, but I wanted to spend more time reading a good book. I moved the hands on our clock ahead 30 minutes when no one was looking. Soon I told my brothers and sister, "Look at the clock. It's time to go to bed." It worked, and they all went to bed a little early, while I got extra reading time.

The next day was Sunday. We all got up and drove to church. But no one was there! Then I had to confess, "I changed the clock last night!" My parents said, "We'll just sit here in the parking lot and wait 30 minutes."

Spiritual Growth

Many different missionaries spoke in our church. They told us about their lives in northern Michigan or Kentucky, while others told us about life in Africa or Japan. Everywhere they went, they told people about Jesus Christ, and it was encouraging to hear about the people who turned to God.

Each Wednesday, we went to the prayer meeting at church. After the pastor or missionary spoke, we divided into smaller groups to pray. The men went into one room and the women into another, while the teenagers and children went into yet another to pray for the requests for our church, sick people, unsaved friends and relatives, and our own needs.

1949
Baptized Again

When I turned 16, I was baptized with my sister Shirley and my father at the First Baptist Church located at 125 Division Street by the city park. Later, the church relocated to a much larger property on the edge of town at 5880 West Frankenmuth Road.

We were baptized by immersion in front of the church and before many witnesses. The church had experienced a fire and was recently renovated. The bathroom floors were painted red. When we went to change into dry clothes, our wet socks picked up red paint from the floor.

Both Dad and I had been baptized as infants. Those baptisms were our parents' decisions. This time the decisions were ours—as a public testimony to follow our wonderful Lord.

Camp Chetek

I attended Camp Chetek, a Baptist Church camp in northwestern Wisconsin when I was 16. The preacher spoke from Romans 6:11–13 about being dead to sin but alive to God. The experience was especially challenging for me, and I yielded myself to God as a living sacrifice. I wanted God to use every part of me—head, hands, feet, and mouth—as instruments of righteousness to bring Him glory.

The camp reinforced the commitment I made in eighth grade. I had gone forward in church to dedicate my life to serving the Lord as a missionary. My public commitment was another vital step toward serving the Lord.

The Little Flute

I liked taking piano lessons. In junior high, I learned to play the flute. The sound was pleasing to me, plus it was easy to carry, not like a big, heavy tuba. I also played the piccolo, a half-sized flute, in the Vassar High School marching band.

We had fun marching on the football field before the games and at halftime. The band would make unique designs like a broken ax when Vassar played against the town of Bad Axe. At times, it was the initials of the city we were playing in. We played the other school's marching song. We even did exercises in the dark with lights attached to our band hats.

Eye Surgery

I had surgery to straighten my "lazy" left eye, but it was always weaker than my good eye. My poor eye would go out of alignment when I was fatigued.

Best Decision Ever (1941)

Working Days

During high school, I worked for Uncle Ed, who owned Erb's Food Shop on the corner of Main Street and Huron Avenue. I would go to the store to work two or three hours after school, and all day on Saturday. I worked as a checkout clerk, stocked the shelves, and marked the prices. I also carried groceries up from the basement storage area. Efficient Uncle Ed would often say, "Take a load both ways." I made 50 cents an hour, which came out to eight dollars a week. I put five dollars in the bank and had three dollars to divide between God's work at church and spending. That doesn't sound like a lot of money, but I could buy a dress for five dollars and a candy bar or an ice cream cone for five cents. I was able to save money for college while working. It only cost 600 dollars for my first year at Bryan College, but college costs have risen dramatically since then. I appreciated a good Christian boss who wouldn't sell liquor or cigarettes.

Driving Lessons Past and Present

Dad told me that my great-great-grandfather Heman Lewis bought a Ford Model T Touring car with a cloth top and a crank starter in 1919. Dad said the crank starter was an arm breaker! Grandpa Temple drove it first, and then Grandpa Lewis got behind the steering wheel. He decided to practice driving on the winding roads in the Silverwood cemetery. When Heman thought he would hit a building, he got scared and yelled, "Whoa!" while pulling back really hard on the steering wheel like he would have on the reins to stop his horses. (He had had a great team of horses before getting his first vehicle.) It was a good thing that he was going slowly when he hit.

Being the oldest child put me in line to be the first to learn to drive and get my license. My dad taught me. After some lessons, Dad took me to Caro, the county seat, to get my first license. I knew I had to take a driver's test, and I was ready. Someone from the department came to me and said, "Come on inside, and I'll issue you your first license." Later, I asked Dad, "Why didn't I take the driver's test?" The man in Caro knew Dad, so I didn't need to take the test. Thankfully, I made Dad look good by being a careful driver.

1951
Gull Lake Bible Conference Center

For many Gideon events, Dad included the whole family, such as at the Gull Lake Bible Conference Center in Hickory Corners. While there, we heard the Bryan University Gospel Team sing and share their testimonies about what God was doing in their lives. Dad commented to our family, "That's a school where I'd like to see our kids go." If we listen to the Lord, God can use events in our lives to guide and direct us. That fall, I enrolled at Bryan. Five of the six Temple children included Bryan in their college training.

We toured the Post Cereal Company nearby in Battle Creek, Michigan, and received samples. On our way home, we ate at a fancy restaurant. A butler with a towel over his arm served us! That impressed all of us.

High School Graduation

My senior year courses at Vassar High School included speech, typing, English literature, government, home and family, economics, and band. Typing and the marching band were my favorite subjects. I was a band member for all four years. I had a problem holding the sheet music while marching, so Dad fastened a mousetrap on my holder. It kept my music in place, but my friends all smiled with amusement at his handiwork.

I'm not a natural public speaker, even though I took elocution, public speaking, and reading lessons as a child. When I was a senior, I didn't want to give a public speech upon graduation. Consequently, I graduated with the third-highest GPA in my class of 53 seniors.

My high school portrait

Our yearbook, *The Echo 1951*, included this quote about me, "See you in church Sunday." I'm thankful the yearbook staff and classmates knew of my spiritual side. I graduated from Vassar High School on June 6th, 1951.

David's Early Life (1936~1953)

What's in a Name?

My parents named me David, which means "beloved," after David in the Bible. My biblical namesake was my hero; I loved to hear about him. He killed Goliath, the giant who had defied and spoken against God, with a single stone from his slingshot. He also killed a lion and a bear with his bare hands while tending his father's sheep as a young boy.

My understanding of life deepened as I approached my teenage years. I then realized David had had an affair with Bathsheba. His life was not always as a superhero, but he was "a man after God's own heart." He did not sin any less than usual, but he had a tender heart with a strong will to obey God.

Mother wrote the following in my baby book: "Great hopes are wrapped up in this wee babe. May he truly be as David of old, a man after God's own heart."

I like the relationship and influence of being named after someone in the Bible. Our four children, Deborah, Daniel, Elizabeth, and Stephen, have biblical names so they can relate to their namesakes.

Cameron is the middle name I was given, just like my father, Luther Cameron Henry. This was my Grandmother Mariah Jane (Cameron) Henry's maiden name and a reminder of the Scottish ancestry on her side of the family. Eventually, my wife and I passed on the same middle name to our older son, Daniel Cameron Henry. Then one of Dan's daughters named their youngest son Ethan Cameron Fields. Names and histories are often passed on with special meaning.

Grandfather Henry

Grandfather Hubert Melanchthon Henry Sr. died long before I was born. His parents gave him the middle name Melanchthon, who was a close associate of Martin Luther. My father was named Luther, and my brother was Martin Luther Henry.

In 1894, Grandfather Henry attended the Bible Institute of the Chicago Evangelical Society, which was later renamed Moody Bible Institute, and took the foreign mission course. He was a Sunday School superintendent, taught Sunday School, and did itinerant missionary work on horseback in northern Iowa with the American Sunday School Union before moving back to North Chicago.

Malaria Fever:

My father and grandmother Henry were sick with malaria, a potentially lethal infectious disease with a high fever. In that era, quinine was a common treatment. As a result of the disease, they burned down their "infected" home in Iowa and returned to North Chicago. Both the disease and the cure left lasting negative health effects on my father.

Life-Changing Spiritual Choices

My father turned to Jesus Christ at age 12, and my mother was a child when she accepted Jesus as her Savior. My parents made a huge spiritual decision for our family before I was born. They switched to Waukegan Bible Church, several miles away from our home. Father grew up in a mainline denomination that offered history and tradition but did not help in his spiritual growth. Mother also left the denomination in which she was raised.

At Waukegan Bible Church, my parents became very active in teaching Sunday School, helping in Daily Vacation Bible School, and serving on the mission board. The church was an old wooden building with a stage. Later, I was intrigued to learn that it was a converted beer parlor and dance hall. People and buildings need to be changed for God's glory.

Tribal Heritage

My maternal grandfather, Martin Bachanz Jr., was Wendish. Other names for the Wends are Lower Sorbians or Lusatian Sorbians. They were a small Slavic people group in East Germany near the city of Cottbus. The Nazi era, under Hitler, put a lot of pressure on them to learn German and assimilate into German culture.

My grandfather Bachanz immigrated with his older stepbrother, Friedrich Schimlick, to the Sheboygan area of Wisconsin in 1887 when he was 16 years old.

Many Wends Turned to Jesus Christ

The majority of the Wends became Christians during the Reformation, which Martin Luther started in 1517, and studied under him in Wittenberg. Within a dozen years, Luther translated the New Testament into modern German, compiled a hymnal, and wrote a catechism. This inspired the Wends to translate these books into Lower Sorbian. The New Testament was translated but not published until 1709, while their hymnal was published in 1574.

My grandmother, Clara (Truttschel) Bachanz, modeled a traditional Wendish costume.

Drehnow, Germany, in 1905, with my mother's relatives in front of their house.

Standing is Great-Uncle Christian, who made a full-size oak china cabinet as a wedding gift when my parents were married on August 9th, 1930. Next are Great-Uncle Matthes, Grandfather Martin Bachanz Jr., and Great-Aunt Anna. Seated are Great-Grandmother Louise (Schimlick) and Great-Grandfather Martin Bachanz Sr. The little boys peering through the window are Uncle August, age eight, and Uncle Carl, age five. The women are dressed in their Wendish dresses and elaborate head coverings. In 1868, Great-Grandfather Bachanz purchased this house and the adjoining barn and stable, which continue to be used by his descendants.

President Ronald Reagan stated, "Freedom is never more than one generation away from extinction. We didn't pass it on to our children in the bloodstream. It must be fought for, protected, and handed on for them to do the same." This is also true for the blessings of our Christian background. It is not passed on to our children in the bloodstream. We each must make a conscious decision to put the Lord first in our lives and keep building upon our Christian heritage. We had no choice about who was in our background, but we can choose our legacy.

1936
Preparation for Living in the Arctic

January 1936 was a frigid month—never above zero—even dipping to minus 30 degrees. (All the temperatures in this book are given in Fahrenheit.) Our family lived in North Chicago, a small town on the shores of Lake Michigan, 50 miles north of downtown Chicago. Little did I know that God was giving me a glimpse of my future life and ministry!

My father, Luther Cameron Henry, tried to repair our family car, a 1934 Plymouth sedan, but it refused to cooperate. The extreme cold had gotten to it! The vehicle had a six-volt battery, weaker than the ones we use today. Dad had to walk to work due to the prolonged cold spell.

Nelda Louise (Bachanz) Henry knew my birth would be very soon. In desperation, at 1:00 a.m. on Sunday, my father made three urgent phone calls: for a taxi, begging the driver to hurry; then to the doctor to meet us at the hospital; and finally to the hospital, stating that we were coming as fast as possible. It was minus 17 degrees!

The taxi sped over the four miles to Victory Memorial Hospital, currently called Vista Medical Center East, in Waukegan, Illinois. We arrived at the hospital with only a few minutes to spare. The night nurses assisted with my birth. I was a healthy eight-pound boy, greeting everybody available at 2:10 a.m., two minutes before the doctor arrived. The doctor charged half his usual fee since he missed my first cry.

At that time, they did not allow young family members into the maternity ward, so my five-year-old brother climbed on top of a high snowbank and peered in the window at me.

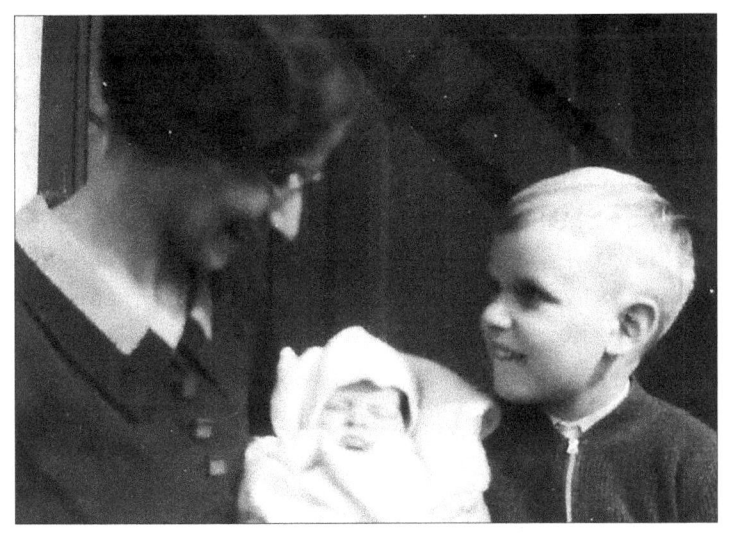

My first picture at six weeks with two particular admirers: my mother and my brother, Martin, who had been born in June 1931

1938
A Dangerous Adventure

When I was two, my mother made a warm, brown-colored snowsuit from an old fur coat, which made me look like a little cinnamon bear. One Sunday afternoon, Mother, Martin, and I walked to Central Grade School six blocks from our home.

The following Monday morning, instead of playing in our yard, I ventured out on my own. My mother and grandmother Henry called, prayed, and looked up and down the alleys and everywhere. Martin came home from school and joined in the search. Then they guessed I may have gone back to the school. Well, I did, and the adjoining railroad track looked safe enough to me. It was almost 5:00 p.m., time for the North Shore Line train running between Chicago and Milwaukee. The sun was setting, and it was getting darker. My cousin Richard Henry found me sitting and playing in the middle of the railroad track. After that, my mother had a strong preference for bright colors.

Later in life, I realized how quickly that express train could have wiped me out and ended my life. The Lord was gracious to me. We all have experiences in life where we can look back and say, "Thank you, God, for allowing me to live longer."

The express train track was turned into the Robert McClory Bike Path. When Kay and I were married, I asked Cousin Richard to be my best man.

The House on Lincoln Street
A 14-room, two-story house at 1616 Lincoln Street in North Chicago with my grandmother was home. It was a grand old house with 12-foot-high ceilings on the first floor and 10-foot-high ceilings on the second floor with one bathroom. My grandparents borrowed 2,000 dollars in 1910 to upgrade to electric wiring and coal heat.

My grandfather, Hubert Melanchthon Henry Sr., died in 1912 when my father was 14. My grandmother Henry and father took in boarders to pay the bills. Mariah Jane Henry had me call her "Aunt Jennie" because she thought being a grandmother aged a person. I never realized she was my grandmother until after she died.

Once, Uncle Henry Gelwicks, a 90-year-old carpenter from Minneapolis, visited us. He was my grandma's brother-in-law. Martin and I saw him tumble from the second floor and roll down the long flight of stairs to the landing, jump up, and announce, "I'm okay!" He came down faster than I ever did!

1939
Riding the 400
My father and I rode on a passenger train called the "400"—taking that many minutes to travel between Minneapolis and Chicago. Uncle Henry showed us five houses on one block that he had built—without any power tools!

Drowning at Third Lake
We enjoyed a gala family reunion with our Bachanz relatives, the first one in 10 years. The joyous Sunday event was held at our Third Lake cottage on August 6th, 1939.

Grandpa Martin Bachanz Jr., David, Grandma Clara Bachanz, and Martin are in our boat.

The following day, Martin, age eight, and I, age three and a half, went fishing with our Grandpa Martin Bachanz Jr. in our 16-foot orange and black flat-bottom rowboat. When we reached our fishing site, Grandpa Bachanz reached for the can of bait worms, lost his balance, and fell overboard. My brother quickly grabbed Grandpa's hand, but it slipped out of his tight grasp.

Thankfully, Martin could steady the boat and hold onto me, even though I lost my fishing pole in the shuffle. He rowed the two of us safely home. My brother's and my shrieking cries alerted our parents that something terrible had happened. Martin screamed, "Grandpa fell overboard!" He pointed at one of our favorite fishing holes.

Later, we heard Grandpa had suffered from a heart attack. The tragedy left a gaping hole in our family.

The Bible verse on our living room wall was a source of comfort to Grandma Bachanz and the rest of our family. "And we know that all things work together for good to them that love God, to them who are the called according to *his* purpose" (Romans 8:28, KJV). This key Bible verse was used at Grandpa's funeral service in Plymouth, Wisconsin. God is vitally interested in every detail of our lives.

Would He Ever Talk?
Friends began to wonder if I would ever talk. My parents knew I was an intelligent, happy little boy. I was shy and started speaking when I was five. Grandpa's drowning had traumatized me.

1940
Kindergarten
Each day for kindergarten, I walked six blocks to Central Grade School. I met the deadline to enroll that fall rather than waiting until the following year. At Christmas, I preserved an apple by completely covering it with cloves. I still have that apple, and I enjoy smelling the cloves.

1941
First Grade
For first grade, I continued at Central Grade School in North Chicago. I would order chocolate milk for every lunch at school instead of drinking bland white milk. At home, Grandmother Henry would drink Sanka coffee every day. I wanted to drink coffee like my grandma but was told, "It'll stunt your growth," and I didn't want that. Later, I found out that Sanka was decaffeinated coffee.

Raised in a Christian Home
We always thanked the Lord for our food. In the evening, my mother would read a Bible story out of *Egermeier's Bible Story Book* or straight from the Bible. Then we would kneel together and pray as a family. We memorized many Bible verses. Mother enjoyed reading other books to us. She was a former school superintendent at Pigeon River Elementary School in Sheboygan, Wisconsin.

My parents made a conscious decision and effort to follow the Lord and the Bible more closely. They followed Joshua's example: "But as for me and my house, we will serve the Lord" (Joshua 24:15b). God knew what was best for me, so He placed me in a home where my parents sought to serve the Lord. Their commitment gave me a tremendous advantage in learning God's way from birth, but I still had to make up my mind to follow the Lord or go my own way. I may have accepted Jesus Christ as my Savior when I was six. At age 13, I made sure. More on that later. Our choices in life have consequences!

David's Early Life (1936~1953)

Martin, David, and their father are enjoying the water at Third Lake. We wore one-piece swimsuits, which looked more like bodysuits.

World War II

My parents would turn out all the lights in our North Chicago home after hearing the siren during the blackout air raid drills. We didn't want the Germans to see our lights and know where to bomb us!

The government introduced rationing because certain things were in short supply during the war and to ensure that the military received adequate supplies. People were issued ration books that limited certain items each person could buy, such as gasoline, tires, meat, sugar, butter, cheese, and other things.

My Favorite Toys

I learned to ride a bicycle without using training wheels. Uncle Allen Henry, my father's half-brother, would run along the sidewalk beside me with his hand under my seat to keep me upright. It was a big day when I could ride without any help and go faster!

I played with various construction-type toys: Lincoln Logs for building houses; an Erector Set for making a variety of movable or stationary toys with nuts and bolts; Tinker Toys (like the Erector Set, but using wooden parts), and building blocks. I enjoyed building birdhouses, boats, houses, and bridges with scraps of wood. I kept busy with stamp collecting, coin collecting, doing puzzles, climbing trees, exploring the woods, and swimming.

1942
Another Grandparent Dies

My grandmother, Mariah Jane (Cameron) Henry, died on March 22nd at age 84. Her coffin was in our house's parlor. I remember secretly sneaking in and prying open one of Grandma's eyelids to see if she was sleeping or not.

"Victory Speed"

President Franklin Roosevelt put the country on daylight saving time, called "War Time," to save fuel, as our national attention was on winning the war. All auto production was switched to military vehicles until the war ended. The maximum "victory speed" was 35 mph to conserve tires and gas. It was an economical speed for better gas mileage.

Picking Cherries: David

Our family liked peaches, so we drove to an orchard near Muskegon, Michigan. The trunk and back seat were loaded with wonderful fruit. There were also pick-your-own cherries, and I had fun climbing in the big trees. I was sick for several days after eating too many cherries that had been sprayed with pesticides.

Moving to Third Lake

Our family moved to Third Lake, 12 miles from North Chicago. My dad spent a lot of evenings and weekends after his regular job making significant improvements so we could live there during the winter. Raising the building and adding a basement was too dangerous for this little boy. I was given the job of straightening bent nails. Many were given a second life, but it may have been busy work to keep me occupied.

One of Dad's improvements was building a massive stone chimney. My parents' car was a 1940 Ford sedan. Father would stop at a farmer's field on his way home from work and load some big stones into the trunk. Dad and Martin would lay out the rocks on the ground to match them for building the chimney, going up three or four feet at a time.

I'm pushing our "power" mower!

The hearth was formed with many pieces of local, broken Indian arrowheads inlaid among other stones. One arrowhead was four inches long and in perfect condition. My father found them while using a spade to dig up our large Victory garden in the backyard.

We had a big basement. Dad built a raised platform, 12-foot by 16-foot, for our Lionel train sets and their long tracks. I liked seeing the headlights coming down the track pulling a string of train cars. Great wintertime fun!

Another Dangerous Adventure

A few days past Martin's birthday, my mother sent me out to play while they baked his cake in our new oven. They called me when the cake was ready, but I was long gone.

Later, a policeman escorted me home. He told my mother, "You'd better watch that little son of yours. A woman picked him up near Gurnee and brought him to our station." Barefoot and clad in only a sunsuit, I had walked six miles. I explained, "I was going to meet Daddy!" Blistered shoulders and sunburn were my punishment as well as the remaining evidence of my adventure.

Swimming Lessons

At Third Lake, I played in the shallow water near our shoreline. As I got bolder, I ventured farther out on the sandy bottom as far as I could. Then, to keep my nose out of the water, I tilted my head back. Later, I realized there were holes out there!

I learned to swim on my own by dog-paddling, but it was pretty slow. After moving to Florida, I took swimming lessons and learned the proper strokes.

Purple Martins

My grandfather Bachanz had built a big birdhouse for purple martins, the most prominent member of the swallow species. The house was two feet square with holes for 12 families—an apartment complex for birds. My father mounted it on a pole at the edge of our large garden, 15 feet high. I liked to hear them sing. While flying, they caught and ate insects such as gnats, flies, and mosquitoes. In the winter, they migrated to South America, but we looked for their return every spring. We wanted only purple martins to nest there, so my brother climbed a tall ladder and cleaned out any nesting materials from other birds looking for a free home.

Flying Paper Airplanes

Grandma Bachanz and Aunt Lillie shared a fourth-floor apartment in Milwaukee, Wisconsin, and we often drove there to visit. Building houses with Lincoln Logs and playing checkers with someone new was fun. Once, I peered out their living room window at the big yard below. I thought it would be nice to throw paper airplanes and see where they landed. So I launched one. Grandma was more cautious, so I stopped watching them fly. She was worried that I could fall out of their window.

Stay out of Trees

We had several fruit trees in our yard, but I especially liked climbing one big apple tree. While climbing, I slipped and fell, hitting the corner of a beehive my father had placed under that tree. Father heard the crash and hurriedly brought me inside. After catching my breath, I ended up with some reddened skin and more scratches.

I enjoyed the summers. All a little boy has to do is jump out of bed, grab a pair of shorts, and play, play, play. I wore loose-fitting shorts. My brother saw me up in a tree and hollered, "You need to wear underwear. People can see everything you've got!" After that, it took longer to get dressed.

Indians

When I was older, I visited Crescent Lake Bible Camp near Rhinelander in northern Wisconsin with my father and later as a camper. We would always look for Native American places, like the Menominee Indian Reservation.

It was fun to wear my costume and headdress as a little boy in North Chicago, but more so when we lived at Third Lake. This was a prime hunting area for the Algonquian Indians.

Martin and I were pretending we were Indians.

Side Jobs

Martin and I had side jobs mowing lawns for people. Depending on the size, the payment was usually a dollar, maybe two for a larger lawn. When you're young, the sweat and hard work of a push mower burn off excess energy.

1943
Home Alone

One time, Martin and I were home alone at Third Lake. He told me to eat a spoonful of sugar. I did, and it was good. Then he encouraged me to eat a spoonful of flour. I didn't want to, but he kept urging me. So I did. It was terrible! Immediately, the flour in my mouth exploded all over! Because I was the little brother, I had to clean up the mess, and we did not want our parents to find out.

Baking a cake is an excellent example of Romans 8:28, where all things work together for good when we love God. By itself, a spoonful of sugar is good, but a spoonful of flour is yucky! The result is outstanding when the cook combines the ingredients and bakes them.

One-Room School

From second to sixth grade, I attended Druce Lake School. It was a simple one-room school with 22 students spread over eight grades, and one teacher, Sarah Fields. I rode my bicycle to school.

Occasionally, the teacher told me to sit in the dunce corner and wear the hat for 10 minutes. Sometimes during recess, the older girls would throw me into a big thistle patch for teasing them. Another time, a boy threw a stone at me, so I told the teacher. The boy's version was, "David threw a stone at me, and it bounced off of me and hit him!" It seemed that the teacher believed his version!

We usually played "war" during recess since World War II was raging at that time. I also had a G.I. Joe crew cut that required no care.

Smoking cigarettes was popularized by soldiers and cowboys. Students would smoke behind the school garage. Smoke is smoke, I thought, so I took a sheet of paper, rolled it lengthwise, and lit it. I blew it out and took a deep breath, inhaling the smoke. Boy, did I ever cough! Some things in life you do only once!

Saving Stamps

We could buy the minuteman stamps at the post office in denominations of 10 cents, 25 cents, 50 cents, or larger. A full book was valued at 18 dollars and 75 cents. By the end of the war, I had 10 of these bonds, which matured to 25 dollars in value in 10 years. All the children purchased defense stamps during that time. We were very patriotic and supported our country's war efforts.

Fast-forward: I cashed in those bonds when we moved to Alaska in 1958 and needed money. They were worth 30 dollars each by that time.

Life at Home

At home, we had a large map on our kitchen wall where I kept track of our military's movements during the war. At church, my favorite song was "Onward, Christian Soldiers."

Since meat was rationed, we raised sheep, chickens, rabbits, squabs (young pigeons), and goats. The rabbits grew and reproduced the best. Taking care of animals kept us boys busy.

We sheared our two sheep and sent the wool to a company that made two warm blankets. We milked the goats and drank their milk. One kid goat would follow me around the yard or along the road.

One time, I was playing in a narrow ditch connecting Third Lake and Druce Lake. The shallow water allowed me to jump in and grab an eight-pound carp 20 inches long. My surprised mother asked, "How did you get it?" Martin answered, "He just jumped into the water and picked it up with his bare hands." I had carried it a quarter of a mile home. I don't like eating carp, but it was fun catching them.

A Sister Is Born

My sister, Miriam Louise Henry, was born in October 1943. She was named after Moses' sister in the Bible. It was a big day in the Henry household when she arrived home from the hospital. Our parents were thrilled to have a girl to help balance our family. Now, I am the middle child. Martin and I were raised on cow's milk in North Chicago, but our sister was raised on goat's milk at Third Lake.

1945
Feeling Inflation

Inflation was one of World War II's consequences. The size of a five-cent candy bar shrank from one and a half ounces to one and a quarter ounces, but the price stayed the same! Today, a five-cent candy bar would be tiny!

Stopped by a Moonshiner

Minnie Joy, a missionary supported by our church, was serving in the hills of southeast Kentucky. Minnie and her co-worker, Elsie Pfister, lived and ministered in Bledsoe, away from major roads.

While driving to visit them, we passed a section of a winding gravel road where one lane had been washed away due to high water. Miriam started singing "Trust and Obey" in her childish voice. I'm glad the Lord prompted her to do that. We needed the reminder.

As we neared our destination, a boy flagged us down, so Daddy stopped to see what he wanted. Then a man walked up to our car. You could smell liquor on his breath. He was trying to sell home-brewed moonshine and reached into the front of my dad's coat, searching for his billfold. Daddy asked him, "Do you know where Minnie Joy lives?" The man took his hand out and pointed down the road. Instantly, Daddy sped off! We thanked the Lord that the man did not shoot at us with the pistol strapped to his side.

Later, Minnie Joy spoke to our Sunday School at Waukegan Bible Church about her service in Kentucky. She kept referring to her "co-worker" doing this and that with her. I thought she was saying "coal worker." I knew they burned coal and mined it in the hills, but why did she need another person to work on coal? My vocabulary and comprehension were growing.

1946
Almost Froze to Death

My father worked as a plant superintendent at Fansteel Metallurgical Corp., a chemical company in North Chicago, for 30 years. He graduated from Waukegan High School as the valedictorian of his class in 1916, then worked as a chemist and eventually became a foreman. His beginning pay was 50 dollars a month and 600 dollars per month when he resigned.

Father commuted 12 miles to the factory each workday. While driving home during a big snowstorm, he got stuck in a drift and almost froze to death. We were concerned because he was later than usual. Once he arrived home, he sat with his feet in a pan of warm water. He informed us with a determined voice, "We are moving to Florida!" That was the straw that broke the camel's back—no more snow and cold for us.

1947
Moving to Florida

The following summer, we moved to Ellenton, Florida, near Bradenton. Packing was challenging while giving or ditching many things. We loaded a two-wheel trailer to pull behind our 1942 Nash Ambassador sedan. Martin pulled a large four-wheel farm trailer with his stick-shift, half-ton 1939 Ford pickup for our 1,300-mile trek south! Goodbye, Third Lake; I'll miss all the fun times!

1948
Down on the Farm

"Tropical Acres" is what my parents called the 40-acre farm and house they purchased for 15,000 dollars. It had many orange and grapefruit trees and lots of sunshine.

My sister, Miriam, has two oranges covering her eyes while I help our parents pack bushels of organic oranges and grapefruit to sell.

The garage had a bank of Delco glass-cased batteries that produced 32 volts of electricity, providing light for our house. Five acres of okra were watered by an irrigation well. Martin powered the pump using an old Model A engine. Our nearest neighbor lived two miles away.

Snakes

Daddy, Martin, and I were making changes to the slabs of concrete used for our sidewalk. I noticed a snake and shouted, "Look at this pretty snake!" It was a beautiful coral snake over a foot long. It had a black nose, and its body had alternating wide red and black bands separated by narrow yellow bands.

It was an eastern or harlequin coral snake, North America's second-most deadly venomous snake. Shortly after that, I found its mate close to our house. I killed it without hesitation. Many other plain-looking non-poisonous snakes in Florida feed on rodents, frogs, toads, eggs, and baby snakes.

A Much Bigger School

Martin and I walked more than a mile on a sandy side road to our bus stop by the idled Fuller's Earth plant. We rode through Ellenton, and after several more miles, I would get off at Palmetto Junior High.

Martin continued farther, crossing the Manatee River to Bradenton and Manatee County High School.

Going from a small one-room country school with one teacher to a consolidated school where I was one of several hundred seventh and eighth graders was a culture shock. When the bell rang, we changed classrooms for each subject.

Our homeroom class started the day by saying the Pledge of Allegiance and the Lord's Prayer, followed by reading the Bible. This was before the government took the Bible and prayer out of public schools through the Supreme Court decision in 1962.

The Southern boys called me "Yankee." I did not appreciate that name at all, but it stuck, even though I tried to fit in and not be different. Many Northern kids attended our schools because their parents spent their winters in Florida.

Fast-forward: I started working in Michigan to pay for college in 1953. I wanted to buy a typewriter. On two occasions, when I asked the clerks some questions, I quickly realized they didn't understand my Southern accent! So I tried to talk like my northern friends.

A Few Difficulties in School

I joined an after-school club, called the "400 Club." Members would write out $400 + 1 = 401$, $401 - 2 = 399$ on the following line, and then continue adding one and subtracting two until arriving at zero. It took time and was part of my discipline!

Because of a problem in eighth grade, my parents were called in to have a little talk with the principal. While waiting in the office, my mother noticed a stack of permission slips to attend Bible classes after school. My parents realized they could teach the Bible in public schools. There has to be a bright side to a special meeting with the principal.

Riding Bareback

Mr. Philips retired his old nag to our farm to live out the rest of her life. She was gentle, calm, and cooperative. He gave us a saddle, but I was young and rode her bareback, which required balance. I let her walk or trot, but she didn't run. She was old and her backbone ridge made me sore. It was like sitting on a two-by-two board or riding on a racing bike with a skinny seat.

After a year, she quietly passed away near a swamp on our property. We left her lying right there while God's cleanup crew of maggots did their job with her body. We didn't smell anything since she was far enough away from our house. The pile of bones marked her final resting spot.

Miracle Camp

I memorized Bible verses with the Bible Memory Association, which later changed its name to Scripture Memory Fellowship. For 10 weeks, I memorized 10 or 12 verses that I had to say perfectly to a helper each week.

Memorizing the verses earned me a week at Miracle Camp in Ringgold, Louisiana. The program's founder, N. A. Woychuk, led the meetings and helped us apply the Scripture to our lives through messages, songs, and testimonies.

Why Memorize Bible Verses?

If it's essential, memorize it. I memorized my Social Security number, date of birth, phone number, and address. You get the idea. Memorizing Bible verses helps us to know, live, and plant the Word of God deeper in our hearts. The Holy Spirit can help us recall verses when we need them.

1949
First Time Hearing Billy Graham

Dad took a friend and me to hear Billy Graham speak on the infield at a baseball park in Saint Petersburg, Florida. Billy was ministering with Youth for Christ, and 50 young people attended. I was impressed to be with other youths listening to the preacher.

Boarding School

My parents felt I would do better at a Christian boarding school. So we drove to Asheville, North Carolina, and I enrolled at Ben Lippen School for my freshman year, 1949-1950. At that time, the school was an all-boys school. It was also expensive, and my parents did not have much money.

Picking up Pop Bottles

I picked up unbroken, discarded glass pop bottles along the road and turned them in at a gas station for a refund of two cents each. At that time, a bottle of pop cost 10 cents, so I only needed to find five good, used ones. The bottling companies would wash, sterilize, and reuse them.

My Heaven or Hell Choice (1949)

A Weak Church

We attended the only local church, the Ellenton Community Church, where the Methodists provided the pastors and some support. My parents enjoyed working with children, so they were asked to take over the youth department. They taught Sunday School, and more children started attending.

My dad taught an adult Sunday School class for a month about eternal security in the Bible, which certain people didn't appreciate. It didn't fit their doctrine. Unfortunately, the church was more interested in social issues than being biblically correct and meeting the spiritual needs of the people.

Leaders in the church also found out my parents were teaching a children's Bible club in a segregated school. Several board members visited my parents and bluntly told them, "You cannot minister to colored children!" (Shortly after this, the term "colored" was changed to "black.") Certain members referred to my parents as "Communists," a derogatory term for anyone with a differing view.

A Strong Church

Our family switched to Calvary Baptist Church in Bradenton. It was an independent Baptist Church with good Bible teaching by Rev. Luttrell and his son, Don. This church was a great place to grow spiritually, where God challenged me to commit my life to serve Him as a missionary.

Heaven or Hell?

"If I died, would I go to heaven or hell?" That haunting question bugged me. I knew Jesus Christ was the only way to heaven, but I wasn't sure if I knew Him. I wrestled with my doubts. Maturing from a boy into a young man brought significant changes besides a deeper voice.

Life began to move faster, and interest in girls was developing—plus more serious, deeper thinking about life.

"Was I saved or not?" I earnestly sought the answer by reading through the entire Bible and asking that one nagging question for a year. I looked up every verse in several good articles to help convince me that I was safe and secure with my Savior.

Finally, the Lord gave me His peace. As He said, "Peace I leave with you, My peace I give you; not as the world gives do I give you. Let not your heart be troubled, neither let it be afraid" (John 14:27). I did not have to fear hell anymore. I relied on and believed in Jesus Christ as my Savior, making Him the Lord of my life. God said it; I believe it, and that settles it.

I could not depend on my parents' faith or on my family's background to save me. I had to make that personal decision myself. Now the Lord gave me forgiveness of all my sins, His inner peace, and a purpose for living.

Being saved is not simply a matter of praying the sinner's prayer. It's much more. Everything in life centers around Jesus Christ. It's a matter of believing in Jesus and following Him. It means He is our Savior, Lord, and Master of our lives. It's our growing relationship with Him. (See Appendix A, How to Know God.) I don't know the exact date I accepted Him as my Savior, but I know I did.

Growing Spiritually Has Its Ups and Downs

When I was 13, I was baptized at Calvary Baptist Church as a public witness before others. Pastor Don Luttrell asked if I had anything to say, so I commented, "The Lord saved me from so much, to so much more."

I put this famous quote by C. T. Studd on my bedroom wall: "Only one life, 'twill soon be past, Only what's done for Christ will last." It helped me realize that my time here on earth is short, so I need to live for Jesus Christ all the time.

I attended the special missionary meetings at Calvary Baptist led by Paul Smith from Canada. At the meeting, I went forward with 11 other young people to dedicate my life to the Lord for missionary service wherever the Lord would lead.

As I was leaving the service, one huge disappointment happened. A high school classmate questioned me, "Did you really mean it?" My sad answer was, "No." But then another verse hit me, "And Jesus said unto him, 'No man, having put his hand to the plow, and looking back, is fit for the kingdom of God'" (Luke 9:62).

The next thing I did was repent and recommit my life to the Lord. I wanted to be fit for the kingdom of God. Later, when I felt weak spiritually or was tempted to turn back, I would think of this verse. Scripture can guide us throughout our lives. I also thought about the time Peter denied Christ three times in a row, then repented and stood firm for the Lord.

A Lifelong Process

As I journey through life, I continually come back to needing and finding forgiveness and peace. I sin less now than I used to, and my sins are mostly in my thoughts or may be considered small things to some people. My life following Christ is not perfect, but I keep my focus on pleasing Him each day.

1950
Hearing Billy Graham Again

I enrolled at Ben Lippen School in North Carolina for my freshman year of high school. The school took all of us students to Columbia, South Carolina, to hear Billy Graham speak at a crusade. I sat in the upper balcony. It was difficult to see Billy at that distance, but many people went forward to receive the Lord. The crowds were impressive!

My Mother's Covenant

On March 21st, 1950, my mother signed the following covenant and pasted it in her Bible:

"Lord, I give up all my own plans and purposes, all my own desires and hopes, and accept Thy will for my life. I give myself, my life, and my all utterly to Thee to be Thine forever. Fill me and seal me with Thy Holy Spirit. Use me as Thou wilt; send me where Thou wilt; work out Thy whole will in my life at any cost, now and forever."

Teaching Boys and Girls

My parents began teaching Bible lessons in Manatee County Public Schools in conjunction with Mrs. Alma Gray in nearby Sarasota. Teaching children gave them the experience, confidence, and desire to join a mission in Kentucky.

My Heaven or Hell Choice (1949)

A Bigger, Scarier Snake

When I was 14, I walked out in front of our house and saw it. A huge rattlesnake was at the bottom of the deep drainage ditch by our driveway. I dashed back to our house, yelling, "There's a big rattlesnake out front!" I grabbed my .22-caliber rifle and some shells and hurried back. I didn't want to miss this one and wonder where he was lurking. Mother and Miriam observed from a safe distance.

The eastern diamondback rattlesnake was watching my every move as he went into a coiled position. I could hear his rattles, which also rattled me. I prayed, took a deep breath, and took careful aim. My shot struck him in the middle of his head! He was six feet long and heavy to hold up. On its tail, there were 12 buttons.

David is holding the rattlesnake.

In the Navy

Martin enlisted in the Navy in August. He started his training in San Diego, California, and they shortened his boot camp to nine weeks, quickly deploying him to Hawaii. He worked 12-hour shifts transferring ammunition to planes headed to South Korea. After one tour of duty, Martin was honorably discharged as an Aviation Ordnanceman Third Class. Uncle Allen Henry served in the Navy during World War I. My father was underage, so he missed being enlisted.

1950 Census

My father worked for six months with the Bureau of Census in Florida. We also lived off of our family savings while my father regained his health.

Local High School

From 1950-1951, I attended Manatee County High School for my sophomore year and rode the bus. I earned a ribbon for my part in a 660-yard relay. But most sports were too far from home for me to participate in.

1951
Back to Boarding School

For my junior year, 1951-1952, I returned to Ben Lippen. They had smaller classrooms and lots of sports. Football was fun. I played right end, but I often sat on the right end of the bench. In soccer, I earned a letter and played halfback.

I enjoyed running, especially the mile. In one race, I passed several people going into the last curve of our quarter-mile oval track, even though I was instructed not to pass on curves. I experienced an exhilarating "second wind" for the first time. It was the final lap, so I passed anyway and won the race.

I ran the 440-yard leg of a relay race with our team of four in a large regional meet in Asheville, North Carolina. I was the second person in our relay team of four boys. When the first runner tried to pass the baton to me, another runner accidentally knocked it to the ground. I had to stop, go back, pick it up, and do my best. I made up for the lost time, but we didn't win the competition.

Fast-forward: Ben Lippen's campus has been dramatically expanded to become Billy Graham's training facility, The Cove. The main building, where we had high school classes and dormitories, burned down. It's a beautiful location continuing to help God's people.

Music Lessons

My parents thought I should take music lessons, so I took violin lessons and learned a little. Unfortunately, playing an instrument wasn't in my genes. But it helped give me foresight and perseverance much later in life while working with the Sakha people in Siberia to produce their hymnal. No, I never did play an instrument, nor did I develop a singing voice, but I have a good ear for appreciating great music. My future wife played the piano, flute, and accordion.

1952
A New Job

My father finally felt he was well enough to go to more difficult living conditions in the hills of Kentucky. Grandmother Jennie Henry had dedicated him to the Lord's service when he was born.

My parents joined Scripture Memory Mountain Mission (SMMM) in the hills of southeastern Kentucky. Their home office is at Camp Nathanael, on a 500-acre campus near Emmalena, a year-round Christian Camping and Conference Center. Later, my father served as camp director for four years. Since similar missions served in neighboring communities, the mission focused on five counties. Their primary goal was to win the unsaved to Christ and encourage them in their Christian walk. At that time, they could freely teach the Bible in public schools and have the kids memorize Bible verses to earn a free week at Camp Nathanael. Father and mother ministered to the children in 16 public schools in Letcher and Harlan counties, reaching 2,500 children every two weeks.

Oven Fork, Kentucky

Between my junior and senior years of high school, we packed for the move to Kentucky. This time, we had fewer possessions than when we moved from Illinois. Our Ford Woody Wagon and a trailer held our belongings.

We were heavily loaded! Near our destination, we went up a long, mountainous road when our vehicle overheated. I had the honor of going to a nearby creek to fill our radiator with water. This happened three times before we finally crossed that mountain.

We lived in a lovely home at the foot of Pine Mountain. The rent was very reasonable.

Our Post Office

The post office, Maggards Cash Store, and a few houses were in Oven Fork, a mile from our home. One time, when I picked up our mail, the postal worker asked me questions about the information on my postcard. I didn't give much of an answer.

Dad and I are standing by our home.

Fresh Meat

My father would buy older goats for 10 dollars each and bring them home. Then we would butcher them and fill our freezer. It provided cheap, quality meat—without the trouble of having to raise or care for them as we did in Illinois.

A Senior at Frenchburg

The only public high school in economically depressed Letcher County was in Whitesburg, 13 miles away. I would have taken the 6:00 a.m. Greyhound bus that crossed over Pine Mountain, with two big switchbacks on a mountainous road, and then returned late in the day.

My parents heard about a Christian boarding school in Frenchburg, Kentucky. The town's population was less than 500. The school was supported by the United Presbyterian Church and operated as a public school in partnership with the State of Kentucky.

My Heaven or Hell Choice (1949)

My family is in front of our 1948 Ford Woody Wagon, including Martin (home on leave), Mother, David, Father, and Miriam.

Hell for Certain

Thirty boarding students lived at Frenchburg High School, and most of us were in the senior class. Two classmates came from Hell for Certain, Kentucky, otherwise known as Dryhill. The town's name is derived from a nearby creek with the same name.

I heard that a man was chasing his mule but couldn't catch him. Finally, in exasperation, he said, "That mule is going to hell for certain!" Thousandsticks, Black Bottom, Wolf Lick, Mud Lick, Big Rock, Talcum, Viper, Raccoon, and Dwarf were just a few of the many unique names in the region.

Jobs

I was a janitor at the United Presbyterian Church. My duties included cleaning, dusting, tending the coal stoker for heating the church, and doing other odd jobs. I received 25 cents an hour, which was the top wage during that era.

When I served as treasurer for our senior class, one of the primary fundraisers was to sell candy bars. This kept us busy counting nickels and dimes.

My Last Grandparent

Grandma Clara (Truttschel) Bachanz, age 83, passed away on November 14th, 1952. This was during my senior year, so I did not attend her funeral service in Plymouth, Wisconsin. She was interred at Farmin Cemetery beside her husband, Martin. Grandma had lived a quiet, Christian life.

1953
Spiritual Challenges

Some of the faculty at Frenchburg urged me to consider attending one of their denomination's five colleges. But my classroom experience pointed me in the opposite direction. One teacher was promoting the theory of evolution as a fact. So I piped up in one of her classes and stated that early "cavemen" in Genesis 4:22 were highly developed and had forged all kinds of tools out of bronze and iron.

Class Activities

I was a member of both the Bible club and the letter club. The Letter Club consisted of athletes who earned a school letter in sports. I played with our basketball team, the only available athletic activity. Basketball was a big thing in rural Kentucky schools, but we didn't have a gym.

I graduated from Frenchburg High School with my class of 25 members on May 21st, 1953. My father gave the invocation for this 40th annual commencement.

My senior picture

Higher Education

Bryan College (1951~1957)

College History: The Monkey Trial

Bryan College, a private Christian liberal arts college with a 128-acre campus, is on top of a hill just outside of Dayton, Tennessee. It was founded in the aftermath of the world-famous 1925 Scopes Monkey Trial to establish a Christian institution of higher education that would teach creation from a biblical worldview. During the trial, live monkeys were on display in Dayton to mock Christians for not believing in the theory of evolution.

William Jennings Bryan, the "Great Commoner," was the prosecuting attorney for the state of Tennessee, showing that John Scopes had broken the law by teaching evolution in a state-funded school. Bryan had been the Democratic nominee in 1896 for president of the U.S. when the Democratic Party represented the working class.

Name Changes

William Jennings Bryan University was chartered in 1930 and admitted its first class in the fall of that year. Its stated purpose is to provide "for the higher education of men and women under auspices distinctly Christian and spiritual."

From here on, we will go with the new name "Bryan College" or simply "Bryan." Since our life experiences merge at this point, we will indicate who is speaking by placing that person's name after the section title.

1951
First Impressions: Kay

During a rainy week in the spring of 1951, while my high school class took their senior trip, I chose to go with my family to visit Bryan College. It was a long drive, 700 miles from Michigan.

The center of the main administrative building was a gaping hole, missing the large, central section where the tower was planned. Not too impressive! The rain was leaking everywhere in the unfinished entrance.

The dean showed us around, stating, "If you've come to see the buildings, we don't have them, but if you want a spiritual, friendly college—this is the place for you." Their motto is: "Christ Above All." The Great Depression and World War II slowed the physical development of the campus.

Catalogs continued to come from many universities and colleges as I tried to decide. One day I told my mother, "Mom, the Lord must want me to go to Bryan; every time I think of another university, Bryan just comes up in front of me." Now I know why. I received an excellent Christian education, and God placed me there to meet David, my future husband.

College Days: Kay

I enrolled at Bryan College as a freshman in the fall of 1951. Everything was new and exciting. It was great to meet other young people who also loved the Lord and were seeking God's will for their lives. Besides studying, we had lots of fun.

1953
Hard Choices: David

It was difficult to choose where God wanted me to go. I filled out an application along with my autobiography for one school, but I did not feel God's peace about going there. I would have felt out of place in the big city at a large school. I also needed to work my way through college.

My parents were serving with Scripture Memory Mountain Mission in Kentucky. Mrs. Franklin, the mission director's wife, told my mother while they were working together in the Camp Nathanael kitchen, "I'd be happy to have all my children (five boys) graduate from Bryan." I finally chose Bryan and enrolled in the fall of 1953 as a freshman, while my future wife was already a junior.

First Sighting: David

During the summer of 1953, I went to Grand Rapids, Michigan, with David Franklin. I quickly realized I was underage at 17, for most jobs require a person to be 18. My friend got a job driving a dump truck for one dollar and 55 cents per hour. The best I could do was work at Eberhard's Supermarket for 75 cents an hour as a carry-out boy. The store manager gave me a pay raise to 85 cents an hour the second week I was there. I would also help stock the shelves, change prices, or clean.

I stayed at the Grand Rapids School of the Bible and Music campus. Not much was happening there since they were on summer break. (The school has since merged with Cornerstone University.) I walked four miles to Eberhard's Store to save money.

One weekend, we took a trip to the Maranatha Bible and Missionary Conference Center in Muskegon, Michigan. David Franklin would graduate from Bryan in 1954, so he knew Kay Temple, my future wife. I even saw him talking to her when she and her family visited Maranatha. But he never introduced me to her! I had to wait until I enrolled at Bryan that fall for my next chance to meet her. I had my first sighting from a close distance.

1954
Working My Way through College: David

During the winter, I did various jobs in the maintenance department at Bryan. To work in the girls' dorm, I had to announce my entry by saying, "Man on the floor." The usual female response was, "Pick him up" or "Don't step on him."

In 1954, Bryan College received a large grant to complete the main administrative building. There was a beehive of activity, with scaffolding all around as crews worked on the outside brickwork and built the tower section. I did electrical wiring for the main entrance and other jobs as needed.

Exciting News: Kay

David and I enjoyed our meals together in the dining hall. Someone informed me that I had a long-distance phone call, which I was expecting about the birth of my new brother or sister. I was so excited that I took off running to talk on the phone. Since I left the dining hall so abruptly, several students thought we had a fight and had broken up!

Sheryl Lynn Temple was born in November 1954. I was thrilled to have a little sister, 21 years younger than I was!

"Oh Henry!": David

On Sadie Hawkins Day in November, the roles were switched, and women asked men out for a date. Kay's roommate, Charlene (Watkins) Smith, asked me for a date, and she presented me with an "Oh Henry!" chocolate candy bar. The dating at college was casual and just a friendly time together. I was in the early stages of dating Kay.

1955
Summer Work: David

During the summer, I worked for my future in-laws at Vassar Electric in Michigan. I mainly worked on rewiring electric motors. It was good work, plus I got better acquainted with Kay. I rented a room from an elderly lady in town.

Toccoa Falls: David

We rode in separate cars with our group of 50 Bryan students and teachers to attend the annual fall missionary conference at Toccoa Falls College in Georgia, a small Christian college similar to Bryan.

One of the exceptional speakers and a college donor was R. G. LeTourneau. He was a business magnate and inventor, especially in earthmoving equipment. His company supplied 70 percent of the heavy equipment used in Europe during World War II, and it was used to build the Alcan (Alaskan-Canadian Highway), which provides access to Alaska. He donated 90 percent of his salary and company profits to God's work. What a testimony of God's provision!

Bright and early one morning during the conference, both Kay and I, along with three others, paddled small boats to an island. We had our devotions as a group and briefly explored the island before returning for breakfast. Our first official date later that fall was at a missionary conference on campus.

High Jumping: Kay

One year, I broke the girls' record in the high jump for intramural sports. But shortly after that, another girl broke my record. I also like to run, but I didn't pursue track at Bryan.

My Major—Christian Education: Kay

I loved children and teaching, so I majored in Christian education. The department head was Kermit Zopfi. We had practical experiences like playing games in class, rather than just studying about Christian education. My studies helped prepare me to train our four children and teach others. My regular classes included Bible, Christian education, education, English, French, history, and piano.

Extra Activities at Bryan: Kay

College classes ran from Tuesday through Saturday, so we would study on Mondays rather than skip church on Sundays. All four years at Bryan, I was involved in a Christian Service Assignment. (Every student at Bryan served in Dayton or nearby.) My assignment was to go to a public school with several other students to teach Child Evangelism Fellowship classes. I used these materials throughout our future ministry time in Alaska and Siberia.

David and I were both active with the Foreign Mission Fellowship. It was a national organization functioning on Christian campuses to associate with fellow Christian students seeking foreign missionary service. There were informative meetings about activities and needs in various world regions.

Africa interested both of us, so we joined this campus prayer group. A good friend pointed out that we were not standing together in the earlier college yearbook, but later group pictures showed us side by side, with like-minded people who enjoyed thinking and working together.

I was in the French club only during my first year. David did not take French, and we were not destined to go to a French-speaking country. I served with the staff preparing *The Commoner*, Bryan's Yearbook, in my junior year.

Eyes for the Blind: Kay

One of the students in David's class was blind. Carol Miesel was a good student, but she needed extra help. I would read textbooks aloud and help her do homework. I would even take tests for her, but she had to supply the answers. If I misspelled a word on a test, it counted against her!

Graduation Day: Kay

On June 6th, 1955, I graduated from Bryan College with a BA degree in Christian education and a minor in Bible. Now I was ready to put into practice all I had learned.

A 1946 Ford: Kay

I bought a used 1946 Ford sedan with low mileage from an older lady in our community. It had six cylinders and served us well later as our first vehicle.

Kay's graduation picture

First Teaching Job: Kay

Seventh grade at Reese Public School in Reese, Michigan, was my first teaching job. It was an easy nine-mile drive from my parents' home in Vassar. Several students were taller than I was, and a few parents thought I was a new student.

Bible reading was allowed in public schools at that time. Being an active Gideon, my father supplied me with a schedule of verses for reading each day. One Catholic student brought hers to school and commented, "My mother told me you wouldn't read out of the Catholic Bible!" I replied, "I'll be glad to use yours." That day, I read from her Bible. One significant difference is that the Catholic Bible includes the books of the Apocrypha added between the Old and New Testaments.

"Dear" Season: David

A big oak tree on Bryan's campus had mistletoe growing up high. I climbed up. The limb was shaking, but I was determined and got some fresh mistletoe for Kay.

In the fall of my junior year, Kay rode the Greyhound bus to Bryan. Her school was on break for Michigan's deer hunting season, and this was our chance to be formally engaged. Yes, it was "dear" season, and I asked her the big question. Of course, she said, "Yes!" This all happened on one of the campus benches at Bryan in the middle of November 1955.

Kay's students were all excited when I came to Michigan the following summer to work. Her seventh graders wanted to see who she was going to marry! College got out sooner than public schools, so I joined them on a field trip to Greenfield Village and the Henry Ford Museum in Dearborn, Michigan.

1956
My Family: Kay

The Temple family before I was married: back row: Thomas, Shirley, and Marilyn Kay; front row: Mother Vera, Father Robert, Sheryl sitting in Father's lap, Nancy, and James.

The Diamond: Kay

My aunt Flossie, Florence Madeline (Temple) Burridge, and her husband, Chester Frank Burridge, had no children. She died in 1953 before we were married and left me her ring as a gift to use for my engagement. We had it resized to fit my finger, and I had an engagement ring with my family history.

Preparation for Our Big Day: David

For our wedding, I purchased a blue suit in Saginaw, Michigan, and a matching pair of blue suede shoes. I felt in style. Elvis Presley had already sung "Blue Suede Shoes" and made the song famous before we were married.

Fast-forward many years. Every year at the Tanana Valley State Fair in Fairbanks, Alaska, an Elvis impersonator sings for "Senior Day." He always includes "Blue Suede Shoes." His song brings back laughs and good memories—even after 65 years of marriage.

Panic Attack: David

My new blue suit was carefully hung in a closet at Uncle Ed and Aunt Leola Erb's house, two blocks from the First Baptist Church in Vassar. Their son, David, borrowed a blue suit for our wedding. We were about the same size. David said that his suit was a little tight, but he would make it do for ushering people at church as well as standing with our wedding party.

The Temple house was a mile away, and I kept checking my watch for the right time to get dressed. I was the only one left in the Erbs' house as I got ready. But the blue suit didn't fit! It wasn't mine! I frantically called Kay. She tried to assure me that no one would hide or take my suit as a joke. I checked every room and closet in their house—but I couldn't find my suit!

Kay arrived at church and informed Aunt Leola. She immediately figured out the problem and called her son, who was already ushering people. He ran home, and we quickly changed suits. I hurried to the church and rushed in, realizing I was a little late.

Even the *Flint Journal*, in Michigan, had a news clip mentioning the bridegroom losing his wedding suit. At most weddings, there always seems to be something that doesn't go as planned.

Our Life Verse

We chose Psalm 48:14 to be our life verse: "For this God is our God forever and ever: He will be our guide even unto death" (KJV). We both made this awesome God of the Bible our God. Now, through the ups and downs of everyday life, our Guide is leading us to the end.

Our BIG Day

God teaches in the Bible that a Christian wedding means: one man + one woman = one lifetime together—"till death do us part."

The Vassar Pioneer Times printed the following article in June 1956 about our wedding:

"Mr. and Mrs. David Henry are traveling to Knoxville, Tennessee, following their marriage Friday evening at Vassar Baptist Church. The bride, the former Marilyn Kay Temple, is a daughter of Mr. and Mrs. Robert Temple of Vassar, and Mr. Henry's parents are Mr. and Mrs. Luther Henry of Oven Fork, Kentucky.

Three hundred guests witnessed the candlelight ceremony, which was performed at 8:00 p.m. by Rev. Peter Ypma before an altar banked with Schefflera and decorated with baskets of white snapdragons and white gladiolus. Mark Blossom was the organist, and Miss Ruth Ann (Kroff) Borduin-Bragg of Marlette was the soloist.

The bride, who was given in marriage by her father, wore a floor-length gown of lace and net over white satin. The fitted bodice featured a sweetheart neckline, and a pearl-covered bandeau held the fingertip veil in place. She carried a white Bible to which a spray of sweetheart roses and stephanotis was fastened.

Miss Shirley Temple, sister of the bride, was the maid of honor. Miss Nancy Temple, sister of the bride, and Miss Diane Erb, cousin of the bride, were bridesmaids, and Miss Miriam Henry, sister of the bridegroom, was the junior bridesmaid. All wore ballerina-length frocks of lace and net in yellow, aqua, pink, and blue, respectively. They had matching gloves and headdresses and they carried colonial bouquets.

Barbara Jean Temple, a cousin of the bride, wore a mint green dress for her duties as flower girl, while Philip Griffin of Reese carried the rings on a satin pillow.

Richard Henry of Chicago was the best man, William Robinson and David Erb, cousins of the bride, were ushers, and Tommy Temple, brother of the bride, was a junior usher.

The bride's mother wore navy blue with white accessories for her daughter's wedding, and Mrs. Henry wore green with white accessories. Both had corsages of pink carnations.

A reception in the church dining room followed the ceremony. The bride's cake was centered on the refreshment table, and sweetheart cakes decorated with white wedding bells were on either side of it.

Mrs. Ed Erb served the wedding cake, and Miss Irene Esckelson of Vassar, Miss Carol Fox of Pontiac, and Miss Janet DeGrow of Silverwood were assistant servers."

Our wedding party at the First Baptist Church in Vassar, Michigan, on June 8th, 1956, included Miriam (Henry) Llewellyn, Diane (Erb) Pedersen, Nancy (Temple) Horwath, Shirley (Temple) Spear, Kay and David, Richard Henry, Bill Robinson, David Erb, Tom Temple, flower girl Barbara (Temple) Johnson, and ring bearer Philip Griffin.

Postponed Honeymoon: Kay

After our wedding, we drove 600 miles to Knoxville, Tennessee. Shortly after we arrived, we received an urgent phone call from my mother. She wondered if we had plunged over a mountain or if something else had happened. Frankly, we were too occupied and didn't think about keeping anybody else in the loop. We didn't have any problems, either.

I enrolled in the required education classes at the University of Tennessee the following Monday. The classes were part of a six-week summer program to give me the proper credits to teach in Dayton, Tennessee when we returned for David's final year at Bryan. At that time, Bryan was not accredited.

David searched for a summer job. He was offered one to ride around on an ice cream truck. That wasn't him! A nearby grocery store offered him a temporary job in their produce department. The choices for short-term jobs were extremely limited. The interviewer at the grocery store gave him the job because he liked Bryan College. It was easy to find an apartment to rent during the summer within walking distance of everything.

Honeymooning on a Shoestring: Kay

As soon as I finished my six weeks at the University of Tennessee, we drove the 80 miles to Bryan to unload our extra baggage. We would be staying in an old trailer in "Trailerville" on the hill just behind the main administration building. It was small, cozy, and had just enough room for two newlyweds.

Then we left for a month to go to California and see a few sights. To conserve our cash, we camped for many nights. David's parents gave us a double-wide sleeping bag. We did not have a tent but were camping out in the open. One night we slept by a fence that was corralling horses. All night, they kept neighing and keeping us awake. We were too close to their fence, but we couldn't move to a new site since it was already dark.

Another night, we camped near a railroad track. We were abruptly awakened by an oncoming train in the middle of the night. At first, we thought it was headed straight toward us! No more camping by train tracks either.

In El Paso, Texas, we crossed the border into Ciudad Juarez to see a bit of Mexico for a day. We bought fresh peaches. The border agent said we couldn't take those into the U.S., and David asked if we could eat them on the spot. That is what we did while waiting on the bridge before crossing back to Texas.

We journeyed through New Mexico and Arizona and saw the Navajo Indians with their beautiful jewelry and handicrafts, as well as their flocks of sheep on the reservation. We took in outstanding sights of God's creation, such as the impressive Grand Canyon, the Petrified Forest, and Carlsbad Caverns with all the bats.

Death Valley: David

People recommended driving through Death Valley at night, but we wanted to see what was there. We had no air conditioning, and neither did anybody else. People hung a canvas water bag over the front of their vehicles to have cool water to drink. Our Ford had little wing windows on our front doors, so we turned them to blow the desert air directly into our faces. At every gas station or store, we bought cold pop. We made it, but it sure was hot!

California: Kay

California was beautiful. We saw David's brother and family, who lived in San Diego. Then we headed north and crossed the Golden Gate Bridge. In San Francisco, we walked together in front of a large hotel in the middle of the day. A bold man approached David, offering sexual services. We kept moving!

Heading to Michigan: David

We headed up through the southeastern corner of Oregon then cross-country to Michigan. One day for lunch, we stopped at a picnic area beside a gas station. The camp stove was on one side of the table while Kay prepared squash. We both sat down on the same side of the picnic table. Suddenly, the table flipped up! We and our stove ended up on the ground! The gawkers at the gas station had a good laugh!

Haircut: David

I had thick, curly hair and realized I needed a haircut before seeing my in-laws in Vassar. We stopped near a graveyard, and Kay proceeded to cut my hair with a pair of hand clippers. When we could afford haircuts, we stopped using those clippers that always pulled my hair!

Visiting Home: Kay

My parents were glad to see us. We had had no contact with them or anybody else while on our honeymoon. We picked up the rest of our wedding presents and headed back to Tennessee for David's senior year.

I wheeled Kay in a wheelbarrow (on the left) along with another couple.

Married Housing: David

Our welcome to "Trailerville" included the old tradition of celebrating marriage with a chivaree as it was getting dark. A loud, ragtag band of students accompanied us in a mock serenade, banging on pots, pans, or anything noisy.

Teaching Second Grade: Kay

I applied to teach in the Dayton School District. To our shock, the courses I had taken at the University of Tennessee left me three credits short of becoming a certified teacher! So I ended up teaching a second-grade class in Dayton as though I did not have a college degree. Consequently, I received a lower salary of 1,000 dollars a month. Nothing else was available. I enjoyed teaching but had to drop out for maternity leave without pay in the spring.

Africa, Here We Come: David

We prepared for our missionary journey during my remaining time at Bryan. We thought we were going to Africa. People who had asked about wedding gifts were told, "Don't give us blankets, we won't need them." It seemed like good advice, but we did not realize that we would be heading north to Alaska!

Looking for Help: David

The wide variety of great speakers included missionaries, Bible teachers, preachers, and other Christians at the daily chapel services. One time, I sought further counsel from a speaker representing a well-known mission working in Africa. The man told me they had certain days when they shut their compound gates and did not allow any Native visitors. I didn't know how to respond, so I thanked him for his time and walked away.

I may not have understood their situation, but we did not want to shut out local people. There are other ways to get away from it all, as Jesus did. I wish I would have told him that was a huge "turn-off" for me. We see all people as being created in the image of God.

1957

A Linguistic Focus: David

A representative from Wycliffe Bible Translators came to Bryan and gave a short workshop on using linguistics for Bible translation. We were impressed and interested. Several missions were doing translation work in Africa with gifted missionaries, but Wycliffe had no work there at that time.

I had taken several linguistics courses at Bryan. Kay was also fascinated with different languages. As a young person, she would study and look at the foreign words offered by various missionaries.

A Baby Girl!: David

We sped 40 miles to the Memorial Hospital in Chattanooga, Tennessee, when the contractions started. I asked if I could watch the delivery. The nurses laughed and calmly said, "No, we don't allow husbands in the delivery room. Just go to the waiting room, and you will be the first one to see your baby."

Of course, Kay was the first to see Debbie. Then a nurse showed me newborn Debbie. I was a little shocked since she was very red and kept stretching out her arms and legs in all directions.

We both experienced various emotions leading up to her birth. An earlier medical checkup revealed an ovarian cyst about the size of a grapefruit. The doctor said it must be removed before giving birth. So Kay had an ovarian cystectomy and oophorectomy, requiring the removal of one of her ovaries and a fallopian tube when she was six months pregnant. The doctor gave her medicine to prevent premature birth. We had more questions than answers, but we waited on the Lord for the right time for Debbie's delivery.

Praise the Lord, our first child, Deborah Kay Henry, born in April, was a healthy baby girl! We named her after Deborah in the Bible in chapters four and five of Judges. Deborah was known for her wisdom, courage, faith, and action as a judge in Israel during troubled times. We gave her Kay's middle name.

Cry Baby: Kay

Debbie was born shortly before David's senior finals. Study time was precious, but Debbie had colic. Sometimes David would put her in our baby buggy and give her a bouncy ride around campus, trying to get her to quit crying. Several girls told me they felt sorry for our new baby, but no one offered to try to comfort her. David made it through his finals as a pleased father.

My Major—Biblical Greek: David

I majored in biblical Greek under Dr. John Anderson. It was challenging! It aids in reading commentaries and authors who explain things using Greek. I enrolled in all the Bible courses. I especially liked and did very well in the inductive Bible study under Dr. Irving Jensen, who authored many books. My other classes included Bible, Christian education, Hebrew, and history.

Bryan College (1951~1957)

I held Debbie as we stood between my parents on graduation day.

We Recommend Bryan

We like Bryan's motto: "Christ Above All." It reflects the college's direction and purpose and guides us in our daily lives. Since we had found a good Christian college, we also urged others to go there.

The following 21 family members and relatives either graduated from or attended Bryan College. Those who graduated are listed with a degree after their names and the person's relationship to us.

1938 Rev. Lewis Llewellyn Sr., BA, David's brother-in-law's father
1954 Rev. John Rathbun, BS, our daughter's father-in-law
1954 Joyce Rathbun, BA, our daughter's mother-in-law
1955 Marilyn Kay Henry, BA in Christian education
1956 Pearl Rathbun, BS, our son-in-law's aunt
1957 David Cameron Henry, BA in Biblical Greek
1957 Shirley Spear, Kay's sister
1962 Nancy Horwath, Kay's sister
1965 Miriam Llewellyn, BA, David's sister
1966 David Llewellyn Jr., BA, David's brother-in-law

1969 James Temple, Kay's brother
1969 Thomas Temple, BS, Kay's brother
1980 Daniel Henry, BA, our son
1980 Deborah Rathbun, BS, our daughter
1980 John Rathbun, BS, our son-in-law
1980 Joel Rathbun, our son-in-law's brother
1985 Denise Phipps, Kay's niece
1995 Christopher Llewellyn, David's nephew
2014 Joshua Rathbun, our grandson
2014 Danielle Reynolds, BA, our granddaughter
2015 Isaac Reynolds, BA, our granddaughter's husband
2017 Isaac Reynolds, MBA, our granddaughter's husband

Beyond College (1957~1958)

Moving to "Big D": David

Upon graduation, we packed up like everybody else. This time we bought a small two-wheel trailer to pull behind our car and moved 800 miles to "Big D" (Dallas, Texas). It was an incredible sight to be driving west across Texas and see the city's skyline emerging out of the flatlands. I enrolled at Dallas Theological Seminary. One of their housing units at 4422 Sycamore Street was within walking distance of the seminary.

Scofield Memorial Church: David

Scofield Memorial Church on the corner of North Carroll and Swiss Avenues was a two-minute walk from our apartment. A former pastor was C. I. Scofield, a well-known Bible teacher famous for the popular Scofield Reference Study Bible. I used one of these Bibles when I was a teenager. Kay worked in the nursery.

The congregation numbered 250 people, with a maximum building capacity of 300. When the church grew, they would start another one in the Dallas area and urge members to join the new church plant within their region to help them grow. We liked this idea rather than building a more prominent building.

Dallas Electric Company: David

Residential work with the Dallas Electric Company kept our crew of eight busy. I enjoyed working with one of their journeymen electricians. To see if a wire was hot, he would wet his fingers and touch the wire. I still jump when I get shocked by 110-volt electricity. I purchased a three-quarter-inch wood bit 18 inches long. I drilled many holes with that bit using a hand drill.

One recurring work order was to add more electrical outlets to a home. Since I was a new employee, I got the "honors" of crawling under a house to run the electrical wires. There was usually enough room to crawl, but there were always lots of cobwebs. I did not like this part of my job, but work was work.

Yes, a Scorpion!: David
We had a lovely baby crib for Debbie set up in our kitchen. One day, we found a two-inch dead scorpion under it! To protect Debbie, we took four tin cans, removed the paper labels, and put a crib leg in each can. Now, no scorpion could climb into her bed.

How did it get there? We guessed I may have picked it up while crawling under a house. Afterward, I would check my pants more carefully before coming home to ensure I didn't bring any unwanted guests.

The Box Factory: David
In the fall, I resigned from my electrical job to attend seminary classes. Then I worked on the evening shift at a box factory during the winter. The work mainly involved sorting, gathering, and bundling printed boxes for various companies. This job cut into my study time but allowed us to keep up with our expenses.

1958
Profs: David
The professors at Dallas Theological Seminary were outstanding. Dr. Merrill Unger taught the Old Testament as well as biblical archaeology. Dr. John Walvoord was the president of the seminary. He was a profound thinker and taught theology. Dr. Charles Ryrie was very concise, clear, and down-to-earth in teaching systematic theology.

Several students switched their majors to get into classes with Dr. Howard Hendricks. He promoted solid Christian leadership with a shepherd's caring heart for more than 60 years. He was just beginning his teaching career when I was there. One of his classes was about preachers who major in minors, going on and on about unimportant information. He stated, "It's like shooting canaries with cannonballs!"

Rev. John G. Mitchell, who founded the Multnomah School of the Bible in 1936, taught the book of Matthew for one whole week. He was an excellent teacher who taught from his heart, not using notes or modern-day aids. Someone asked if he had memorized the book of Matthew, to which he responded, "No, I just read through the book of Matthew once a day to familiarize myself with God's Word. It becomes part of my life." He was a real "Lion of God" and taught the Bible for more than 60 years.

More or Less Study

The seminary offered a basic four-year course. I liked the classes but decided I didn't need all four years. Since I majored in biblical Greek at Bryan and took all the Bible courses there, I chose to end my seminary training after one year. No, I did not cram four years into one year either.

Linguistics Courses

We enrolled in the first-year course with the Summer Institute of Linguistics (SIL) at the University of North Dakota in Grand Forks during the summer of 1958. The intense eleven-week course specialized in the linguistic sounds of any language in the world. When a new, strange-to-American-ears sound was introduced in class, you could hear it repeated almost anywhere on campus.

The last two weeks included practical training with local Sioux Indians. We worked one-on-one with them, collected language data, and processed the information. The course taught us how to reduce spoken language to written language and develop a literacy program with written materials.

More than 100 students were training to join or were already members of SIL. First-year students like ourselves were the largest class. Others were taking second-year or advanced courses working with language material they had gathered. A few people from other missions were training with us.

Junior Members

We moved back to Vassar and were accepted as Junior Members of Wycliffe Bible Translators, Inc. (WBT) and SIL on September 26th, 1958. The next step was to build our team of prayer partners and financial supporters. Both are vital to the success of the Lord's calling. We needed to move quickly to participate in the Arctic Training Camp in Nenana, Alaska, which was only available that winter. We kept up the momentum, packing, praying, and trusting the Lord to fill all of our needs in His timing. Kay was also expecting our second child.

Danny Boy: David

In October, Kay went into labor. We drove 21 miles from Vassar to Saginaw General Hospital (now part of the Covenant Health Care System). Our son Daniel Cameron Henry was born at 11:42 a.m., a healthy baby boy weighing seven pounds and 13 ounces.

We named him after Daniel in the Bible, a man who faithfully followed the Lord all of his life, often under extreme pressure. He was an excellent example to follow. We gave him my middle name, Cameron, the family name on my grandmother's side of our family.

The doctor advised waiting until Danny was six weeks old before taking him on our long trip to Alaska in the dead of winter. We had many things to keep us busy as we stepped out in faith into the unknown.

Commissioned to Go

The First Baptist Church in Vassar, Michigan, had an exceptional commissioning service on October 23rd before sending us to Alaska. The not-to-be-forgotten service included both of our fathers fulfilling an important role. Rev. Peter Ypma led the service.

David's father challenged the congregation to pray for and support us. Pastor Ypma gave us the charge to follow through with God's calling on our lives. Then they laid hands on both of us. Kay's father and Pastor Ypma prayed for David as he kneeled. Then Kay kneeled, and David's father and Pastor Ypma prayed for her.

Generous Giving: David

An incredible reception followed the service. God's people gave us various gifts for our work in Alaska. The "love" offering that night was over 800 dollars. That was back in 1958! No wonder I was overwhelmed as I told Kay, "That's more than I earned in a whole year working at Bryan!"

The Adult Sunday School classes gave both of us beautiful Eddie Bauer down parkas rated down to 40 degrees below zero. The hoods on these parkas had outer ruffs of gray wolf fur and inner ruffs of darker wolverine fur next to one's face. Wolverine fur is more frost-resistant than most furs and doesn't hold the snow in its hair. They served us well for many years, even at temperatures lower than minus 40 degrees. A couple of layers of clothing underneath added extra warmth.

Church members Leon and Helen Siver owned the Western Auto Store in Vassar. Leon took me to his store basement and offered me any gun I wanted. I chose a 30.06 pump-action rifle with a clip holding four shells. What a blessing! I've shot many moose plus caribou and bears, with that gun over the years to feed our family.

Our Home Churches

Kay's home church, First Baptist Church, was in Vassar, Michigan. She was baptized when she was 16, joined the church, and grew spiritually. Calvary Baptist Church, in Bradenton, Florida, was David's home church. At age 13, he was baptized and grew in the Lord.

After we moved to Alaska as missionaries, we planned to live here for the rest of our lives. Fairbanks was our base for serving the Koyukon people in Alaska and later in Siberia, and we joined Bethel Baptist Church in Fairbanks as our home church.

These churches generously supported us with their prayers, finances, and encouragement. They functioned as our "sending" churches, working closely with us and providing a level of accountability and direction in doing the Lord's work. As we aged, the local Fairbanks Native Bible Church became our final home church.

More Outstanding Backers

We deeply appreciated the same encouragement and support from other churches that had become family to us over the years. Churches in Michigan included Dayton Center Church in Silverwood, First Baptist Church in Flushing, Flint Baptist Temple in Flint, New Hope Community Church in Williamsburg, Rich Bible Church in Mayville, and Tuscola Community Church in Tuscola.

Outside of Michigan, Faith Bible Baptist Church in Egg Harbor Township, New Jersey, supported our work. Several other churches supplied our unique needs and projects. God also raised faithful individual encouragers and supporters, which rounded out our special backers.

North to Alaska
The Koyukon People

Arctic Training Camp: Nenana (1958~1959)

North to Alaska!

We drove a brand new 1959 Chevy Bel Air to Alaska for a dealer from Detroit but had to relinquish it to the Chevrolet dealer in Fairbanks upon arrival. The company in Detroit paid us a 100-dollar delivery fee for gas and one oil change en route. The gas and oil change cost 110 dollars, so the Lord provided inexpensive moving costs with another 100 dollars spent on motels and roadhouses. We added 4,400 miles to the odometer. The shipping expenses to Alaska were much higher. Later, the car companies equalized the new vehicle transportation costs to Alaska or anywhere in the US.

The Chevy Bel Air sedan was full-size with lots of room. Our personal effects fit in the back on the floor, so Debbie and Danny had the whole back seat area in which to sleep, play, and move around. No seatbelts in those days! The trunk was spacious, and we tied a small table and baby crib on top. We liked the wing-shaped tailfins, a distinct mark of that year. Now, north to Alaska!

We left Vassar on November 19th, 1958. It was warm in Michigan, but we hit minus 35 degrees at one motel in northern British Columbia. We had no problems starting the following day, even without plugging in the car for the night since we got up early. (Cars in the north have since added electric heaters to preheat the engine for the extreme cold.) The roads were very smooth along the Alcan Highway, with enough packed snow to cover the gravel and fill in the potholes. This trip was before they paved the road.

Arctic Training Camp: Nenana (1958~1959)

At Arctic Camp

We arrived in Nenana (neh-NAN-uh), Alaska, 60 miles south of Fairbanks, on November 29th, 1958. We crossed the frozen Tanana River on an "ice bridge," a safe area of thick ice indicated by 55-gallon drums. Vehicles were ferried across the river during the summer. The current bridge that spans the river and speeds up transportation was built in 1968.

Nenana had a mixed population of predominantly Native Indians plus many non-Native Alaskans, numbering about 275. It is situated at the confluence of the Nenana and the Tanana Rivers. Wycliffe's Arctic Training Camp was held in simple, unimpressive cabins belonging to the Yutana Barge Company on the Nenana River. Their barges were dry-docked for the winter. The Alaska Railroad runs via Nenana on the route between Anchorage and Fairbanks.

The Arctic Training Camp started in October, but it continued throughout the winter months. George Fletcher, the camp director, with his wife, Alda, and their daughter, welcomed us.

Camp participants who went to Eskimo groups were:

Dave and Mitzi Shinen and their children went to the Siberian Yupik, in Gamble.

Don and Thelma Webster, to the North Slope Iñupiaq, in Wainwright.

Wilfred and Donna Zibell and their children to the Kobuk River Iñupiaq, in Ambler.

Camp participants who went to the Athabascan Indian and Tlingit groups were:

David and Kay Henry and their children went to the Koyukon, in Kokrines.

Paul and Trudy Milanowski and their children went to the Upper Tanana, in Tetlin.

Dick and Susan Mueller, to the Gwich'in, in Arctic Village.

Three single women: Con Naish, Gil Story, and Shirley (Gerbaulet) Frye to the Tlingit Indians, in Angoon.

Herb Zimmerman worked with the Dogrib Athabascans in Canada.

Our Activities

At Arctic Training Camp, we spent our mornings in classes and afternoons in outdoor activities. We learned basic Arctic survival skills, how to run a dog team, and walk in snowshoes. We had general help for language learning with Athabascan languages. *My Life with the Eskimo* by Vilhjalmur Stefansson was high on our reading list and generated many discussions. The author included information about his time with the Eskimos and eating every part of an animal to survive. We prayed for each other, especially regarding where members should start their language work. The mothers took turns caring for babies and younger children.

1959
Alaska—the 49th State
The US purchased Alaska from Russia in 1867. On January 3rd, 1959, President Dwight Eisenhower signed a proclamation admitting Alaska to the Union as the 49th state. Celebrations were held in Fairbanks, Anchorage, and other cities, but we were too busy adjusting to our new life. One gas station in Fairbanks sold gasoline for 49 cents a gallon.

First Aid: Kay
We discussed accidents reported in the news, deciding the best course of action to take if we were in such an emergency. We practiced first-aid methods. Since I was the smallest, I was carried home on a stretcher with a "broken" leg.

Frostbite, a real problem in Alaska, and how to avoid or deal with it, were discussed. Many advocated putting snow on the frozen skin, but we learned to warm the frozen area slowly with the palms of our hands.

Staying Warm: David :
Nenana is in a cold region of Alaska. Local people thought we were "arctic testing" the Eddie Bauer parkas, but we were just trying to stay warm. I went on an overnight camping trip with the men when it dropped to negative 50 degrees. Five of us headed across the Nenana River on snowshoes and disappeared into the woods. We were heavily loaded with our backpacks, sleeping bags, and supplies.

Each of us made our own shelter. We cleared away most of the snow and put down a generous layer of small spruce boughs on the ground under a lean-to shelter of willows with larger spruce branches overhead. I had a five-pound down sleeping bag but soon realized the only part of me that stayed warm in the sleeping bag was covered with my down parka lying on top. In the middle of the night, several of us got up and walked around to get warm. We were learning survival skills for living in the northern climates wherever God placed us.

Women's Overnight Campout: Kay

I went on a women's trip to an island in the Nenana River. I carefully arranged split firewood, all ready to light the following morning. It was warmer on the women's trip, but we woke up to eight inches of fresh snow. A mound of snow showed me where I had carefully set up for my morning campfire!

Donna Zibell and I started back ahead of the rest of the women to nurse our babies. We couldn't find our well-worn path crossing the river with all the new snow! But the other ladies helped us find the trail.

Target Practice: David

We learned about guns—a necessity for hunting and living in the "bush" of Alaska. I had hunting experience growing up in the country and on a farm, but Kay did not. She took her turn hitting the target.

During the winter, Don May, Don Sauer, and John Miller from Denali Bible Chapel in Fairbanks shared meat from their successful fall moose hunt. We received a steak from the hindquarter, and it was an inch and a half thick, 18 inches by 10 inches!

Baking Without All the Ingredients: Kay

The local store was out of eggs. I substituted one teaspoon of baking powder plus a quarter cup of water for an egg. It worked but it lacked natural nutrients.

Ice Fishing: David

I set my bait, a piece of fresh fish, on a hook under the ice and tied the line to a stick to leave overnight. The following day, I reopened the hole and pulled out a 15-inch burbot. The temperature was cold, and I suffered frostbite on one earlobe. My ear was super sensitive to the cold for the rest of that winter.

Housing: David

Our cabins were formerly single-standing Army prefabs, 15 feet by 15 feet, with an entry door and a potbelly stove in the middle for burning coal or wood. When we stored things against the wall, they would freeze to it. The nail heads in the prefab wall would frost two or even three feet up from the floor during an extreme cold spell. The frosted nail heads worked like a thermometer, indicating how cold it was outdoors.

One time, I walked to Nenana about a mile away, and Kay got cold. She started to panic and threw in several pieces of cardboard to get the stove going. Then Kay told herself, "Settle down. Don't panic. The kids won't freeze, and you will get this fire going." And she did.

An Athabascan Believer

It was thrilling to meet a local Christian believer, an Athabascan teenager, Paul Starr. He had accepted the Lord shortly before we arrived in Nenana. Paul would often come by our cabins, driving his team of four dogs.

Paul gave both of us our first opportunities to drive his dog team. Thankfully, his dogs were friendly. We hollered, "Gee" (turn right) and "Haw" (turn left) to the lead dog. But shouting "Whoa!" and holding down the brake on the back of the sled was more necessary.

Paul is running with four dogs. Our cabins are pictured in the background.

Meeting Paul enabled us to meet his mother, Elizabeth Starr, and sister, Anna. We tried to learn a few Athabascan words with Paul's mother. Walking around Nenana with them, we learned about the Native way of life and culture.

Arctic Training Camp: Nenana (1958~1959)

Nenana Bible Church

We attended the Nenana Bible Church, a small mission church started by Arctic Missions, later renamed InterAct Ministries, Inc. Pastor Mel Jensen and his wife, Pat, were serving there.

The Old Red Truck: Kay

Our mission purchased a well-used red truck from Ralph and Ruth Daily, former missionaries in Nenana with Arctic Missions. Several of us drove into Fairbanks and looked for a property for Wycliffe's headquarters in Alaska.

When we returned from Fairbanks, the ice road was barricaded. We parked the truck and started to walk toward the train trestle to cross the river. Another truck approached us from behind. It was with the Federal Aviation Administration, and they told us, "Go ahead and drive on the ice road; we'll follow you." We plunged into water that came just over the running boards and made it!

In the spring, the water got pretty deep along the edges of the river. The ice in the main part of the river would rise with the runoff from side streams and melting snow. At first, the anchor ice held the main river ice to the shore as it bent, making a concave area with a foot or so of water before the river ice finally broke loose with the rising water.

The Tripod: David

Nenana is home to the annual Nenana Ice Classic, a money pool contest where people guess the exact time the tripod set in the Tanana River will move. I have never won, but then again, I have never lost any money either. The highly visible black-and-white tripod moved as the ice broke at 11:26 a.m. on May 8th.

Cleanup at Camp: David

In the process of cleaning camp, we found a can of chocolate-coated roasted grasshoppers. We weren't anxious to taste them or see if we liked them. So I said, "Let's send them to my parents as a joke." We included them in a small package. My parents remarked, "If Dave and Kay eat them, we will give them a try."

First Glimpse of the Mighty Yukon River: David

A friend at Denali Bible Chapel, Al Wendt, offered to fly me to Kokrines, a small Athabascan village about 165 miles from Nenana. He had a two-passenger plane on skis with plenty of power. We picked a lovely spring day during the first week of April. I was amazed—so much land, so few people! We passed only one village named Tanana (TAN-nan-naw), with 250 people at the confluence of the Tanana and Yukon Rivers.

Kokrines

We landed on the smooth, snowy landing strip on the frozen Yukon River, directly in front of Kokrines. Small green spruce trees marked the designated area. The abandoned Bureau of Indian Affairs' two-story white school stood out among the village cabins. We were welcomed by Larry and Mickey Scripter, who served with Arctic Missions, and their three children.

Kokrines was a beautiful location for a village, nestled in front of the Kokrine Hills on the north side of the Yukon River. It was on high ground, so there was no concern about flooding. A little creek by the Scripters' house provided drinking water. Only 15 miles upriver was the Novi River, with many moose and other wild game. Kokrines has a long Native history.

The Scripters planned to move to Talkeetna, Alaska, and serve there. Many in the village had already moved downriver to Ruby for schooling and employment opportunities. The school had been closed for several years.

Renting on a Shoestring

The Scripters rented their cabin to us for 15 dollars a month, plus they gave us good deals on buying their furniture. Larry snagged 30 drift logs coming down the Yukon right after the breakup. We appreciated the firewood.

Making the Move

We decided to move to Kokrines on Monday, June 1st, 1959. Wien Air Alaska was very cooperative and picked us up with our freight at 7:30 a.m. in Nenana. We flew directly to Ruby.

The twin-engine Beechcraft 18 had plenty of room for the four of us plus our baggage. When we landed in Ruby, the baggage ties came loose, dumping our things on the floor and blocking the exit door. The pilot leaned out his window and asked people on the ground to open our door and move the baggage out of the way so we could get out.

Ruby

We met Russ and Freda Arnold and their children, who were serving under Arctic Missions. They were knowledgeable about Kokrines since they had temporarily stayed there while building their home in Ruby. We saw more of them in the years to come.

Native Hospitality

Claude and Bertha Demoski invited us for lunch in their cabin. It was good to meet them and see their interest in us and the Koyukon language. One always looks forward to seeing people again, especially while living in the same region.

Thirty Miles to Go

Larry Scripter was ready to go as we climbed into his boat with our baggage. The 25-horsepower outboard motor propelled the small boat quickly along the 30-mile journey upriver. We were anxious to meet the people. Everything was new and exciting for our family.

God gives each of us One Lifetime

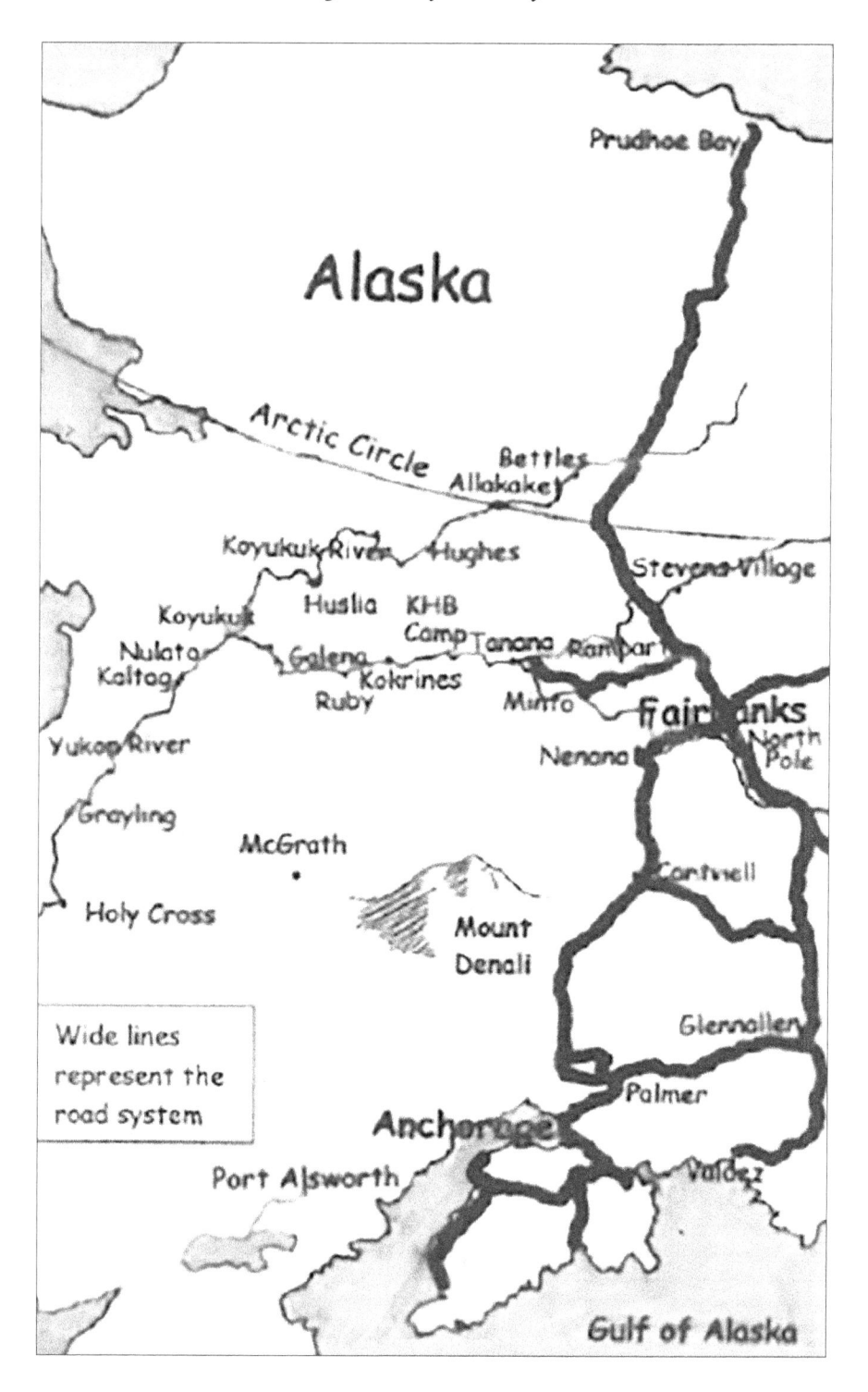

Wide lines represent the road system

84

Yukon River: Kokrines (1959~1960)

Welcome

Everybody in Kokrines stood on the bank, watching our little boat come up the Yukon River. They knew that our family was coming and would be living in the Scripters' one-story log house, measuring 20 feet by 30 feet, with a basement half that size for storing things.

The village ran on "Indian Time!" Life was event-oriented rather than going by the clock or "White man's time." Today's event was the Henrys' arrival. June was a busy time to prepare for the king salmon run, with plenty of good eating plus preserving them for the coming winter.

New Friends

The Natives usually came to visit us during the afternoons and evenings. Mornings were quiet, but people in their boats and occasional barges traveled up and down the Yukon River any time of the day or night.

Our Prayer

"Lord God, thank You for a safe move to Kokrines. Help us point people to You and live for You. You are our Creator and our only hope to help us as we live in this world full of sin. Amen."

Koyukon 101: David

In Kokrines, most adults were bilingual in both Koyukon and English, but the middle-aged and elders preferred Koyukon. I asked Mary Jane Philips, "How do you say, 'Come in?'" She responded, "Onee', bedeenee." So that's what I told the next person who knocked on our door. She came in and smiled. Soon somebody informed me only to say, "Onee'," meaning "Come in." "Bedeenee" means "Tell him or her."

Transportation

We purchased the 16-foot runabout boat and outboard motor from the Scripters. The boat was fiberglass with a few dings, but it was light and fast, riding almost on top of the water for more speed. Thankfully, they let us buy things "on time," but we did pay them off by the end of the summer.

Summer Mail: David

Larry Scripter had made weekly mail runs to Ruby. He asked if I would like to become the mail carrier. Like the Scripters, we would receive more first-class mail than the other residents. The payment was 10 dollars per trip. One could also shop at the Northern Commercial Store on Ruby's waterfront or do whatever business needed to be accomplished.

I thanked Larry for the offer but told him, "I would rather see a Native person have the job." Gilbert Cleaver ended up running the summer mail. He enjoyed getting out of Ruby and seeing whoever was along the Yukon River. He also brought news from Ruby.

During the summer, we received weekly service. Gilbert brought a mailbag, plus any boxes to our house and then picked up outgoing letters. Locals would stop by for their mail. It was a good opportunity to visit everybody who hoped to get a piece of mail. We provided a free service.

Volleyball and Rhubarb: David

In a small village, everyone wanted to play volleyball and was needed to make it more competitive. I noticed people kept going into the tall grass and getting something to eat during the game. The wild rhubarb, or *ggooł*, was ready to eat. The wild version tasted similar to the traditional one and was sweeter. One picked it when the green stalks were a foot or so tall. Then you peeled back the base of the stem and got a couple of inches of green, sweet rhubarb. But it was only good in the early spring before it got tough and went to seed. There was no local store to buy any snacks, so we all enjoyed the natural food.

A Child Leaves Us: David

Louise Albert was a frequent guest. She was a friendly teenager, taking care of her father and stepmother's little four-month-old baby girl named Kitty.

Unfortunately, the baby had convulsions. On June 16th, Louise cried to Kay, "I think Kitty died!" Kay hurried over and saw the baby, who was already stiff.

Louise's father, Philip Albert Sr., and his second wife, Justine, were upriver cutting firewood and making a raft of seasoned wood to take down to Ruby.

The other men were not available, so I took Frank Titus as my river guide to find the Alberts. We located them and passed on the sad news. I was new to the Yukon River and using an outboard motor, but God protected us.

The Kings Are Running!: David

When the king salmon are running in the Yukon River, everyone gets excited. On July 2nd, Gilbert and Zeta Cleaver asked, "Would you like to stay at our fish camp for a couple of days? We're going down to Ruby."

"Yes, sure, but we've never been to a Native fish camp!" The next day, we took our boat and gear to "Big Eddy," five miles downriver. Gil and Zeta left a half-hour after we arrived. There was not enough time for many questions. We learned the Native way: asking only a few questions but learning more by observing how they cut fish. Then practice, keep practicing, use common sense, and do the best you can.

The 30-foot king salmon gill net had an eight-inch mesh, so we caught only the big ones and let the baby salmon go and grow. The mesh size was standard for king salmon nets in 1959. Most of the kings caught were at least three feet long.

I would go out in our boat and harvest the kings. We would typically get 10 salmon at a time. As soon as I landed, I could see the floaters bobbing again, indicating that we had caught another one in the net. But that king would have to wait. I was too busy cutting fish and all there was to do at the camp. What a generous blessing from God, our Creator. Almost more kings than we could handle. Not much time to sleep.

We had a big pressure cooker/canner, tin cans, and a sealer. We would prepare 10 one-quart cans at a time. Later, we found out we had not sealed every can. That winter, we stored the canned salmon in our basement. Soon, we heard a can explode! There were bits of salmon all over and a strong, sour, fishy smell. Then I checked all the cans for bulged ends and threw them out.

We cut many salmon and hung them in the smokehouse to cure. We tried cutting strips, but it required extra time and patience. This was the only time in our lives we got tired of eating king salmon!

Kay washed cloth diapers by hand with a scrub board since it was before the age of disposable ones. We kept the smokehouse fire going, plus we fed and watered their dogs. A loose dog pulled clothes off the clothesline, and we lost a pair of Debbie's yellow pajama bottoms. It was a long, super busy week, but we would be glad to do it again.

Debbie pointing out the kings

A Fish Wheel: David

Larry Scripter rescued a fish wheel that came down the Yukon River with the ice during the breakup shortly before we arrived. One basket was damaged, but the other was good.

I partnered with Al Hardluck to repair and set the wheel more than a mile below Kokrines. It was fun to see the water current turn the wheel and watch fish drop into our box.

The first time Kay went with me to check for fish, we took Debbie and Danny along with us. She was so anxious and curious to see everything that she stepped out of the boat onto the platform with me. Suddenly I yelled, "Grab the boat!" We hadn't tied it to the fish wheel. Later, she had dreams of what could have happened to our two precious children in the boat alone, drifting downriver. We thanked God for His protection.

Yukon River: Kokrines (1959~1960)

Keeping a Diary: Kay

I started keeping a diary of daily events until life got too busy. Josephine (Adamon) "Koda" Corning kept a diary on her calendar. She could not read or write. Instead, she drew a picture of a boat on the river, meaning a visitor came; a rabbit, meaning she snared a rabbit; or a grouse, showing that she was given a grouse on a certain day.

Native Name for Kokrines

The Koyukon name for Kokrines was *Bek'edeneekk'eze Denh* (the place of red ochre). The earth-tone red clay pigment was used to stain wood.

Fast-forward: George Albert became a master craftsman, using birch to make genuine Native snowshoes. He would also color them with red ochre upon request.

Gardens: David

The Scripters had two garden plots. One was on the bank overlooking the river, and a bigger garden was behind their house. They had planted a variety of vegetables, which were doing well by the time we arrived. I worked in the garden pulling weeds until 11:00 p.m., forgetting about the time because of the long hours of sunlight. June 21st, the longest day, gave us almost 22 hours of direct sunlight, and the twilight was not dark. One surprise was a frost that nipped our zucchini squash on the Fourth of July.

An Elder Passes: David

On July 14th, I went to Ruby to get a few supplies. I traveled with Gilbert and Dave Corning, a village elder. Dave waited overnight in Ruby to connect with Thursday's mail plane to fly to the Public Health Services Hospital in Tanana. He passed away 11 days later in Tanana.

Hardy Peters, Dave's stepson, and his wife, Helen, came by boat, 90 miles from Tanana. They brought Dave's body to be buried in Kokrines. Franklin Albert and the new Episcopal priest, Curtis Edwards, came with them for the burial rites.

Lost Two Grandparents: Kay

I lost two grandparents in July 1959—my paternal grandfather, Lewis Peter Temple, and my maternal grandmother, Edith (Maier) Wall.

Both lived near me while I was growing up. When we left Michigan, I realized there was the possibility of never seeing my grandparents again. I grieved along with my parents.

We need to spend time with and enjoy one another while we can. Only God knows when we will take our last breath. I'm thankful both of them had asked Jesus into their lives and were living for Him. That gives me hope of seeing them again in heaven.

100 Pounds of Popcorn

What do you do with a 100-pound bag of popcorn? You eat it, one kernel at a time! It took our family five years to eat that much! When ordering and comparing prices, we found that the 100-pound bag cost less than smaller quantities in nice little packages, so we ordered the big bag. We still like popcorn!

Moose around Kokrines: David

Every morning, I would go out in front of our house and look up and down the river. We could see a long way, and I wanted to see what was new. Sometimes a moose or black bear would swim across the Yukon. The Yutana Barge Lines regularly passed by with a full load headed to Galena Air Force Base. In the fall, I took Allen Hardluck up to the Nowitna (Novi) River to hunt for moose in my boat.

Telegraph Station

Kokrines served as a telegraph relay station shortly after 1900, installed by the U.S. Army Signal Corps. Technology keeps changing, and the system was abandoned in the 1930s. They left hundreds of miles of heavy-duty wire, which has helped many locals, including us, who used it to make a sturdy cable to anchor fish wheels.

A former, ambitious missionary had used the cable to string an antenna more than a quarter of a mile long behind the village. We tied our battery-powered Zenith transoceanic radio to the antenna. We could even hear HCJB, a Christian shortwave radio station in Quito, Ecuador. The radio worked well, but it used eight size-D batteries.

First Black Bear: David

I went upriver a couple of miles in our boat to get a load of firewood. As I was cutting driftwood, I noticed something coming across the Yukon River straight toward me. It looked like a log swimming, but I knew it wasn't.

The black bear landed 50 feet from me and headed straight for the woods. I took him down in one shot with my 30.06 rifle. The fat on his back was three inches thick, so I saved it. His stomach was loaded with blueberries and smelled sweet. The meat was tender and tasty.

We cut the fat into small pieces, which Kay rendered, filling several gallon containers. She substituted bear fat for Crisco when baking a pie crust. Once Kay went with Sophie Peters to check her traps. She had bear lard, which she spread like butter on her crackers.

Tundra Topics

Every evening at 9:20 p.m., except on Sundays, we would turn our battery-powered radio to KFAR in Fairbanks and listen to *Tundra Topics* for 15 minutes of vital messages. This was the only way to pass information about a baby being born or how a person was doing in the hospital in Fairbanks. Listening unified the interior of Alaska as everybody heard the same program.

Let There Be Light: David

Village generators, or even personal ones, were not available, so we used kerosene-burning lamps with Aladdin mantles in the winter. The mantles were fragile, so we had to be careful not to bump or touch them. They burned quietly and lasted a long time. A Coleman lantern with Blazo white gas was brighter but noisy, and the gas pressure needed to be pumped periodically.

Let Your Light Shine

We did not agree with a Wycliffe policy discouraging members from sharing the gospel until one could share it in the Native language. The Koyukon language was challenging to learn, plus the young people knew English much better than their language. Historically, the school system, churches, missions, and the government strongly discouraged the use of their Native language.

People knew we were Christians by our actions and how we lived. We continued to have our personal Bible study and family devotions with our two young children. One of us would read aloud a Bible story and comment on it, and then we kneeled by one child's bed and prayed together. All the other members from Arctic Training Camp went to villages with an existing church where they could fellowship.

Moving Away

Gilbert and Zeta Cleaver had moved to Ruby early that spring with their four children. Grandma Bob, Zeta's mother, moved with them. Later in the summer, Philip and Justine Albert and their four children moved. Living in Ruby allowed people to be on call for work in Fairbanks, Galena, or elsewhere.

The Bureau of Indian Affairs closed the Kokrines Grade School in the mid-1950s. Consequently, the school-age children were sent to Wrangell Institute in Wrangell, Alaska, and the high school students to Mount Edgecumbe High School in Sitka, Alaska. Both schools were far away. This experience proved to be harsh and brutal for many of the students.

A Trip to the Hot Springs: David

Frank Titus and I walked 16 miles to Melozi Hot Springs in the fall. He led, walking at his fast pace, and then he would stop turn around and wait for me. As soon as I caught up, he would take off again. There was no time to rest! I was younger than Frank, but he had done a lot of hiking in his lifetime and was in great physical shape.

We came across a black spruce tree two inches in diameter that a grizzly bear had broken down. We could see telltale pieces of brown hair left on the tree.

The hot spring facilities were "ancient history." The old cabins that Natives and reindeer herders used in the past were decrepit and falling in.

According to official records, the collection pool was too hot at 131 degrees to dip our tired feet into the mineral-rich water. Inside one cabin was a huge cast iron, old-fashioned bathtub on stubby legs. We found a five-gallon gas can and carried water from the superheated pool to put in the tub. Then we let it cool before soaking our tired feet. It felt great!

Last Trip to Ruby: David

We needed a few more food supplies before the freeze-up. Henry Titus loaned us his lumber scow, which was bigger than our boat. Sophie Peters also needed to go to Ruby. I mounted my outboard motor on Henry's boat, and away we went.

Sophie's eyesight was poor. In the store in Ruby, she sat down on a chair. She was small, but we heard a big crunch! She accidentally sat on a bag of potato chips, which would have made a good TV commercial. We purchased all the supplies we needed at the Northern Commercial Store.

On our way home, we stopped at Paul and Lena Cleaver's fish camp. Baby Danny, who was lying on the deck, was soaking wet. Then we realized the boat was leaking with our heavier load. We were riding lower in the water, just below the sideboards that needed to be chinked better. There was always something to deal with! Kay and Sophie helped bail the water out of the boat using tin cans.

Pulling up the Big Boats: David

Henry Titus had a houseboat with an inboard engine, which needed to be stored safely on the bank for the winter. I helped the men use the windlass to pull it up. They slid a wooden sled called a go-devil under his boat. A steel cable was wrapped around the shaft and attached to the go-devil. We had to turn the windlass round and round by hand. The main shaft was tied to an underground cable, and a "dead man" was buried in a perpendicular position to hold everything securely.

Freeze-Up

Gilbert's last mail run ended just before the ice started running in the Yukon River. First, the water froze along the edges of the river, and then larger pieces of ice flowed. When the ice stopped running and froze solid—everything changed!

A week after the ice froze, people walked or drove their dog teams across the river. We cut a hole in the ice and dipped all the clean water we wanted for drinking or washing and hauled it up the 30-foot bank to our house. The Yukon River water was clear in the winter. If any bugs were in the water, we didn't see them!

A Good Surprise: David

Our nearest neighbors were Henry and Agnes Titus. Most cabins in the village were small, with one big room. People would put up curtained room dividers if they had older children or guests. I knocked on Henry's door in the middle of the day. He called out, "Onee," and I walked in. He was lying in bed on top of his covers, reading the Gideon New Testament. Now that was a good surprise!

A Son

Henry and Agnes adopted their nephew, Allen Captain when he was 15 months old. Our children enjoyed having a playmate their age. They could have been called the Three Musketeers!

A Koyukon Alphabet: David

Frank was a great language helper. He was single, 20 years older than I was, a hard worker, available most of the time, and patient. He was bilingual and outstanding in both Koyukon and English. He told me, "Learning Koyukon is like a dog learning to bark. They just open their mouths and start barking." That worked for him, as did learning from his parents and others who spoke Koyukon fluently in his generation.

One of the first things I did was establish an alphabet for reading and writing. I wrote a linguistic paper about the Koyukon sounds. I asked Frank about related possibilities for their sounds. He laughed and jokingly accused me of trying to "change his language."

Kay and I also learned from other Native people. I always carried a tiny three-by-five pad of paper and pen with me. We tried to listen, learn a little, and practice a lot with everyone.

How Many Pancakes Do You Want?: David

Once, when I was visiting Frank, he asked me, "How many pancakes do you want?" As a hungry young man in my mid-20s, I answered, "Four or five." Frank was amazed and responded, "I eat a lot, but only one pancake is good for me."

Then I understood as he pulled out his 10-inch cast iron frying pan. He poured in the sourdough mixture and filled the frying pan with a thick, tasty pancake. Yes, one was enough! I had answered from my own experience with small, thin pancakes. Culture and language must be learned together.

A Pizza Party: Kay

One evening, we invited the villagers over for a pizza party. I served pizza with little slices of pepperoni sausage. Sophie asked, "What kind of meat is this?"

Her brother, Al Hardluck, responded, "Deeltsaa'e! (Mouse!)" That made everybody laugh! It looked like it could have been round slices of mouse meat.

Yukon River: Kokrines (1959~1960)

David holding Debbie and Kay holding Danny in front of our house.
The icicles were pretty, but they showed that the roof was not well insulated.

Winter Mail: David

We had to wait two months until the river froze solid to commence the winter mail run. Our first winter mail arrived by dog sled rather than pony express, and it contained a backlog of reading material and letters to answer. Nobody had heard from us for two months either! The dog sled driver then informed Wien Air Alaska in Galena that Kokrines was ready for their plane.

A single-engine mail plane with skis landed on the safe sandbar across the mile-wide Yukon River. Fortunately, the Yukon River froze over smoothly that winter. We took turns snowshoeing to pack down and mark the official "landing strip" on the river, much closer to the village. They wanted 3,000 feet, so we guessed at it.

Thursday was mail day. Someone would yell, "Mail plane!" and then everybody would run to meet the plane. The last mail plane arrived at the end of April, even though water had started to appear along the shore by then.

Winter Trek to Melozi Hot Springs: David

Frank and I returned to the hot springs with our dog teams. This was easier than walking with heavy backpacks. Frank led the way again. We had to cut down small black spruce trees at the top of a steep hill. Then we tied a rope from the master rings on our towlines to the tops of the trees. The trees provided us enough drag to have control going down that steep trail.

We stayed overnight in an old cabin with only a few gaps in the roof, so we made a fire and kept warm. I rolled out my sleeping bag and planned on getting a good night's rest. In the middle of the night, a mouse tried to join me. I gave him a violent send-off from my bag. It was difficult to fall back to sleep after his unwanted visit.

First Moose: David

Moose season opened in November. The state law allowed me to get a hunting license for only 25 cents since our income was low. We had finally met the one-year residency requirement.

I hunted alone and decided I wanted to shoot a bull. I found one that was two miles away. That was a lot of meat, and the rack measured 48 inches! It was easy to break the trail and bring our dog team to haul the meat home.

Dog Team Adventures: Kay

Our seven dogs and I happily started down the sloping riverbank in Kokrines to cross the river. I was glad I now had a pair of *kkaakene*, Native-made fur boots with a thick piece of tanned moose skin for the soles. Previously, when I had worn my store-bought boots, my foot slipped off the brake, and I fell going down the bank. It embarrassed me to know that people watched and asked, "Are you all right?"

I headed home after loading the moose head minus the antlers. The dogs were excited to be going back. My foot missed the brake on one incline, but I hung on as the loaded sled and I plowed through my dog team, landing in the middle of them. I went to Red, our all-important lead dog, and hollered at him. I finally got them all straightened out, jumped on the runners, and headed home.

We soon realized why our Native friends were shooting cows in November. Our meat was tough, so it had to be ground into hamburger or pressure-cooked. A giant moose rack looked nice, but you can't eat it! Live and learn.

This was a better-tasting young moose. David provided the transportation, and we all shared the meat. That's Simon Pilot (who was blind) and Bergman Cleaver.

The Nurse Is Here

Barbara, a registered nurse with the Public Health Service Hospital in Tanana, arrived in Kokrines. She stayed overnight with us, and the local people came to her for their medical needs.

Washing Clothes: David

Our Maytag washing machine, which has a gas engine, became part of our kitchen-living room furnishings. In the winter, we ran a flexible exhaust hose outside through a hole in the wall and quickly washed our clothes. It was noisy! During the winter, I would haul cans of water in our dog sled.

In the summer, I moved it outdoors. I used a wooden yoke with long wire hooks to carry two five-gallon gas cans of water. The Yukon ran muddy, so we let the water settle. Then I poured the cleaner water off the top for washing and rinsing our clothes.

Christmas

We celebrated Christmas Day by inviting the whole village to our house for dinner and giving each person a present. We played a recording of "O Come, All Ye Faithful" in Athabascan. The tune was good, but they understood only a few words because it was in a related dialect from Nenana.

1960
Overnight Guests

Don Stickman Sr. and Jessie Johnson stopped on their way to the Novi River. Don talked extensively about various Native beliefs. He also told how he used to fly a small plane to hunt wolves with a gunner using a shotgun with slugs. There was a 50-dollar bounty on wolves at that time. Wolves had been overkilling moose and the deep snow made them easy prey. The moose population was recovering.

Indian Night School: David

I hooked up our seven-dog team and went to visit Paul and Lena Cleaver, halfway between Ruby and Kokrines. It was easy going on the sled dog trail on the smooth Yukon River ice. To keep my feet toasty warm, I had to get off the runners and jog.

After supper, when it was dark, Paul blew out the kerosene lantern as we went to bed. He liked to sing, and he had made a memorial song for Dave Corning. He told me many Koyukon riddles:

"What am I? I wear my cap in the winter." Answer: Snowcap on top of a tree stump.

"What am I? I drag my shovel on the trail." Answer: A beaver dragging his tail on the trail.

"What am I? We come upstream in red canoes." Answer: Red salmon migrating upriver.

"What am I? I'm sweeping my body like a broom." Answer: The tops of long grass swishing in the snow.

"What am I? I walk before my brother." Answer: A cane.

I liked this kind of "Indian Night School" where you lie in bed, you can even shut your eyes but keep your ears wide open, and try your best to remember what you are learning. There was no chance for notetaking either. Good for stretching one's memory! I've heard similar riddles from different people with minor variations.

Yukon River: Kokrines (1959~1960)

Paul and Lena Cleaver standing in front of their cabin

Native American Citizenship: David

Paul related about the time when Indians couldn't go to a bar and buy a drink in Ruby or even vote because they were not equal to white people. The issue was real to him as an old-timer, but all new to me since it happened before my time.

Flashback: On June 2nd, 1924, President Calvin Coolidge, a Republican, signed the Indian Citizenship Act. It had been a bumpy road for Natives to be fully accepted. This act conferred citizenship on all Natives as American citizens. History is important!

Koyukuk River: David

Kokrines was a fantastic place to live but with a declining population. We realized that we needed to move. We kept praying and asking God to show us His will for our future.

In the spring, I flew to Huslia to find out where to move. I stayed with a teenager, Samson Henry. One night, the young boys in town came to visit. They asked me to name many different animals and birds in Koyukon, and I knew most of them. Fascinating test!

Another person told me, "We didn't put you with Olin because he's a medicine man." That would have been interesting, too. I got to meet him in person. They were traditional healers with special powers and connections to the spirit world. Olin was the last publicly recognized one in the Koyukon area. Medicine people also need to hear about Jesus Christ, who is much stronger than any other power.

Hughes: David

I stopped in Hughes, the next village up the Koyukuk River, and stayed with Deeł'o (Little Arthur) and Salokh (Julia Old Man), his grandmother. Deeł'o took me around town to meet everybody in his village of less than 100. It was fortunate that the population was small, as I enjoyed a cup of fresh coffee at every home. I was floating by the end of the day!

Allakaket: David

Next, I flew to Allakaket on the Arctic Circle. I heard one young girl playing with her doll on the front steps of a home and talking to it in Koyukon. I met and spoke with several people and left our request to move there for the village council to consider.

Their Answer

Here's their brief answer, which we received in the mail from the Allakaket Village Council dated April 18th, 1960:

Dear Mr. Henry

As there is no place for you to live it would be a good idea for you to move here

Signed by: Frank Sam, Chief; Johnson Moses, First Councilman; Moses Henzie, Second Councilman

We were disappointed, but they were right. There was no available housing for our family at that time. We rightly assumed they left out the word "not." Now we changed to plan B.

The Ice Is Moving!: David

"The ice is moving!" The Yukon River breakup in Kokrines was spectacular, and the high bank was a safe place to be. The first significant change was to see the "landing strip" move 500 feet downriver before stopping for several hours. After that, it didn't take long to move again and break up. The bigger pieces of ice were jostled, tossed noisily, and viciously ground into smaller chunks. It sounded almost like a train passing by.

The breakup brought drift logs and trees. I snagged a few of them with our boat. Large 55-gallon gas drums went by. The empties floated high in the water, but the full ones floated very low. Were they full of gas or water? A couple of 100-pound propane tanks drifted by. An unfortunate person upriver lost an open riverboat, but there was no way to rescue it from the ice. If the Indians didn't collect these "treasures," they would keep going downriver to the Eskimo people and out to the Bering Sea.

Getting Ready for Spring

Spring was always an exciting time after a long winter. People worked on getting their boats and motors ready for another summer. Henry Titus nailed a new white canvas on his handmade canoe frame to go muskrat hunting.

A Common Story

Many people related a story about a Catholic priest who left Kokrines by boat, upset with the people in 1937. He dramatically cursed the village and said something like, "May this place turn to dust."

The Curse Turned into a Blessing

Can God turn a curse into a blessing? Yes! God did it for the Jewish people when a king hired a false prophet named Balaam to curse God's people. "Nevertheless, the Lord your God would not listen to Balaam, but the Lord your God turned the curse into a blessing for you, because the Lord your God loves you" (Deuteronomy 23:5).

God also loves the people in Kokrines. In the past, through the Catholic Church, people heard about God and Jesus Christ. They were taught religious practices but not about knowing Jesus Christ personally. God wants to forgive our sins. He wants to have a direct relationship with each of us. The change happens inside our hearts, and then others can see a difference on the outside by the way we live, think, and do things.

Missionaries by Boat

Billy McCarty Sr. was the village chief who also managed the Northern Commercial Company store and ran the post office from 1937 to 1947. Several missionaries came down the Yukon River by boat in 1946. Billy encouraged them to use the four empty Quonset hut buildings that the Army had evacuated after World War II. They taught the people about Jesus Christ, the Bible, and how to live the Christian life.

Fifteen Evangelical missionaries have served in Kokrines before us. This list is not necessarily in order: Independent missionaries: Al and Doris Franz; Louise Robison; Don Blood; Harvey and Joyce Kennison; Slavic Gospel Mission: Olga Erickson: Arctic Missions: Ed and Lila Smoggie, Don and Rose Nabinger, Russ and Freda Arnold, Larry and Mickey Scripter.

One of Billy McCarty's daughters, Clara Honea, accepted the Lord when she was 15 years old in Kokrines. She continues to live for the Lord. The era of steady witness and teaching about Jesus Christ and the Bible is still bearing fruit. Koyukon people living in Ruby, Tanana, and Fairbanks can trace their Christian legacy back to Kokrines.

Fast-forward to 1982: Three of Henry and Agnes Titus' daughters, Dolly Titus-Yaeger, Vivian Titus, and Mary Jane (Titus) Philips, by our log house in Fairbanks

Athabascan Workshop

At the end of May, we flew to Fairbanks for our Wycliffe Conference and an Athabascan Language Workshop. It was beneficial to compare notes with other members working in Athabascan languages around the state. We enjoyed listening as well as sharing.

Eunice Pike, an expert linguist, was down-to-earth and practical. She advised us, "Don't use any big words, since you want people to understand you." We also told her we had a "kł" consonant cluster in Koyukon. She kindly informed us, "That is extremely rare in any language in the world. Check that again." She was correct, it was a "tł" consonant cluster.

Yukon River: Koyukuk (1960~1964)

God Is Leading: David

On June 6th, 1960, the Wien Air Alaska single-engine plane settled below the treetops onto the sod landing strip in the Yukon River village of Koyukuk (KOY-yuh-kuck). Many adults and children came to see who was arriving or just passing through. George Fletcher and I flew 295 air miles west of Fairbanks. We wanted to determine if this would be a good place for us to live and continue our work.

The Native village is situated at the mouth of the Koyukuk River, where its cleaner water mixes with the Yukon River. The village had two rows of cabins along the riverbank. A large trading post, liquor store, and post office run by Dominic and Ella Vernetti were in the center. The grade school was at the upriver end, along with a small storage facility for Chevron gas.

One hundred and twenty Native people called this home. Everybody lived close to the river with a steep 15-foot riverbank. People would cut a few steps into the bank to make it easier to get to their boats. River travel was their primary means of summer transportation. There were no roads or vehicles.

George and I visited various people, including Benedict Jones, the village chief. All the pieces of the puzzle were nicely falling into place.

A Cabin for a Handshake: David

I spoke with Edward Pitka Sr. about buying a small, vacant log cabin between the store and the school. The house was old, so I would need to redo the sod roof and add a second window pane to the front window to keep us warmer.

The well-used cabin was purchased for 30 dollars down and a handshake. The total price was 85 dollars. The house came with a cache, an elevated storage shed six feet by eight feet, and an outhouse. Edward and I both thought it was a good deal. No paperwork, no real estate deed, no signatures, or even property boundaries. Just a simple verbal agreement between two new friends. I paid off my remaining debt of 55 dollars before Christmas.

Rafting the Yukon: David

Now I was anxious to return to Kokrines and pack. I asked George, "What do you think if I build a raft and move here?"

He gave me a half-serious answer, "If you don't mind a slow trip by floating!" George did not think we would travel down the Yukon with everything we owned.

Back in Kokrines, we were ready and listening for the low rumble of the Taku Chief's diesel engine pushing a barge down the Yukon. Our belongings were on the beach. Everything was coming together, just as planned.

I raced out to meet the barge in my boat so they could land and pick up our freight. I could see that they were heavily loaded, and the barge's deck was riding six inches above the waterline. The captain on the barge told me, "We're overloaded, no room, but we'll be back in six weeks!"

I was shocked and wondered what to do next. I headed back home to tell Kay about the change in plans.

What Should We Do?: David

I prayed, "What should we do?" It didn't take long to decide to build a raft and float down the Yukon to Koyukuk. I told Kay. She listened, but with reluctance, said, "I think we need to pray about it." I was one step ahead of her and responded, "I've already prayed about it." Besides, I was anxious to start building a raft.

I gathered 10 of my best drift logs and cut off the ends to make them 40 feet long. Now, I knew why I had snagged those logs floating down the river. Since there was no electricity, I built the raft using hand tools: a wood auger with a two-inch bit, an axe, and a crosscut saw.

Being in my mid-20s, I had energy and enthusiasm. I drilled several holes, made short wooden pegs, and pounded them into the crosspieces, thus stabilizing the raft. I placed 10 short logs crosswise, making a little platform to keep our belongings dry.

Yes, the raft was beginning to look like a river-worthy craft!

Needing a Maternity Smock: Kay

I was pregnant and growing. My maternity clothes were loaned to my sister Shirley in Michigan, but Agnes Titus came to the rescue. She made a smock out of an old skirt of mine. She was fast, and it fit! I was thankful Agnes was a seamstress since I needed a pattern to guide me.

Launching Our Raft: David

We slept as well as our family could under the midnight sun and the excitement of moving. I tied our boat to the raft's stern so I could steer it. We placed our few boxes of personal belongings plus a small table and a propane kitchen stove on the platform, trusting everything would be secure and dry.

Kay and our two young children, Debbie, three, and Danny, almost two, jumped into our boat. Kay was six months pregnant with Elizabeth when we moved. We prayed, committing our raft and the trip to our Lord. I pushed off and motored out into the Yukon River. Our outboard motor was way too small to push the raft as the big barges do, so we quietly drifted with the current. I used the outboard motor to land or move into the river's main channel.

The first stop was Ruby, 30 miles downriver. The river was calm for drifting. The beautiful scenery passed by slowly while we were confined to our little boat.

Soon Danny let us know that he wasn't feeling well. We thought he might be coming down with measles, so in Ruby, we contacted Russ and Freda Arnold, who cared for local medical needs. Kay, Debbie, and Danny stayed one night with them while I continued floating downriver. Thankfully, no measles!

The next town, Galena, was 50 miles from Ruby. One windstorm with rough water caused a little concern, but everything stayed dry and safe. There were plenty of "just drifting" hours, and I guessed the Yukon River flowed about five miles an hour. I watched, waited, and prayed as I stayed in the main channel.

Just Checking!: David

A speedboat passed, then circled back and stopped beside me. The man didn't recognize me, but I recognized him. He was the Catholic priest, James Plamondon. Someone had pointed him out to me at a distance when I had briefly passed through Galena on a previous trip. I knew I would get to meet him sometime!

He asked, "Where are you going?" I didn't want to answer since I knew he would not appreciate it. After a slight hesitation, I replied, "Koyukuk."

His instant reaction was, "That's my village!" and he sped off downriver. Well, here I was cruising at a snail's pace in the middle of the Yukon! I had already put money down on a cabin, and I was confident I was following the Lord's leading for us to move to Koyukuk.

On to Galena: David

In Galena, I picked up Kay, Debbie, and Danny as they had flown from Ruby. We continued the final leg of our trip together. Someone kindly warned me about the shallow water just below Galena, and we missed that sandbar.

But farther downriver around Green Cup, I hit a hidden sandbar and suddenly stopped. I ran out onto the raft with a pole to find out how deep the water was and locate the edge of the sandbar. The bar was directly under the raft, and we were stuck. I decided to ferry part of our freight about a mile downriver to a suitable landing location to lighten our load in hopes of sliding off the sandbar. After shoving with my pole, we were moving again. That unscheduled stop took four hours!

Painfully Slow: David

In the old days, the steamboat would come directly down the Yukon toward the village of Koyukuk. By the time we arrived, a large sandbar had grown into an island in front of the town. One had to go downriver almost a mile before coming upriver to the village. The large diesel-powered barges with Yutana Barge Lines would do this with freight for the village store and gas for the Chevron warehouse each spring and fall.

I followed the same route. It was painfully slow, pushing our raft upriver and around the boats tied along the shoreline. Everybody came out to the bank and peered at us as we inched our way to the landing spot in front of our cabin.

It was quite a trip—110 miles, three and a half days along the Yukon River. I landed and tied up each night, even though it was light enough to see. I needed the break and the safety of resting on solid ground.

Mission Accomplished!: David

Chief Benedict spoke in Koyukon, and instantly, the men pitched in, carrying our belongings back to our cabin. Later, I cut up the raft, and we had several cords of firewood—getting a jump on the coming winter. I packed that wood to our place.

Our Prayer

"O Lord, thank You for a safe raft trip and for providing a home for us. We want to see Native Koyukon people trust in You and live for You. You are our God and our only hope. Thank you most of all for sending Jesus Christ—for all of us. Amen."

The Welcoming Committee

It was good to be in a village with so many people. Visitors came by to check us out, and our two children made new friends.

One local, who was inebriated, visiting us on our first day, stated, "I want to join your church." We didn't even have a church, and we didn't understand his request!

Later, we found out that three men from Anchorage had been in Koyukuk the previous summer. They held meetings, talked about Jesus, and taught Christian songs. The longer we lived there, we realized the results of those meetings were minimal. They told the people, "Everybody who loves Jesus, raise your hand." Most raised their hands.

Yukon River: Koyukuk (1960~1964)

We would also have put up our hands. We desired to see faith in Jesus that changed people's lives.

Adjusting

Our cabin was much smaller, only 14 feet by 18 feet, compared to the big one in Kokrines, but it was easier to heat in the winter. It didn't take long to unpack and get settled.

The king salmon run had just started. But that summer, we didn't catch kings at fish camp since we didn't have a place to set our nets. We were given delicious king salmon, which we enjoyed. We got to know the people and learned more about life in Koyukuk.

The ~~Drinking~~ Dry Cabin: David

We knew our cabin used to be the drinking cabin. Visitors traveling by river would often stop in from downriver villages looking for a drink. But we served only coffee or tea! We were glad to convert a "wet" cabin into a "dry" one for the Lord.

One time a man from Kaltag stopped by. When he saw I was a White man, he explained the way of salvation to me. Of course, I listened intently!

Later, missionaries Don and Rose Nabinger with Arctic Missions in Kaltag were concerned if they were making any difference for the Lord. I relayed the story of the man who visited us from Kaltag. Yes, the people were listening!

Dominic's Store

Dominic and Ella Vernetti ran their well-stocked store in the center of town. Natives from neighboring villages even came here to purchase items since they offered more choices. The post office, where Ella served as the postmaster, occupied one room inside the store. They also ran a small, heavily padlocked liquor store.

Dominic's store was a two-story building with bedrooms upstairs, while the kitchen, living room, and main store were on the first floor. In 1960, we could buy two apples or oranges for a quarter, which was reasonable.

To Drink or Not to Drink: David

Every time I went to the store, Dominic would call out to me, "David, come, drink coffee with me." Sometimes, I turned him down. I didn't like coffee at that stage of my life. Then I reconsidered, thinking that if I didn't drink coffee, he might think I was not normal! Maybe he'll think Christians don't enjoy coffee. So I replied, "OK." And then I started enjoying his excellent coffee. Now I'm hooked!

A Jar of Gold Nuggets: David

Dominic was an Italian and a former gold prospector on the Koyukuk River near Hughes. He commented, "I prospected and even slept on the very place where gold was discovered by Hog River." At that time, a gold dredge was operating there. He showed me a pint jar of nuggets he had collected. It was heavy and impressive!

Bush Furniture

The store operated a Chevron gas distribution center where they sold gasoline, kerosene, Blazo white gas, and aviation gas in five-gallon cans packed with two of them in a nice wooden box. Blazo boxes were in high demand since they served many practical purposes. We used them as extra chairs, cupboards, and shelving.

Mail Day: Kay

Three mail planes a week stopped in Koyukuk! That was a significant upgrade from only one a week. I wrote lots of letters. Being ready for the next mail day spurred me on.

Another Untimely Death

A man ran to inform us that four-year-old Judy Huntington had been shot and died. Immediately, we went to Jennie's house to see what we could do. Her six-year-old son had been playing with a gun and accidentally shot Judy. Life includes some sad events!

Working as a Deckhand: David

Dominic hired me to be a deckhand for his barge. Stanislaus Ambrose of Nulato and I loaded most of the freight on the barge for our trip downriver. The cargo of canned food and a wide variety of dry goods filled all the available space. It included twelve cases of malt syrup to make homebrew to be traded for bundles of sun-dried salmon! Madeline Solomon was the cook. Helpers included Madeline's daughters, Jenny Pelkola and Naddy Bahr. Francis McGinty Sr. was the pilot. We all pitched in and worked together on whatever needed to be done.

We slept on the boat in tight quarters. Major engine problems developed in the long stretch below Kaltag, so Dominic sent for Leo Demoski in Nulato. He repaired the diesel engine in a few hours.

As we neared the new village site of Grayling, I was impressed with a monstrous two-story smokehouse loaded with salmon, which belonged to John Deacon. We kept traveling downriver as far as Holy Cross. I met Pious Savage and a few other Native people living there. It was a busy week.

Grandma Elia: Kay

To check my blood, I had flown to Tanana a month before Elizabeth was born. I wanted to make sure I was not building up antibodies, thus requiring a blood transfusion after the birth of our child. This was a major concern since I had lost a cousin shortly after her birth due to this blood problem.

Grandma Elia, who was 91, was my host this time and again in September when I waited for our daughter to be born. She was a good cook and great company. I enjoyed meeting the people in Tanana. They were surprised when I said that my husband was David Henry. A Native man with the same name lived in Tanana, so I had to explain that one. Life is interesting! After Elizabeth was born, Grandma Elia made her the cutest pair of Indian slippers with beadwork on them.

It was good to attend the Tanana Bible Church services. Mel and Pat Jensen, with Arctic Missions, served as missionaries.

It's a Girl!: Kay

Elizabeth Ann Henry was born in September at the Public Health Services Hospital in Tanana with no issues or complications. I was concerned about her right foot turning in too much, which may have been spastic from the birthing process. The Lord took care of our concerns with no need for further medical help or issues with our blood types. Thank you, Lord, for watching over our daughter and easing our minds.

We named her after Elizabeth in the Bible found in Luke chapter one. She was righteous before God and lived a consistent Christian life as the mother of John the Baptist. Her name, Elizabeth Ann, has a nice ring to it.

Hospital Roommates: Kay

Four of us shared one hospital room. I met Bertha Moses from Allakaket, who gave birth to a son named Rudy. We laughed together because she was an Eskimo, I'm white, and we communicated certain things in Koyukon. Another roommate was Olga Solomon from Kaltag. She had a son named Terry Lee. We have continued our lifelong friendships.

Home Again: Kay

I was so happy to return home to Koyukuk. Debbie and Danny were excited to have a baby sister. Life changed when we added another member to our family.

When I laid Beth Ann on our bed to change her, Debbie said, "We've got our baby that cries!" And her eyes were just sparkling with joy. Danny was very quiet and shy at first.

Beth Ann enjoyed her first Christmas with Danny and Debbie.

The Hospital Bill: Kay

We were grateful that the Native hospital in Tanana allowed non-Natives, depending on the space available and the urgency. Being pregnant was obvious. My bill could be 15 dollars per day. When I checked out, their office was closed. I wrote back and asked about my bill. The head doctor answered, "No charge. You were no trouble. Your appreciation for us and what we do was enough." I didn't tell him we were on the lower end of the income scale!

A Needed Addition: David

Kay stayed in Tanana for 10 days. My main project was to cross the Yukon River to a sandbar with piles of driftwood. I searched for small logs, about eight inches in diameter, to build a 10-foot-by-14-foot addition on the back of our little cabin. It was challenging to take the kids with me in our boat on every trip. Debbie and Danny enjoyed playing in the sand and among the driftwood.

The extra room gave us space and a separate bedroom. We put cardboard on the walls for warmth and pasted heavy building paper with a pretty design from Montgomery Wards over it to brighten things up.

A Tragic Accident: David

I was in Dominic's Store the night of September 25th when a group of five travelers stopped in to warm up and buy a few snacks for their trip. They were traveling at night from Nulato back home to Ruby.

Early the next morning, at 2:00 a.m., Koyukuk village heard the tragic news. Those five people—Claude and Bertha Demoski, Leo Demoski, Emmett Nollner, and Harry Pitka—were in a boating accident below Galena near Green Cup.

The loss of life was a major shock to everyone! We had visited Claude and Bertha in their home in Ruby a year ago when we moved to Kokrines. We expected to see them again and continue our friendship. We never know when it will be our last day or the last time we get to see a friend on this earth!

Moose Nose Soup

Saunders and Evelyn Cleaver, with their two children, Josephine and Joseph, lived in front of us. Saunders was a good hunter and he shot a moose. One of the best parts of a moose was the nose meat, which was soft and sweet-tasting. They made delicious moose nose soup and Josephine, their young teenager, brought some to us. That was a first for us.

Learning Koyukon: David

The process of learning the unwritten Koyukon language was going slowly. It took much of our time just to exist! Everybody was busy getting ready for winter. I made a large bulletin board on our wall with pictures and Native words, which helped us see and learn them. The school printed a paper, and we submitted a few stories with words written in Koyukon.

1961
My Grizzly Bear: David

Looking for a moose, I hiked behind the village on the last day of September, the end of hunting season. It was easy walking, with only a sprinkling of new snow on the ground. We needed meat, and I knew Saunders had shot a moose back there.

As I came out into an opening, I noticed an unusual pile of fresh dirt three feet high, just 30 feet in front of me. Suddenly, a grizzly bear popped up on the far side of that mound, put his front paws on the pile, and stared at me.

I instantly stopped. I didn't have a shell in the chamber of my 30.06 rifle! Fumbling around, I pumped a shell from the clip into the chamber. I also prayed, "Lord, help me." A lot can happen in a brief instant! The bear continued looking at me as I loaded my gun. I raised my rifle. I was nervous and shaking, but I hit him with the first shot as he waited for me. After that, I always carried a shell in my chamber and put my gun on safety when I walked alone in the woods.

That dirt pile was a cache for the grizzly. It was a mixture of Saunder's moose and dirt. I cut open his stomach, which was swollen and more than a foot long. It was full of fresh moose meat. Being stuffed helped him wait for me to shoot. He was eight feet long and had four-inch claws.

Everybody was happy I had shot the bear. Young people had been there roasting moose meat on a stick, but it had not harmed anyone.

Yukon River: Koyukuk (1960~1964)

The Rabbit Chokers Club: Kay

I joined the Rabbit Chokers club, the local women's club Ella had formed. It was effective in bringing all the village women together. One of their annual projects was to draw the names of all the village children, knit socks, or make something for each child to receive at Christmas.

The club's initiation requirement was to snare and thus choke a rabbit. Effie Nelson helped me make a snare out of stranded picture hanging wire. I fenced off a small line of branches and brush with an opening. I tied the wire to a strong stick over the space and set the snare, just big enough for a rabbit's head to go through. The next day I checked my snare line and choked my rabbit. I joined the club.

A Small Dictionary

We self-published a short dictionary of 11 pages in Koyukon. It showed that the Koyukon language could be written in a modern alphabet. We wanted to stimulate more interest in seeing Koyukon words in written form.

A Sled Load of Wood

Mary Vernetti, Rita Dayton, Annie Dayton, Madeline Solomon, and William Dayton "Dougal" with a sled load of firewood.

Out of Commission: David

I was anxious to get a moose before the second hunting season ended on November 30th. I put on my snowshoes and walked around the big island across the Koyukuk River. It was seven miles upstream on the Koyukuk River, then another 10 miles following the slough back to the Yukon River, and finally back home on the third part of the triangular island. No moose, just a few old tracks, but I was exhausted!

The next day, I was laid up in bed, hardly moving, and very stiff and sore. The following day, I felt like rigor mortis had set in, and I needed help to roll over in bed. The public health nurse from Tanana was in town and advised me to get to Fairbanks as soon as possible.

It was a major struggle to get on the plane. Then I had to change planes in Galena. The doctor in Fairbanks told me, "You shouldn't be walking. You have an iron deficiency and rheumatoid arthritis." That was not what a young man wanted to hear! He gave me iron pills and prescribed a drug. A few days later, another doctor told me about the side effects of that drug and to get off of it as soon as possible. I took extra vitamins for the next six months and got plenty of rest. After a year, I was back to normal, with only minor arthritic pain.

During my downtime, people helped us, especially with our wood supply. One time, Joe Nelson brought his double-bladed ax to split wood for us. He took a manly swing and caught our clothesline, causing the ax to hit the top of his head. We felt terrible! Thankfully, it was only a superficial wound!

Remembering the Dead: Kay

I was invited to walk up to the Koyukuk graveyard with Jenny Huntington, several of her children, and two other women. She wanted to go to the gravesite of her daughter, who died last summer.

I watched Jenny "burn food." She believed the person's spirit would be able to "eat" the food that was burned until she became accustomed to eating spirit food. We all sat and ate snacks together.

We needed to know what people thought and did, but we do not burn food for the dead. The Bible says that death is when one's spirit leaves the body. See James 2:26. When Jesus hung on the cross between two thieves, he said to the one who believed in him, "Truly I tell you, today you will be with me in paradise" (Luke 23:43). There is nothing in the Bible about a spirit state between being alive or dead. The person's spirit is in heaven immediately.

The Bible teaches that when a person dies, they either go to heaven to be with the Lord Jesus or to hell. In this case, Jenny's daughter went to heaven since she was an innocent little child before God, only four years old.

We go to the cemetery and visit the gravesites of our relatives and friends. We do this to remember them, grieve as we miss them, and thank the Lord for the time He gave us to be with them.

Trapping Season

In the '60s, every capable man and a few women went out trapping to make a living. The fur prices were reasonable. They would catch marten, otter, lynx, ermine, and fox, plus a few wolves and wolverines. The lynx was another source of meat to eat. It is white meat and tastes similar to chicken.

Madeline Mary Isaac stayed with us to go to school while her parents, Hughey and Eleanor Kriska, went beaver trapping. Our kids enjoyed having an older "sister."

Merry Christmas

Two big helicopters swooped into Koyukuk, bringing a live Santa Claus on Friday, December 23rd. We called them "banana helicopters" because of their shape with twin rotors that could hold 40 men each. The Galena Air Force Base, 30 miles upriver, helped us celebrate Christmas. They had to make two trips to bring oranges, apples, and fresh milk, plus toys for all the village children. They even brought a huge cake. Everyone was thrilled!

On December 24th, the school put on a party that included the nativity scene with Mary, Joseph, and baby Jesus, Christmas carols, and Santa. Each child received a piece of clothing made by the women's club. Debbie received a lovely pair of rabbit skin mitts, Danny a pair of handmade socks, and Beth Ann a little trundle bundle or sleeping bag.

Christmas Day was Sunday. The entire village went to the women's club, where we exchanged gifts between men and women. Then Dominic and Ella invited our family, the teachers, and the nurse for a Christmas feast at their store.

Reading Stories: Kay

I enjoyed reading or telling stories to our children. I read both Bible stories and other good stories. I was always glad to include village children in my reading audience. I also read a Christmas story during the annual celebration at the school.

1962
Sewing Projects

Ella Vernetti would sell beadwork and crafts on consignment. One popular item was a Native doll dressed in Indian clothing. Debbie was delighted when Evelyn Cleaver gave her one.

Summer School: David

In the middle of April, we left Koyukuk and flew to Seattle. We took the train and enjoyed the beautiful scenery and open spaces out west. Kay's parents met us at the train depot in Detroit. Then we headed to their home in Vassar, Michigan. Later we drove to Canada to see my parents. This was the first time both sets of grandparents had seen their newest granddaughter, Beth Ann.

My parents, Luther and Nelda Henry, had moved to Picton, Ontario. They joined the Canadian Sunday School Mission and taught the Bible in the public schools as they had done in Kentucky.

VW bug with Luther and Nelda holding a flannelgraph board for telling Bible lessons. They returned to Kentucky after Miriam graduated from the 13th grade.

Yukon River: Koyukuk (1960~1964)

Our summer school with SIL in Norman, Oklahoma strongly advised us not to bring children under a year old to the courses. So we left Beth Ann with Kay's parents. The classes were an intensive 11-week course that required lots of study and homework. This was our second session of SIL, and we were able to ask specific questions related to the Koyukon language. It was hot, and we had no king salmon that summer!

Communists!: Kay

After returning to Koyukuk in the fall, we welcomed the new school teachers, Mike and Susan Lockwood, to the village. They were eager to teach. I baked bread weekly for my family, and I gave them a loaf of warm bread. After we got acquainted, they heard that we were Communists! They soon realized those reports were not accurate.

Ella Vernetti is holding 10 white ermines, and Dominic is displaying a wolverine skin with Sue and Mike Lockwood.

Windup Tape Recorder/Player: David

Our tape recorder/player had a big spring and a windup handle. I had to keep the spring wound up for it to run at normal speed. It worked well for recording Native songs and stories.

One time, a man asked me, "How much does that recorder cost?" That was a fair question, so I asked him, "How much does it cost to get drunk?" He replied, "40 dollars." So I informed him, "Well, this recorder costs three and a half drunks!"

Young Toby: David

Another time, Young Toby, our 80-year-old neighbor, returned from hunting camp with his son, Stephen Toby. He asked me to record a song he had composed while at camp. Then all day, different people would come by to visit and ask to hear Young Toby's song.

Later, Young Toby was not feeling well and was admitted to the Public Health Services Hospital in Tanana. The prognosis was grim, and they could not do anything to help him, so he decided to come home and die with his people.

Saunders and Evelyn provided a bed for Young Toby in their home. He quit eating his last week but enjoyed all the visitors, including me. I wanted to read Scripture to him in the Koyukon language, but we did not have any at that time. I've always felt bad about that.

One morning, we were awakened at 5:00 a.m. by a knock on our door. Saunders informed us of the obvious about Young Toby. I helped with building his coffin and digging his grave.

Good Language Helpers: David

We began working with Eliza (Peters) Jones. She was well-trained since childhood by her elders in the Koyukon language in Huslia. She had a strong desire to write, record, and preserve her Native language.

Ella Vernetti was very helpful and knowledgeable in Koyukon. She was a great organizer, which she learned by running a store, post office, a women's club, plus whatever else needed to get done.

Madeline Solomon was great, too. She was super busy fishing, sewing, and raising a large family. Later, she recorded the Gospel of Mark in central Koyukon.

Doria Lolnitz told old Indian folktales. She was patient as we both listened, and I wrote down the tales. We worked with her on adding the English translation. Then Doria would go home and tell her children the same stories. But we realized she was giving us a more "sanitized" version. Life was interesting!

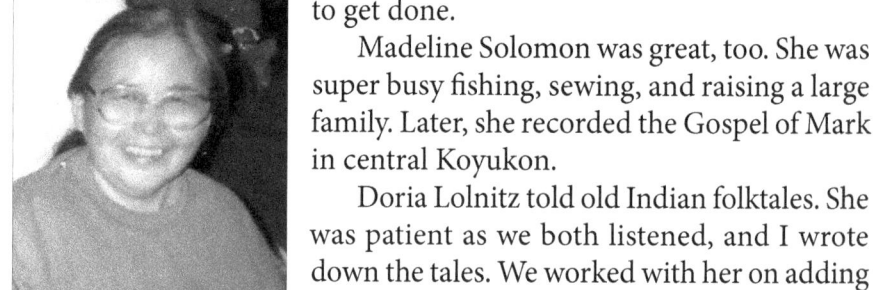

Eliza Jones

Yukon River: Koyukuk (1960~1964)

In God's Time: Kay

Susan Lockwood and I flew into Fairbanks. They had stored their Jeep in Fairbanks, so we had transportation. Since Susan enjoyed music, I invited her to the Nazarene Church. The pastor gave an altar call. Susan wanted to go forward, so I went with her to support her decision. She turned to Jesus Christ and accepted Him into her life. What an exciting evening!

Debbie Had a Birthday

Since we had three children, we always had young visitors. It was a lively birthday party for Debbie at our cabin.

Back row: Paula Pitka, Martha Nelson, Margie Dayton, Violet Dayton, Patricia Esmailka; front row: Joyce Pitka, Danny Henry, Kathleen Edwin, Beth Ann Henry in white, Debbie Henry, Josie Jones, and Marilyn Demoski.

Practical Help: David

The builder at our Alaska office in Fairbanks, Chuck Hoch, came to assist with projects. Chuck and I put plywood on the floor and then linoleum to be warmer. He built our main door with tongue and groove lumber, making it sturdier. We removed the curtains from our Blazo box cupboards and replaced them with plywood doors on hinges.

He built a couch combo for our living room. First, he made a wooden frame and covered it with plywood and hinges. Then he cut the plywood to open the large box. It made a nice storage space for extra blankets or anything big. I purchased single-cot mattresses in Fairbanks. One lay flat, which served as the seat; another was doubled over for the back. We had a couch that was used daily.

Memorial Potlatch: Kay
A big memorial potlatch for Young Toby and Mary Kriska was held in Koyukuk on September 17th, 1962. It was a time for honoring those who had recently died, remembering them with songs and speeches. They served a wide variety of food and distributed many gifts. Saunders gave us moose meat to cook for the potlatch. I baked cakes, made cranberry sauce, and soup. I also baked for Sally Pilot since her oven was too small.

Is the World Going to End Soon?: Kay
Kids ask the best questions. A young girl came to us and asked me, "Is the world going to end soon?" We know it will end, but we don't know when. Yet, we all need to be ready before it comes to an end.

I told her what the Bible says and asked her, "Would you like to ask Jesus to give you a clean heart?" She prayed like this, "I know I have sinned, and I believe Jesus died for my sins. I believe Jesus is my Savior, and I want to live for him. Thank you. Amen."

Then she said, "I want my girlfriends to hear about Jesus, too." She brought along several other girls. And I read to them from the Bible. Everyone needs a "clean heart" before God, and only He can do that.

Other Athabascans in Alaska: David
Irvine Davis and I took a survey trip along the Kuskokwim River. We gathered word lists from Nikolai, McGrath, Takotna, Crooked Creek, Sleetmute, and Stony River to help decide how much they used their Native languages and if Wycliffe should send translators there. I could understand many of their words, spoken with some variation from Koyukon.

More Languages in Alaska: David
George Ryon, the radio operator, and I took several language surveys in southwestern Alaska. Around Lake Iliamna, we went to

Athabascan and southern Eskimo villages, then to Aleut towns on the Aleutian Chain.

We flew with Leon "Babe" Alsworth, from Port Alsworth, in his Taylorcraft plane. He had removed the battery and anything else to allow him to carry more weight. When the two of us were packed in, he leaned out his window and gave the prop a flip to start the plane. He was a fantastic pilot and guide. He had recently turned to Jesus Christ and was very thankful that the Lord had forgiven his sins. He wanted others, especially in his region, to know the Lord.

Simple Light: David

We had one four-foot-long, 40-watt fluorescent light tube, which provided enough light for our little cabin. It was powered by a 12-volt car battery. I rigged a vehicle alternator to be belt-driven by a two-horsepower gas engine on a small sled. I pulled the little sled outside, filled the tank with gas, started the engine, and plugged it into an extension cord running through the wall.

Then we enjoyed the bright light for two hours. When the engine ran out of gas, the battery would keep it going for more than an hour. When the light dimmed, we turned it off and went to bed. The battery recovered overnight and had enough charge to run for another 20 minutes in the morning.

The Old Fashioned Revival Hour: **David**

Edward Pitka Sr. used to trap around the Dulbi River near Huslia when he was younger. He enjoyed listening to Charles Fuller preach every Sunday on *The Old Fashioned Revival Hour* using a big battery-powered radio. When I was in high school, my parents drove to southern California to visit my brother, and we heard Charles Fuller preach in person.

Floods Change Everything (1963)

1963
A Baby Boy: Kay

Waiting for our fourth baby, I flew to Fairbanks and got a ride to Faith Hospital in Glennallen, Alaska. Central Alaska Missions, which later merged with Send International, ran the hospital. I stayed with Vera Kelly for more than a week and enjoyed the Christian fellowship.

Our second son, Stephen David Henry, was born in May. We named him in honor of Stephen in Acts chapter seven of the Bible, who was faithful to God and stood for Him to the very end while being stoned to death. His middle name is after David in the Bible and his father's name.

On my return flight in a small plane from Galena to Koyukuk, I noticed all the water along the edge of the Yukon River. The plane still had winter skis, but the pilot said, "That's OK, we'll land on the wet grass on the runway, and it'll slide." I was apprehensive, but it was a smooth landing.

"How High's the Water, Mama?": Kay

A breakup of the ice on any river is exciting but can be a little scary. Would it be high water up to the top of the bank or over the top? The old-timers reminisced about the last big flood in 1937. Young boys kept track of the rising river by putting a stick in the sloping ground to measure how much the water rose. At one point, it rose three feet in less than an hour! This would have been the time to sing Johnny Cash's song: "Five Feet High and Rising." Everybody kept asking the question, "How high's the water, Mama?" as in his song. The answer instantly spread all around the village.

Stephen was barely a week old when the floodwaters began rushing over the bank on Sunday, May 19th, 1963. I moved to the school at 5:00

a.m. to await evacuation with our newborn and three small children, along with other villagers.

The US Air Force stationed in Galena sent two Chinook CH-47 helicopters to evacuate women, children, and older people at 9:00 a.m. While we were sitting on the floor in the chopper, they wrapped a long seatbelt around several of us. In Galena, they tore up sheets to make diapers for the babies as we waited for a bigger plane to Fairbanks. David and 15 men stayed in Koyukuk to save what they could.

The view is from our roof looking upriver.

One of the men on the roof next door was John Dayton Jr. The lighter-colored building with white trim farthest away is the school, elevated five feet above the ground.

Sally Pilot was in the last group to be evacuated. Already several feet of water had flooded the village. David watched the helicopter let down a basket, and she climbed in. Strangely, the basket kept spinning round and round. She was scared and she screamed all the way up! The Air Force soon grabbed the basket and brought her safely inside.

A Pair of Boots: David

The ice jammed downriver between Nulato and Koyukuk. Then the water in the Yukon backed up, causing the slough in front of the village to flow upstream. It was odd to see a river flowing backward! The river changed directions several times. Someone had discarded an old pair of boots—and we saw them as they went back and forth on a chunk of ice.

Two moose stood on a big block of ice floating downriver, too scared to jump off and swim to safety. Several hours later, people in Nulato saw them still riding that piece of ice.

Our Little Butterfly: Kay

Villagers were flown to Fairbanks, where they stayed at the Barnette School gym. Rows of cots were divided by hanging sheets for each family. The Red Cross and Civil Defense fed and clothed 100 of us. I was busy with baby Stephen and keeping track of three active kids.

At one point, Beth Ann took off on her own. The police saw her but didn't think she was with our group since she was white. A Native lady found her and brought her back. She was then named *Nedenlebedze* (butterfly) because she fluttered here and there like a busy three-year-old.

Buildings Were Floating: David

Back in Koyukuk, all the outhouses floated away first. The community hall, a 40-by-40-foot log building, floated out into the slough. Chief Benedict and other men pushed it back into town with their boats. Eighteen houses shifted or moved. Only our house and three others stayed in place. Before the flood, I threw several cords of firewood onto our roof, adding more weight. Someone "borrowed" a big coil of rope from Dominic's warehouse. The rope was strung along the front of the village to keep houses from floating away.

Several nights were spent on the flat roof of the Catholic Church with most of the other men. I also slept cramped in our boat with items I rescued before the flood. As I watched, I kept praying and wondering about the future.

The Fairbanks newspaper reported that the water was up to the top of the school, which had a high, peaked roof, and that people were sitting on the school roof. Villagers at Barnette School were glad when Kay gave them my accurate report that nobody was on the school roof. I gave daily updates on our single sideband radio from our boat powered by a 12-volt battery.

We were thankful for the shortwave radio, as it was beneficial during this flood. I checked in with Fairbanks for 15 minutes every morning except on Sundays.

Floods Change Everything (1963)

Ice Jam Broke: David

When the ice jam broke, the water started rampaging through town. Standing on our roof, I watched the fast-moving action. A huge chunk of ice, 20 feet across, hit the well-built two-story log house of Leo and Mary Kriska. The house jumped several feet in the air and fell back down, but not exactly square on the lower logs. There was nothing anybody could do to stop it.

Another block of ice hit the house of Arthur and Tecla Malamute. Only their roof had been sticking out of the water. The ice went through their home, and it was gone in an instant!

"How High's the Water, Papa?": David

The water came up to the eaves of our cabin. Most of the things I put up high inside our house got wet. Dominic stayed in his two-story building. Several men tried to get him to move to a safer place, but he said, "The captain goes down with the ship." He stayed put, even though the water was creeping over a low corner of his second floor.

Going Back Home: Kay

We spent thirteen days in Fairbanks. The women and children returned to Koyukuk with boxes and boxes of donated clothing. The people of Fairbanks were generous.

What remained after the floodwaters receded? An inch or more of gooey mud was left on our floor and coated our walls and belongings. We threw out many things, but all our wet clothes were washed. We tried drying the books but realized they would be warped permanently, so out they went. Cleaning up took weeks. This flood prompted the villages of Galena, Koyukuk, and Nulato to move to higher ground.

Material things, like our houses, beds, dishes, and food, can be replaced. Our spiritual possessions in Christ are safe from any flood! Real valuables such as having our sins forgiven, being a member of God's family, having God's peace when very little is peaceful around us, freedom from fears in life, and much more are permanent for followers of Jesus Christ. We were thankful that no lives were lost or injured in the flood.

Yukon River "Coffee": David

Shortly after the flood, I took our neighbor Andrew Esmailka and his wife, Toghondeeno, out on the Yukon River in our boat. Andrew was thirsty, but I had to tell him, "Sorry, I don't have any drinking water." Andrew said, "That's alright, lots of water in the Yukon." Then he took his cup, dipped it in the river, and brought up a cupful. The color looked like coffee with cream. And he drank it! He was happy and suffered no ill effects!

Family Comes to Koyukuk: Kay

Shortly after the flood, my family visited Koyukuk to see their newest grandson. My mother was a school teacher, and received a pay raise after finishing more classwork, giving them extra money to travel.

Herbert and Madeline Solomon let us use their new cabin. It had a double bed, and the basics left after the flood. My parents, Bob and Vera Temple, my 17-year-old brother Jim, and my nine-year-old sister Sheryl met our Indian friends. We were still cleaning up, but my family was a big encouragement.

Vera and Bob Temple held baby Stephen with Beth Ann, Danny, and Debbie.

Everybody slept under a mosquito net to keep the pesky bugs away. We burned Buhach during the day when we were inside. It was popular because it was a safe, all-natural, powdered insecticide that repelled mosquitoes. The container recommended not to burn it indoors, but we had to—there were too many mosquitoes! Later that winter, I sent my mother a pillow I had made. She commented, "It smells like Buhach!"

When my family visited, we cooked salmon as often as we could. My mother told me, "We like salmon, but could we eat something else?" My parents had had their fill!

Fresh Raspberry Pie: Kay

Debbie and Danny were excited to find raspberries growing near our house. They picked a cupful and shared it with us. I told them, "If you pick another cupful, we can bake a raspberry pie for all of us." In no time, they were back with enough berries. After making the pie, I set it on top of the kid's toy box to cool. Someone wanted a toy, so they lifted the cover. The beautiful pie ended up on the floor!

New Village Site: David

The State of Alaska barged in a D9G caterpillar to clear the area five miles downriver at the big bend in the Yukon River. The new village site is on higher ground.

Planning too far ahead, I went up the Koyukuk River with my family. Raphael and Doria and some of their family went to the same location. I cut 40 spruce logs 28 feet long. They were easy to peel in the spring when it was wet under the bark. They were also very slippery, so our families worked together to drag them through the woods and into the water. Then I tied them together and rafted our house logs to the new village site.

Eventually, the village moved behind their old site to higher ground rather than choosing the new site, so our logs sat there, drying. Someone downriver needed them, so that was the end of our dream for a new cabin.

First Grade

Debbie entered first grade at the Koyukuk School for the school year 1963-64. She was excited and ready to learn. There were 49 students in grades one through eight.

First and Only TV

Dominic and Ella had the first and only TV. People crowded into their living room to watch shows broadcast from the Galena Air Force

Base. The pictures were poor quality, black and white with lots of "snow" on a small screen. But it was the best available at that time.

Fast-forward: After the pipeline days in the mid-'70s, everybody in Koyukuk and other villages erected satellite dishes and could receive hundreds of stations on their widescreen color TVs.

Only in Alaska: Kay

Only in Alaska do you hear a woman tell another woman, "Be sure and bring your gun," as you were getting ready to go berry picking. If you saw fresh bear signs, you kept moving to a different berry patch. When I went blueberry or cranberry picking with the ladies, somebody always brought along a rifle.

One time, my neighbor Toghondeeno said, "I'll ask Andrew to bring his rifle and go with us." We picked a good spot and filled our berry buckets. Then we discovered our "guard" had fallen asleep!

God Protected Me: David

In the late fall, I went hunting by myself five miles downriver by the big bend in the river. The ice was running, but the Yukon was frozen along the edges extending out into the river 15 feet.

Taking the easy road back home, I walked along the ice on the river's edge. Suddenly, I fell through the ice. I raised my gun to keep it dry, kicked as hard as I could, and prayed—all at the same time. I got wet only up to my waist, even though I didn't hit the river bottom. I prayed, "Lord, help me! I want to see my wife and my family!" It's incredible how much can happen in just a few seconds!

I crawled along the ice, back onto the bank of the river, and started walking home. My clothes began to freeze. Halfway home, I had trouble bending my knees, but I kept awkwardly moving anyway. I had difficulty getting my partially frozen pants and boots off at home. God helps us even when we do foolish things!

1964
Only 28 Once: David

In January, I turned 28, and Kay made two birthday cakes—one for our family, and one to share. I invited all the men to our house that evening to celebrate with me by eating cake and homemade ice cream. Three of our kids went to bed in the back room as planned, but not

Beth Ann. Every time a man came in the door, she would sing "Happy Birthday."

One-Dog Race

The village held dog races on the slough in front of the town, where everyone could watch and cheer them on. Debbie qualified by age, but we didn't have a dog. Dominic loaned her his pet collie dog. Dominic stood at the finish line and called, "Lassie." She ran straight to him, and Debbie won the race.

The Good Friday Earthquake: David

Arnold Huntington was visiting us on Good Friday, March 27th, 1964, when we felt one big shock wave. Instantly, he yelled, "Earthquake!" A megathrust earthquake with a magnitude of 9.2 shook Anchorage for four minutes. It is still the most powerful recorded earthquake in North America.

Koyukuk is 350 miles from Anchorage, the way the crow flies, yet we felt a strong jolt like many others in Alaska. I turned on the radio, but everything from Anchorage was silent until the next day.

Goodbye, Koyukuk

We had outgrown our small two-room cabin. As we prepared to leave, we sold the cabin and outbuildings to Joe and Amelia Nelson. They needed a home and did not have much money. Our deal was 20 dollars in cash and a friendly handshake for everything.

The Lord gave us four years to live and serve there. The Lord also blessed us with two more children, Beth Ann and Stephen, while we were living in Koyukuk.

We left in April, flew into Fairbanks, and then on to Seattle. There we purchased a used station wagon, spent the summer traveling, and eventually drove to Michigan.

We compared our study notes.

Summer School Again

First, we drove to Norman, Oklahoma, to enroll in the third-year course at the Summer Institute of Linguistics (SIL). The studies were intense! This time, we concentrated on Koyukon projects.

These courses were held in partnership with the University of Oklahoma. We enjoyed their campus facilities. Summer was hot, with 90 and 100 degrees considered normal. To cool off, the whole family went to a children's pool and splashed about. We bought Beth Ann an umbrella. Every day she stated, "I think it's going to rain." We wished she was right.

Working with Verbs: Kay

I wrote a paper titled *Koyukon Classificatory Verbs*. Native people divide these common verbs into twelve categories, which classify objects according to size, shape, texture, and number of items. This system is part of their complex worldview and way of organizing everyday life.

Here are seven examples dealing with sugar:
1. A single sugar cube is there.
2. Cubes of sugar are there.
3. Cubes of sugar are scattered.
4. A box of sugar is there.
5. A bowl of sugar is there.
6. Loose, granulated sugar is there.
7. Wet, sticky sugar is there.

The word "sugar" remained the same in each sentence, but the verb form changed to precisely describe the situation. The article was published in *Anthropological Linguistics* at Indiana University in 1965. We were told, "If you can't write about it and explain it, then you probably don't fully understand it."

Which Way is the River Flowing?: David

I wrote about *Koyukon Locationals/Directionals,* which was published in *Anthropological Linguistics* at Indiana University in 1969. Much of Indian life, livelihood, and culture are closely tied to this complex system. The most common words are defined relative to the direction the water is flowing in the river.

When we see a map, we usually ask, "Which way is north?" And then we'll know which direction is east, south, and west. Old-time Indians didn't use maps. They would look at the rivers and observe which way they flowed. The 11 Koyukon villages in Alaska are on either the Yukon or Koyukuk Rivers. As a result, directions are all river-oriented: upriver, downriver, across the river, and back from the river. Indians follow this system, whether speaking in Koyukon or English.

One time, I brought an Indian friend to a house in the country near Fairbanks. I asked him, "Do you know where we are?" And he replied, "Where's the river?" He did not ask which way was north.

A Huge Computer

The University of Oklahoma allowed SIL members to submit language data on their computer systems. We entered over 100 pages of Native stories on notched IBM punch cards. Their dedicated room was 20 feet by 30 feet, lined with nine-inch reel-to-reel tape recorders and a master IBM computer. The computer took 10 hours to process our work. Then they gave us a printout of the concordance of all the Koyukon words. It was 10 inches thick! Their computer was "modern" and fast—back in 1964!

Maranatha Bible Camp

Summer school ended on Friday, August 28th, and our annual North America Branch Conference began the following week at Maranatha Bible Camp in Maxwell, Nebraska. It was good to spend two weeks together hearing various field reports. The break from the intense studies of summer school was necessary. Our family stayed in the camp's facilities until the end of October to attend another special study session concentrated on Athabascan languages.

Off to Grade School

Both Debbie and Dan went to grade school in the small town of Maxwell, Nebraska. Debbie entered second grade and Dan entered first grade. But Dan had to prove himself to be accepted. His teacher stated, "He's not ready for the first grade, since he did not go to kindergarten." Fortunately, he had learned at home and was ready and eager to learn more. Beth Ann and Stephen were happy "homeschoolers" in nursery school.

In the Archives: David

The archives at Gonzaga University in Spokane, Washington, became my main focal point for five weeks. The Catholic priest in charge invited me to look over their collection of manuscripts. I was especially interested in the Koyukon dictionary by a Jesuit priest, Jules Jetté (1864-1927). He had spent a considerable amount of time in Nulato, Koyukuk, Kokrines, and Tanana.

Fast-forward to the year 2000: The Alaska Native Language Center and the University of Alaska Fairbanks published the unabridged *Koyukon Athabaskan Dictionary* by Jules Jetté and Eliza Jones, with James Kari as editor-in-chief. This dictionary is the new standard for everything written in Koyukon.

In the Thumb of Michigan: Vassar (1964~1965)

The House Next Door: Kay

It was fun to have grandparents close by. We rented the house next door in Vassar, Michigan, early in November. My brother Jim was available when we needed a babysitter.

Debbie was in the second grade at Vassar Public School, where her grandmother, Vera Temple, taught the same grade. But the system put her in the other classroom with Mrs. Parks, a Christian from our church. Dan was enrolled in first grade. Our children walked the mile to school. Debbie took piano lessons, as both of us had done many years ago. Beth Ann and Stephen stayed home.

Fast-forward: Years later, McDonald's purchased my parents' home and the house next door, where we lived. Once, we took my father to eat at McDonald's, and he exclaimed, "We're eating where our living room used to be."

Good Fellowship: Kay

We enjoyed attending my home church, which had biblical preaching and wonderful fellowship. During the winter, we visited other supporting churches in the thumb area of Michigan, sharing about the work in Alaska and showing our slides.

Alaskan Treats

A package from Koyukuk with unique Native food arrived just before Christmas. When we opened the box, we breathed in the aroma of smoked king salmon strips. A two-pound package of Sailor Boy Pilot Bread crackers was included, the staple and favorite of Alaskans. These four-inch unsalted crackers have a long shelf life. The six of us enjoyed an Alaskan meal in front of our fireplace, as we reminisced about life on the Yukon River.

1965
A Winner!: Kay

My youngest brother, Jim, loved and played basketball with the local Vassar High School Vulcans team. He was short, but he was wiry and fast. We watched Jim play in the Michigan regional games in Clio, Michigan, for the state championship.

Our family joined in the cheering section with my parents and sister Sheryl, along with my other brother, Tom, and his wife, Mary, Uncle Ed and Aunt Leola Erb, Uncle John and Aunt Jean Temple, my sister Nancy and her husband, Louie Horwath, and others.

The game was tied with just a few seconds left to play when Jim was fouled. The referee gave him two free throws. There was absolute silence as he took his first shot. He made it! He also made the second free throw! We were all ecstatic! The Vassar Vulcans on the court put Jim on their shoulders and carried him around! We won the championship!

Thinking about Alaska

It was so easy to accumulate things. We started the major job of sorting, packing, and getting rid of items. One box of winter clothes was to be mailed to us in the fall, and another of summer clothing came with us. A pile of things had been stored in Michigan. And a "suspense" pile needed a final decision. We don't like packing!

Bishop Gordon, the Episcopal bishop over the work in Alaska, invited us to move to a village with an Episcopal Church, such as Hughes or Allakaket. Elliott Canonge, our Alaska field director, contacted him in our absence and explained our situation. We continued to pray about housing, when and where to move.

Kentucky Relatives: David

We drove to Kentucky to visit my parents, Luther and Nelda Henry. We enjoyed sassafras tea and had fun together.

In the Thumb of Michigan: Vassar(1964~1965)

Luther and Nelda with Debbie, Stephen, Dan, and Beth Ann

Betty Crocker Coupons: Kay

I asked friends and churches to save Betty Crocker coupons. It took more than six months to collect enough. After gathering 30,000 coupons, I got my Singer zigzag sewing machine for free from Betty Crocker. The pile of coupons equaled the size of the sewing machine!

North to Alaska, Again: David

We said our goodbyes in person to as many people as possible. We didn't know when we'd return to see family in Michigan and Kentucky. It was a genuine joy to have wonderful memories of all the love and warmth of dear friends and family. It was also very encouraging to have people praying for us.

This time, the Lord provided us with a dealer's new three-quarter-ton truck with a camper. And the dealer paid for the gas to get it to Fairbanks! The gas expenses amounted to 175 dollars. We celebrated our ninth wedding anniversary with four lively kids in a camper, leaving on June 8th, 1965.

The only restriction was that we could not use the stove in the camper. Most of our meals were cooked outdoors, using facilities at government campsites along the way. After I woke up early in the morning, I could drive another hundred miles while the kids continued sleeping in the camper with Kay. This gave them time to wake up and get dressed before breakfast. I hooked up an intercom to talk with those riding in the camper, and I could turn it off so they could sing or play at their enthusiasm level and still not bother the driver.

Sights along the Way

One iconic sight was the gigantic Paul Bunyan statue with Babe the Blue Ox in Bemidji, Minnesota. The giant 18-foot-high lumberjack symbolizes strength and vitality as an American folk hero.

We stopped in Watson Lake, Yukon Territory, at mile 635 to see the Sign Post Forest. The number of signs—80,000+—kept growing as people added their hometowns. We looked at and read a few, but did not add to the collection.

End of the Road

We were happy but tired of steadily traveling through Canada after 10 long days. We quickly unloaded our belongings at the Wycliffe Center on a 40-acre plot five miles north of Fairbanks, located at one mile on Chena Hot Springs Road.

Fast-forward: The property is now part of Camp Li-Wa, (short for Living-Water), operating as a year-round Christian Retreat and Adventure Center.

On the Arctic Circle: Allakaket (1965~1969)

The Big Question: David

"Where does the Lord want us to live?" This was our main question. A move to Hughes was out since there was no housing. Allakaket, pronounced (al-uh-KACK-ut), had only one building available, an abandoned store with living quarters. There was no time to build this summer as it was already July, and we had limited finances.

For three days, we combed Fairbanks, looking for the owner's daughter, Jeannie (Evans) Stevens, to find out how to contact her father, Wilfred Evans Sr. He was working in Bornite, Alaska. I sent him a message via teletype, asking for a one-year lease on his building in Allakaket. We prayed and waited for his answer.

Answered Prayer: David

Wilfred answered almost immediately. He added a note with written permission in case I needed to show anybody, stating, "It is alright for David Henry to lease my former store for five years. This deal was in exchange for repairing and taking care of the building." God was very generous and gave us more than we asked for!

About an hour later, when I returned from running errands in Fairbanks, I dashed inside, telling Kay, "I'm flying to Allakaket at 6:00 tonight!" That meant I had to leave at 5:00 p.m. to catch my flight. It was already 4:30 p.m., but I made it. I flew to Bettles on a Wien Air Alaska propjet and waited there overnight before going to Allakaket on a smaller plane the following day.

The Big House: David

The former living quarters and store were huge! The spacious two-story log building was 24 by 60 feet. The Natives called it "Big House," but we named it "The Evans' Mansion." I decided to fix up the living quarters, which occupied only half of the building. The condition of the place was much worse than I thought it would be! Debris was strewn everywhere! One could walk through the walls, but the studs were still there for the room partitions. All the windows were broken, and we needed to buy replacement windows and paint. It took months of hard work to repair everything, but, after that, we enjoyed living there for four years.

Acceptance with Open Arms: David
People are more important than living quarters. The residents of Allakaket welcomed us with open arms. One question several asked me was, "Is your wife learning our language, too?" Yes, we work as a team.

God's Timing: David
Timing is everything! Five days after Wilfred agreed to let us lease his building, the U.S. Geophysical Institute contacted him about renting it. Wilfred told them that he had already decided to let us have the building. It was terrific to deal with people who kept their word and had integrity without a written contract. Wilfred and I had never met in person, and we knew very little about each other.

Moving Day: David
We moved to Allakaket on July 12th, 1965. Our family left Fairbanks at 6:00 a.m. on Wien Air Alaska and flew three hours to Bettles. Then, we took a short flight to Allakaket. Our big house was on the opposite end of town from the runway, and it had started to sprinkle while flying there. Velma (Williams) Simon Schafer offered us a place to sleep at her house for our first night, making our adjustment much more manageable since our house was not yet livable.

Location Is Everything
Allakaket is an Athabascan Indian village with a population of 150. On the north bank of the Koyukuk River is the Iñupiaq (Kobuk River) Eskimo village of Alatna with 30 people. Allakaket is at the junction of the Alatna, and Koyukuk Rivers. Its Native name is *Aalaa Kkaakk'et* (Mouth of the Alatna River). It is 190 air miles northwest of Fairbanks, and there are two mail planes per week.

On the Arctic Circle: Allakaket (1965~1969)

Freighting Down the Koyukuk River

Our only oversized item was a double bed mattress. It was no problem shipping it on the larger plane from Fairbanks to Bettles. But how would we get it from Bettles? It was high on our prayer list.

Answered Prayer

God used the U.S. Geophysical Institute to answer our prayer. They offered to bring our double bed mattress from Bettles as part of a deal to use the store side of the building. It was 90 miles by river from Bettles to Allakaket. They had a bigger boat than ours, and our boat was still in Galena on the Yukon River.

They gave us enough electricity to light our side of the "duplex" and run our washing machine from their small generator located by the school. A single man, Gene Philips, kept their system operating for measuring the ionosphere and lived in the other half of the building on the first floor for a year.

Northern Lights

Allakaket is a prime viewing site for the northern lights, known by their scientific name as the aurora borealis. The extreme cold seems to bring them out, with one or two large bands of colored lights filling and dancing in the sky. They can whistle or sound like radio static for short periods. Our Creator's spectacular light show was fantastic!

Single Sideband Shortwave Radio

We continued using a shortwave radio for our primary contact with Fairbanks six days a week. A weakness of the radio system was that others in Allakaket could also hear. They could listen to their village health aide's contact with the regional hospital in Tanana. It got more interesting if the aide revealed the nature of a medical problem. A little competition with soap operas! Our village health aide was very careful in what she said over the radio.

Clickety-Clack

The boardwalk extended from the Episcopal Mission House past the church toward the airfield. It was eight feet wide, made with four-inch-wide split black spruce poles, and tied down along the sides of the walkway. Every step was clickety-clack, clickety-clack all summer long. It kept the dust away!

Pitching Our Tent: David

We brought a green Army tent, which we pitched on the second floor inside our house. The roof of our building didn't leak, but we needed a dark place to sleep. Living with 24 hours of daylight means it doesn't get dark during the summer. We also needed a haven from all the mosquitoes. After I installed the new windows, we took the tent down.

That first night in our tent, Debbie prayed, "Thank you, God, for the nice house with lots of rooms." The building gave our family plenty of room to spread out and enjoy life.

Remodel 101: David

I built bed frames for all of us. Under the children's beds, I made big drawers for storing their clothing and toys. I constructed simple closets in which to hang our clothes and painted the inside walls and window frames. One young man came to see how the remodeling job was progressing. He commented, "That's a good way to pass the time!" I didn't know how to answer. I was trying to make the house livable, not merely passing the time.

Kay cooked all summer on our two-burner Coleman camp stove. We purchased a second-hand 30-inch propane cook stove in Fairbanks, but it was too large to send on the regular mail planes. We had to wait until fall when Wien Air Alaska flew larger aircraft, like the Pilatus Porter with large doors.

Caught in the Wringer: Kay

In Allakaket, we used an electric Maytag washing machine with a wringer on top. I had to be careful when I fed laundry into the wringer, trying to keep the wet clothes from bunching up. Our children were good helpers as we all worked together.

One time, Beth Ann was standing on a chair helping me but she got her hand stuck in with the wet clothes going through the wringer. I immediately hit the release plate as I prayed. Thankfully, her small hands were more flexible than mine, and there were no broken bones or skinned hands. We were all more careful after that.

Here's the big house with our kids and others sliding down to the river. We lived in the left half of the building.

Saint John's-in-the- Wilderness

It was pleasant to hear the church bell ringing on Sunday mornings, calling everyone to worship. Saint John's-in-the-Wilderness Episcopal Church was established in 1906. Different Natives led the services from the prayer book. The lay ministers were Simon Ned, Elliott Koyukuk, and Kenneth Bergman. They read lots of Scripture. We appreciated the reverence everyone showed while in church.

Heading to Galena: David

I flew to Galena to pick up our boat and motor. Edward and Laura Pitka with their family had moved from Koyukuk to Galena, and he was taking care of our boat. Yutana Barge delivered our new outboard motor. I thought it would be easy to mount the motor and head to Koyukuk. No sweat!

Problems: David

I brought the new Johnson 18 horsepower outboard to the Yukon River and mounted it on the stern. The gas and oil were mixed and the gas tank was filled. I pulled the starting cord several times, but the engine would not fire.

Vincent Yaska Sr. saw that I had a problem and came to help. He was also interested in seeing my new motor and boat. He pulled the starting cord several times, but nothing happened. He removed the engine cover, looking for loose wires.

Vincent suggested, "Let's check the magneto to see if there's a spark." He had much more experience than I and found the problem right away—no spark.

Next, to get to it, the flywheel had to be removed. He remarked, "Don't worry, I can get it off." He went home and got his single-bit ax, which had seen a lot of use. He said, "Just step back," as he took a hardy swing and hit the flywheel, but it didn't move.

Vincent announced, "Your new motor is not broken in yet! One more try." I silently and fervently prayed. There was nothing else I could do. I felt uptight, but I tried not to let it show!

He had good aim as he swung and hit the flywheel again. This time it broke loose. "Yes!"

The next problem was finding a working magneto that would fit. This is why people living in the bush stockpile old motors rather than throwing them away. Vincent ran off and came back shortly with a mag from somebody's motor and installed it. The engine started right up, and I never had to replace that "old part."

I am still learning to leave my problems in the Lord's hands, allowing Him to answer my prayers in His way and in His time. The Bible states, "A man's heart plans his way, but the Lord directs his steps" (Proverbs 16:9). I was glad I had not interfered and stopped Vincent from taking that second swing with his ax, although he gave me an anxiety attack! God intervened and did it His way—the best way.

The little ding didn't hurt anything. The dent was a reminder to me of God's having sent the right person at the right time when I needed assistance.

The Long Trip Home: David

Leaving Galena, I headed down the Yukon River to pick up the rest of our belongings in Koyukuk. Then I started up the long Koyukuk River to Allakaket.

The first night, 80 miles up the river, I stopped at an abandoned cabin, secured my boat, and then went inside with my sleeping bag

On the Arctic Circle: Allakaket (1965~1969)

and 30.06 rifle. I shut the door as securely as possible, laid my sleeping bag out on an old set of rusty bedsprings, and fell asleep. I was tired!

Something walking on the metal roof in the middle of the night awakened me. I grabbed my gun, wondering what was going on. A black bear swung his paw down through a small opening in the roof. It was just like he was waving at me! I clutched my rifle and prayed he wouldn't join me inside the cabin. Eventually, he took off, but I lay there wide awake, listening for a long time!

The next morning, I saw the bear's paw prints going to my boat, but, thankfully, he hadn't jumped in. King salmon strips were under the canvas, and I did not want to share them with him.

Back Home: David
My trip to Galena, Koyukuk, and back to Allakaket took 10 days. I told Kay how God used Vincent to answer my prayer. I had no motor problems on the long trip.

We chose a sunny fall day for our family to go by boat 20 miles upriver to pick cranberries. We walked a little way to a berry patch. Dan commented, "First, I was tired of riding in the boat. Then I was tired of walking. Now, I'm tired of picking berries." He sounded worn out. But he became an excellent berry picker. Together we harvested three gallons of cranberries.

New School Teachers: David
Bill and Syl Brown were the new school teachers. Debbie enrolled in the third grade, and Dan in the second. The school had two classrooms. One room had first through fourth, and the other had fifth through eighth. The high school students were sent off to boarding schools.

The Bishop's Encouragement: David
Bishop Gordon encouraged me to minister in the local church since there was no resident priest. I wanted to supplement the Native lay readers and not compete with them in any way.

Great Language Helper: David
Edward Bergman Sr. assisted me with the translation for use in the church. In the past, a priest spoke in church, and Edward translated the sermon into Koyukon. I asked Edward, "How did you translate words you didn't understand?" He responded, "I did the best I could, but a few times

Edward showed me his fish trap.

had to skip over a word I didn't understand." This is a common occurrence with translators. But in translating the Bible, we must find the right word or choose one with a similar meaning.

Bible Passages in Church: David

On Sunday, starting in October, I gave my first Bible lesson using flashcards. I started by reading, "In the beginning, God created the heavens and the earth" (Genesis 1:1). We had made seven big circles out of construction paper. I showed what God created on each of the six days of creation. The seventh day was a day of rest. I shared in Indian and then in English, sentence by sentence.

Lay reader Ken Bergman asked me to speak right after taking the offering. Following the service, Ken thanked me: "You helped make church longer and more interesting." People said they understood my Indian, but I hoped they were not just being polite. I knew I had a White man's accent, and I needed my cue cards for assistance.

I continued throughout the winter to prepare and present Bible truths in both Indian and English. I normally used flannelgraph figures that I placed on the board as I shared. We wanted people to apply Bible lessons to their personal lives. We were grateful for the eager listeners and this opportunity to serve. Most of the people in the village attended church. Afterward, we all went outside and visited.

Pumping Away: Kay

I knew how to play the piano, so I was destined to use the old pump organ. No one else played. The air bellows had leaks, so I had to keep pumping to keep the proper sound. This organ gave my legs a good workout! Everyone was glad I was willing. Music helped the singing. It was better for me when I chose the hymns; otherwise, I needed to practice before the service in an unheated church building.

On the Arctic Circle: Allakaket (1965~1969)

Kay was on the riverbank with Cesa Bergman, Kathleen and Joe Williams Sr.

First Christmas in Allakaket: Kay

On Christmas Eve 1965, we bundled up in our fur boots, big parkas with fur ruffs, and warm mittens for the walk to the community hall. Everybody was there, eagerly waiting for Santa. He arrived, even though it was negative 57 degrees. This Santa was the funniest we'd ever seen— way too skinny! He carried a tall cane and hopped around the hall. When a child's name was called, he gave them their present and chased them back to their seats, thumping his cane on the wooden floor. The younger children were scared of him, but they wanted their presents and stepped forward. Hard candy, nuts, cookies, oranges, apples, and Cracker Jack were distributed to everybody. It was a happy village celebration!

At home, we read the Christmas story in Luke chapter two. Our children hung up their Christmas stockings, hoping to find a few gifts in the morning.

The next day, I cooked a turkey for Christmas dinner. We invited Edward and Elizabeth Bergman and their five children, ages nine through 20, Cesa Bergman, Edward's mother, and Gene, our neighbor. This was the second time Edward and his family had eaten turkey, and they enjoyed it. We especially liked the opportunity to celebrate the birth of Jesus with friends.

David and Gene took our kids and a few others around town to sing Christmas carols. They also sang to Eva David, who had been bedridden for years. It took longer than we wanted, but there was no school the next day.

1966
Koyukon Calendar: Kay

The Bible verse printed on the calendar was 1 Timothy 1:15, "Christ Jesus came into the world to save sinners." We prayed this would send a clear message to the Koyukon people, stating why Jesus came to our world.

We planned to have reading and literacy classes in the future. Having a verse printed in both Koyukon and English gave the verse prestige. People could look at the calendars on their walls and consider the meaning of the verse. One woman asked me several times, "How do you say this Bible verse?" Once she had it memorized, she could think more about it. We continued to make calendars for the Koyukon people, each year with a new Bible verse.

Sewing Club: Kay

I hosted a sewing club at our place every Monday night. The first time, only two women came, but between 10 and 15 women came during the winter months. This was an excellent opportunity for me to hear and practice Koyukon. It was also a good time to complete sewing projects. Most of the ladies were sewing fancy beadwork.

Caroline (Koyukuk) Bergman brought all the materials needed to make the outside shell for a parka. She cut out the pattern and sewed it together, including the sleeves, pockets, and rickrack. It was after midnight when Caroline and Lydia Bergman went home, but she had finished her project in one evening!

Our electric Singer zigzag sewing machine was getting good use. The women were afraid to use it at first since they were accustomed to hand-crank ones. We were thankful for all the people who donated Betty Crocker coupons so we could get the machine.

Valentine's Day: Kay

Debbie was chosen as the Valentine Queen for the first through fourth grades. She was so pleased that she wore red leotards and a red dress with white butterflies and lace. The teacher made paper crowns for the King, Queen, Prince, and Princess, who handed out valentines for their party.

A Valentine's party for the adults was held at our house. We played games, such as popping the balloon while sitting on it and passing the lifesaver on toothpicks. At first, many were bashful, but all had a good time and laughed and laughed after they got started. It was the first time they had played games like that.

Angel food cake, ice cream, coffee, and tea were served. Two ladies brought cakes. We churned our three-gallon freezer of ice cream and dished it up ahead of time in paper cups. It was a houseful, with 40 adults and no kids.

New Clergy in Town: David

Bishop Gordon flew in from Fairbanks with Jim Bills, the new Episcopal priest, and his wife, Harriett. They moved into the two-story mission house next to the church, which was near the center of town.

Jim asked me to have a personal Bible study with him weekly. He suggested the Gospel of John, and we met together each Thursday. It was encouraging to study the Word of God and help him adjust to life in the bush.

Duck Hunting: David

Joe Williams Jr. invited me to go hunting ducks up the Alatna River. I had a double-barrel 12-gauge shotgun. My method was to load both barrels but I put only one finger on one of the triggers. Joe told me, "Better to have both fingers on both triggers, but only pull one at a time." A duck flew over, and I shot it. It dropped right in front of us! Then more ducks flew overhead, and Joe yelled, "Shoot!" I pulled the trigger, but nothing happened. The first duck was full of lead since I had pulled both triggers.

Fun Under the Midnight Sun

Have you ever played volleyball at midnight? The village people gathered every night during the summer, ending around two or three o'clock in the morning! Everybody joined in and had fun. On the Arctic Circle, the sun barely goes below the mountain peaks during June and July, so there is plenty of natural light. Fewer mosquitoes were out late at night when it was cooler.

"AK Is OK"

Someone came up with the idea to make T-shirts with "AK Is OK". "AK" is an abbreviation for Alaska, but in this part of the world, "AK" stands for Allakaket. Every adult had one of these, and the fad lasted until the shirts sold out. Years later, people still say this catchy phrase.

Steamboat!

People hollered "Steamboat!" when they saw the diesel-powered *Taku Chief* tugboat, coming from Nenana with a barge load of freight. It arrived in June. The barge was the most economical way to transport freight to Allakaket. Our family received two 100-pound tanks of propane.

Bible Conference

We flew into Fairbanks and spent a few days at the Wycliffe Center. Then we drove to the Bible and Missionary Conference at Victory Bible Camp in Glacier View, Alaska. Most mission attendees were with Arctic Missions. The singing, speakers, and Bible teachers were outstanding. We were looking forward to meeting Indian and Eskimo believers from other parts of Alaska. All four children had special activities for kids their age. We felt refreshed. We continued to attend these annual conferences since we appreciated the fellowship and encouragement.

On our way home, we stopped at the Wrangell Mountain Bible Conference in the Copper River Valley area near Glennallen, Alaska. A Native man presented an object lesson, and another preached. People went forward seeking help to live closer to the Lord. They were so friendly and welcoming!

Broken Windows—Again

Returning to Allakaket, we were disappointed to see five broken window panes in our house. We vividly remembered all the shattered windows when we first moved in! We didn't dwell on the problem or get angry but we prayed and wondered what to do. There was no local place to buy window panes.

Soon, Bertha Moses came over with a package of five window panes. People had pooled their money and ordered the glass for us. How thoughtful of these caring people! We thanked the Lord for His solution.

A Helper: David

Larry Neiswender, a single fellow, came to Fairbanks to help Wycliffe members. He flew to Allakaket for two weeks and painted outside around our windows and our front porch. Larry also helped get firewood for the coming winter.

Fish Camp

Lydia Bergman invited us to their fish camp while her husband, Lindberg, worked at the Hog River gold dredge near Huslia. Her fish camp was 11 miles downriver. Lydia put in gill nets, so we had fresh white fish to eat and others to cut for drying.

Life was much simpler at camp. We lived in our wall tent and cooked on a rectangular Yukon Camp Stove with folding legs. The kids enjoyed swimming in the shallow, warmer water by a sandbar. We even found blueberries nearby. All ages liked singing and doing the motions with "O Be Careful, Little Eyes, What You See." It was a delightful two weeks.

House Fire

The interior of Joe and Rhea Williams' house was blackened by fire in September. Everybody in the village worked together by lining up and passing buckets of water from the river to save their house. Rhea and her family stayed with us the first night before moving closer to their home.

Fall Moose: David

I took Grafton Koyukuk upriver to look for moose. Grafton knew the best places to hunt, and we split a moose. We ate good, natural, organic meat all winter.

Three Students

In the fall of 1966, three of our children began school. Beth Ann entered the first grade with nine students, Dan was one of four third graders, and Debbie's fourth-grade class had eight students. Only Stephen was too young to go to school.

Two new teachers arrived. Mr. Hill, an older man, taught the lower grades, including all of our children. He was super conservative. He felt he was saving the state money by running the school generator at an unusually slow speed. The lights in the school were dim. Soon, the generator quit and the lights went out, so they had to use candles and gas lights.

Stephen turned five, so we invited seven village boys to celebrate and eat his birthday cake.

Hog River: David

Jim Bills asked me to accompany him on a ministry trip to Hog River in October. On our way, we stopped in Hughes. I told the story found in Mark 2:1-12 about Jesus forgiving a paralyzed man of his sins and healing him. To aid in telling the story, I used a 10-inch-square box to show a house with a flat roof and a stairway on the outside. About 50 attended the service in Hughes.

The next day, we flew to the gold dredge at Hog River (Hogatza). We had two services since 20 men were working on two different shifts. Twelve men were from Allakaket, plus others from Hughes and Huslia. They fed us well!

On the Arctic Circle: Allakaket (1965~1969)

Caribou in Town: David

A short distance up the Alatna River, I shot two caribou. I saw a group of 15, but my gun jammed. I brought my two caribou home and repaired my rifle.

While I sat down for a tea break at home, someone ran by our living room window with a rifle. I watched three men shoot several caribou on the frozen river in front of the village. We lived in caribou country!

Bear Party: David

The bear parties for men only were held outside the village.

Oscar Nictune Sr. tended two big pots of bear soup.
The paws, kidneys, and heart were cooked on sticks for the elders.

Another time, I went bear hunting with Bergman Sam from Huslia and several Allakaket men up the Alatna River after the first snow. We found fresh bear tracks leading to a den, so we knew it was home. One man took a long stick and poked the bear in its den. Another man shot the black bear as it emerged.

We had a bear party at the kill site. The hunters cut foot-long sections of the small intestines encased in lots of fat. They held one end tightly while using their other hand to squeeze out the unwanted contents. Then, they roasted the empty entrails on a stick over the fire. I took a piece, cleaned it, and ate mine well done.

New Snow Machine: David

I purchased a Ski-Daddler snow machine with a 10-horsepower engine, during its first year of production. Everyone else in town had single-track Ski-Doos, and they were zipping everywhere.

I asked Johnson Moses if he wanted to take a ride up the Alatna River. He said, "Sure, Let's go! Do you have enough gas?"

I responded, "Yes, the manual says we can go 50 miles on one tank." I had just filled the tank.

So we took off. Soon the engine sputtered, and we ran out of gas! We had to walk back to Allakaket. I heard stories about running out of gas with my Ski-Daddler for a long time after that. The good mileage was based on level land and one rider.

Firewood: David

It took many cords of firewood to heat our big house. Our "iron dog," or snow machine, was our primary source for hauling firewood since we didn't have a dog team. The front door of our home was four feet wide, so I took advantage of it. I brought the Ski-Daddler into our living room and turned it around on our linoleum floor, so it was ready to shove out into the cold world. The machine had a bigger and softer cushion seat than the Ski-Doos running around town, which doubled for extra seating. Our house served as a heated garage.

January and February 1966 were frigid, hovering between minus 45 and 55 degrees. We ran out of dry wood, so I was forced to cut green birch. I could make only one trip before the snow machine track became too stiff to operate. Every day, I made one trip for firewood. Green birch worked in our barrel stove, as long as I didn't let the fire die down too low.

Ice in Our Wash Basin: David

The three bedrooms on the second floor were warmer since the heat rose, but the first floor could get cold. Unfortunately, the fire sometimes died out at night, even though I threw in extra wood when it was extremely cold.

Ice formed on the standing water in our wash basin. Nobody wanted to go downstairs until I started the wood stove. I made a tiny room with a bucket toilet under our stairs. One of my regular chores was to empty it.

Bath Time: David

We thawed the snow on top of our stove. The snow melted down considerably, so preparation for a bath required several trips to get enough water. Occasionally, a couple of rabbit nuggets were mixed in with the snow. They were clean, but we got rid of them.

Several round galvanized steel washtubs two feet in diameter were used as tubs for washing clothes. They doubled as a bathtub when our children were small.

We purchased an oval tub measuring three and a half feet long by two feet wide. When our kids were young, we had the girls take a bath together, followed by the two boys, to conserve water. It was barely big enough for adults to sit down and bathe.

Dan, Debbie, and Beth Ann filled and carried round tubs filled with snow.

Race Time: David

I entered the old men's 10-mile dogsled race since this fun race was set up for those who have slower dogs or didn't compete in other races. I had to borrow dogs. Kay entered the women's dog race, and both Debbie and Dan joined the kids' races with their sleds. Dan fell off and lost his dogs, while Debbie held on for dear life to the finish line. We all had fun! The start and finish lines were on the river in front of our house, making it easy to keep up with the excitement and visitors. Many people came to our home to warm up or grab a cup of coffee while watching.

Christmas: David

Christmas dinner with Simon and Pauline Ned, plus their two boys and two grandsons, was enjoyable. We had turkey, which people referred to as a "big willow grouse" since both have white breast meat.

1967
Eskimos Friends

Kobuk River Eskimos lived across the river in Alatna. The older ones were trilingual, speaking Eskimo, Koyukon, and English. Four Eskimo women married Indians and lived on the Allakaket side. The two villages worked together well.

Our translation colleagues, Wilfred and Donna Zibell, lived in Ambler. They sent an Eskimo primer, a *Life of Jesus* booklet, and hymns, which we gave to Oscar Nictune to read and share with his people. Grandpa Oscar had 160 descendants living in this area.

Beaver Trapping: David

Edward Bergman Sr. and I went beaver trapping. We drove 20 miles on my snow machine and camped in his tent. We checked one bear den on the way, but nobody was home.

The next day, he showed me how to set the bait by the snare, using fresh willows and birch poles long enough to push into the muddy bottom of a beaver pond. The following day, we checked the snare; it had caught one large beaver. At home, I stretched and nailed the skin to a piece of plywood. Kay cooked the beaver meat and tanned the skin, following a friend's advice.

Thankful for Caribou: David

Eleven caribou walked across a large frozen lake where Edward and I were camping. He had a 30.30 rifle, and I had my 30.06, but my gun froze up. Edward shot three while the rest ran to the opposite side of the lake. Then he shot over them into the woods, and they ran back toward us. He shot the rest of them, scattered across the frozen lake. (North of the Arctic Circle, there was no game limit.) We gutted all the caribou and turned them with their stomach side down, so they were not so tempting to the ravens. Then we hauled them home over the next couple of days. The heads made caribou soup for Grandma Annie Koyukuk's potlatch.

Edward told me about hunting up the Alatna River when there were very few caribou and little food at home. He saw five caribou, so he shot them! Then, "I dropped to my knees and thanked God for giving me the meat we needed." No one saw or heard him, but God did. The Bible reminds us, "In everything give thanks; for this is the will of God in Christ Jesus for you" (1 Thessalonians 5:18). Edward challenged me. I usually thank God silently, but I need to do it out loud—whether anybody else hears me or not.

Fighting Fire with Fire: David

I went firefighting with 25 men from Hughes and Allakaket traveling to the Tetlin area. We set fires to create a backfire, thus fighting fire with fire. Much of our work was mopping up the smoldering remains.

It was fascinating that the men on the crew spoke more Koyukon while working together than they did in town. I had to be careful not to laugh when I did not want to laugh because they sometimes used language puns with a double meaning or with a sexual twist to them.

I brought back some leftover GI rations. Our kids enjoyed opening all the little cans and sampling what I ate while firefighting.

Adding a Letter to the Alphabet: David

While firefighting, I was made aware of another Koyukon sound in their language that needed to be written to distinguish between two identical words. This voiceless "n" or whispered sound was written as "nh." For example, there is a difference in the final sound in "ten" (trail) and "tenh" (thick ice). "Ten hoolaanh" means "There's a trail," and "Tenh deedaakk" means "The ice is thick."

First Rabbit: Kay

Dan set snares and caught a rabbit. He was beaming. But he had forgotten to take a stick with him to kill the rabbit, so he took it out of the snare alive. He brought it home, thinking it was dead until the rabbit jumped out of his hands and hopped under our porch. Finally, the kids got it out. I hit it with a piece of wood, and we had fresh rabbit for supper.

Ice Trail Melting: Kay

All the sunshine and beautiful spring weather were enjoyable after another long winter. Beth Ann had an Eskimo friend, Sally Sam, who lived across the Koyukuk River in Alatna. She wanted to visit Sally on her birthday, though we thought the ice trail was too dangerous. There were dark spots of varying sizes on the winter trail. If you stepped on those spots, you could get a wet foot or go deeper.

We walked down to the airfield along the front row of houses, looking for her. Some people told us they had seen her heading to Alatna. We continued walking to the ice trail, but there was no Beth Ann. She had seen us coming and walked along the back row of houses to our home. We thanked the Lord for His protection, even when we were not there to help her.

A Baby Boy: David

Jim and Harriet Bills flew to Anchorage to adopt a boy named Tommy. One elder was amazed when they returned and said, "I didn't know Harriet was pregnant!" She was observant, but nobody told her about the adoption part. Jim asked me to lead three church services during his absence.

Initial Translation: David

We were thankful for the initial translation help with Edward Bergman Sr. This enabled us to get a working draft of selected stories from the Gospel of John plus other stories in the Bible.

After we completed the first draft of John 3:1-18, I took the opportunity to ask him deeper questions about his spiritual condition. Edward said, "I do believe in Jesus Christ with my heart." We often hear that if you live a good, honest life, you will go to heaven. But we also need a clean heart that God sees. I wanted to make sure—for his sake. We both learned spiritual truths from translating the Bible.

Field Director's Visit

Elliott Canonge visited us in Allakaket. Jim Bills and Elliott enjoyed discussing theology. After he returned to Fairbanks, Jim spoke at church saying that Jesus Christ died on the cross in our place to pay the ultimate price for our sins as a perfect, sinless person.

The following Sunday, he preached how everyone in Allakaket knew a lot of facts about the Bible and Jesus Christ, but now they needed to know Jesus Christ as their personal Savior.

On the Arctic Circle: Allakaket (1965~1969)

More of the Mundane: David

I helped people fill out their income tax forms. It was straightforward in those days. I also served as an ex-officio advisor for the village council. They were trying to get Head Start and the Neighborhood Youth Program going in Allakaket. More paperwork was required to obtain a village generator.

A Scary Sight: David

Our front door had a large glass window measuring two feet by two feet. Elliott Koyukuk was using his chainsaw to cut firewood on the beach near our house. When he knocked on our front door, his bloody face scared me! The chain had come off and it slapped him in the face. I brought him inside and gave him a clean cloth to wipe away the blood and push the torn flesh back into place. We sent for Bertha Moses, the village health aide, who immediately cared for him.

Family Entertainment: David

We had a stereo record turntable for our entertainment. We played the big 12-inch vinyl LP (long-playing) records. Each one gave us 23 minutes of listening pleasure of classical, modern, or Christian music. Most popular were the records with children's songs and stories. Many village children came to listen to our stories and songs.

One children's record was about a family in the early days of our country. They were heading west in their covered wagon with a group of pioneer families for safety from attacks by outlaws or Indians. In one story, they shot and killed Indians but were thankful none of their family was struck by an arrow. It was an interesting story, but we, as parents, decided it sent the wrong message. We did not want to kill Indians. We wanted them to have the same opportunities we had to hear about Jesus Christ. So I "tested" the record by fire in the wood stove, and it was gone. At certain times, we have to censor what we allow our children to hear.

Taking the Next Step: David

The next step in our translation work was to use a translation checker. Eddie Bergman Jr. was a good fit to help. At 24 years old, he was fluent in both Koyukon and English. He lived with his grandma, Cesa Bergman, who preferred her Native tongue.

Eddie flew to Fairbanks for government training to become an accountant with the new Head Start program. After his training was over, he and I went to the Wycliffe Center for four days. Some concepts were difficult to translate, such as the part about "should not perish, but have everlasting life" in John 3:16. We finally settled on "will never die, but will always live in heaven." We continued to check and recheck for accuracy.

David, Eddie, and Elliott

Fishnet: David

Back in Allakaket, I put a fishnet in a small eddy upriver. It wasn't the greatest location, but we got a few fish. Then Beulah Moses set her fishnet in the same area, and she and Kay went out together to check the nets each day. The Koyukuk River did not have an abundance of salmon like the Yukon River, but we had enough for our families.

A Paint Job: David

A state maintenance supervisor arrived to work on the school's projects. He needed extra assistance with the electrical work and asked me to help during his last week in Allakaket. Oscar Nictune Sr. was painting around the windows on the outside of the building. I jokingly said to Oscar, "Oops! You got some paint on the window panes." Oscar replied, "That's how I know where I've painted."

Meeting the Tall Lady: David

At the Fairbanks airport, we met "The Stilt Lady," an Alaska Airlines representative who greeted people while on stilts. All of us were impressed and thrilled to see her.

Shortly after returning to Allakaket, we were awakened at 5:00 a.m. by Debbie's hammering. She had found two poles and nailed footrests on them. When we got up, she was walking on her stilts. Soon, several of the village kids made stilts and strutted around on them.

Winter Is Coming

With the long hours of summer daylight, you could go outdoors anytime and see what you were doing. Stephen woke up about midnight in the dark and said, "It's funny in here. Why is it dark?" It was August, and he had forgotten that it was getting dark at night as we lost an hour of light weekly.

Sheep Hunting: David

Sheep hunting 200 miles up the Alatna River with Oscar Nictune Sr., Robert Williams, and Jim Bills was an adventure. In some places, the water was so low that we had to raise the motor and push the boat over the shallow riffles using poles.

Oscar wanted me to go on foot with him 20 miles back from the river to look for sheep, while Jim and Robert stayed at our camp on the Alatna River. We saw a massive stone called the Great Horned Owl. It was obvious because of its shape, as we could see both ears on the head. A gold prospector had lived along the stream during a winter of deep snow, and all the trees nearby were cut about four feet up from the ground. We saw sheep, but they were too small.

While traveling back on the Alatna River, we saw one lone caribou, which we shot. A little later, Robert spotted a grizzly bear. We all stood on the river bank and shot simultaneously, like a firing squad. The bear dropped instantly.

At another site, Oscar informed us where to tie up to see all the lowbush cranberries. It was cranberry heaven! I have never seen so many, and so thick, in one place. Oscar was a great guide for our 10-day hunt.

More Cranberries: David
While I was out hunting, Kay and our children picked cranberries. They went five miles up the Alatna River in Beulah's boat, along with another boatload of 18 women and children. Kay and the kids filled a 25-pound flour sack. It provided good eating during the winter months.

New School Teachers: Kay
The new school teachers, Howie and Barb Van Ness, transferred from Tanana. Howie taught the upper grades. Unfortunately, he was sick with strep throat so I was his substitute teacher for a week. I was hesitant because of my disappointing experience with fifth graders in Michigan and their lack of discipline. But the students here were respectful, and I was glad for a positive experience.

Village Power: David
The Bureau of Indian Affairs (BIA) sent two used generators to Allakaket. This was the first time the village had electricity. Previously, they had used kerosene, Blazo lanterns, or candles. Everyone was anxious for brighter lights. Most homes were small and required only two overhead bulbs and one wall outlet. I helped with the wiring and the new power system. Now everyone had electric lights available 24 hours a day! The charge was 15 dollars a month. At this point, it was pretty straightforward, but it didn't take long before people purchased electrical appliances, like coffee makers and more, which challenged the system.

Two Moose and a Raft of Wood: David
That fall, I shot two moose—a cow, and a young bull. The meat was delicious, so our family was well-fed. On another trip, Dan helped me build a raft to bring the five cords of firewood we cut to our house.

New Neighbors: David
Lee and Sarah Simon moved back to Allakaket and lived in the other half of our "duplex." The walls were thin, so we knew they were there.

They hosted a birthday party for one of their grandchildren and invited us. We sat on the floor, with the food spread out on a cloth. The dinner included a porcupine. They had burned the quills over an open fire before it was gutted. Then they scraped the blackened skin to get rid of the quills. The meat had a unique burnt taste. Sarah stated, "This is the Native way to cook a porcupine." Previously, I had skinned one and roasted it over an open fire.

"Turn Your Radio On": David

The lyrics of the gospel song "Turn Your Radio On" should be valid anywhere in Alaska. When the reception was better in the evenings, we could choose between two Christian radio stations. One was KICY in Nome, which started in 1960. The other was KCAM from Glennallen, broadcasting since 1964, but their signal was fainter. We appreciated the opportunity to hear Christian music, Bible teaching, and sermons, even with static.

Fast-forward: In 1985, KIAM, "The Gospel Voice in the Wilderness," was started in Nenana. Later, their programming was broadcast directly to Allakaket.

Christmas and New Year: David

Christmas dinner at our house included Grandma Cesa and three others from her family. The village celebrated with three days of dog races: men's, old men's, women's, and kids', plus snowshoe races, and "iron dogs" (snow machines). I competed in the Iron Dog race. My machine didn't catch their fast ones, and no one challenged me for the last place.

1968
Move to the Big City: Kay

On January 5th, we temporarily moved to the Wycliffe Center in Fairbanks to start adjusting to life in the big city. Our children appreciated living here with endless running water from a faucet and oil heat without cutting, splitting, or burning firewood. Chipping a hole in the ice to fill buckets with river water had been one of their chores in the village.

Our three older children caught the school bus at 8:00 a.m. and returned home at 4:30 p.m. They were in the second, fourth, and fifth grades at Joy Elementary. School in the city was a cultural shock for our children.

Stephen stayed home since there was no local Head Start program. He told us, "I like Fairbanks, thanks for bringing me." He was especially enthralled by all the vehicles since snow machines were the only ones in Allakaket.

He Turned to Jesus: Kay

Mike and Susan Lockwood, former school teachers from Koyukuk, lived in Fairbanks and invited us to their home. Our children enjoyed playing with their daughter, Michelle, and her toys. Mike was away, so Susan told us, "Mike turned to Jesus Christ and is saved!" His recent decision was exciting news!

New Church for Us

We attended a new Conservative Baptist church in Fairbanks called Bethel Baptist. It was small, but the people believed in the Bible and Jesus Christ. Two families in the church, the Holmgrens and the Bartletts, both had children the same age as ours.

Kokrines Again: David

I invited Frank Titus to Allakaket before we left, but he declined. He wanted me to come to Kokrines instead. On January 22nd, I flew to Tanana, then chartered a plane to Kokrines. Only two residents, Frank Titus and Josephine "Koda" Corning, resided there. They lived in a small cabin, 10 feet by 14 feet. I brought some food for them. They had a big moose head thawing out, and we ate all the edible parts over several days.

Frank was anxious to see the translation from the Gospel of John. So I started reading to him. He stopped me and said, "David, that's my language. I'll read it!" He did, too. He read it with minimal hesitation. Frank's brother, Henry, could also read Koyukon without any training. What a wonderful gift and surprise. Frank was a superb language checker. He had a lot of perseverance and worked diligently until we finished the Gospel of John in Central Koyukon.

It was wiser to talk louder rather than keep repeating myself. Both of them had significant hearing loss and wore no hearing aids. Hearing "What?" was not my favorite word.

Frank and Josephine standing in front of their cabin in the summer

Don Honea Sr., from Ruby, stopped in while traveling to trap along the Novi River. Don said, "David, I'm not deaf like those two people." So I lowered my volume until he continued his journey upriver.

"Uncle Sam" Died: Kay

My maternal grandfather, Amos Sanford Wall, passed away in Michigan on February 15th, 1968. He worked as a truck driver for much of his life, but in his later years, he was the friendly janitor at my school. He trusted in Jesus and lived for the Lord. I missed his funeral, but I plan to see him in heaven when my time comes.

Back in AK Again

We moved back to Allakaket and began the daily chores of hauling water and cutting firewood on the first of March. The break spent in Fairbanks was enjoyable, and our kids got a different perspective on life in the city school.

Victory High School

The Gospel Team of four Native students plus a staff member and pilot, Ken Hughes Sr., flew into Allakaket. Two meetings were held in the village hall, with 100 attending each night. Adam Aposik played the accordion, and their group sang several songs surrounded by children eager to hear them. Most were sung in English, but three students sang hymns in Yupik Eskimo. People were impressed with these teenagers. Several adults told us, "I wish our high school students would live for God."

Grandpa Joe Williams Sr.: David

In March, Grandpa Joe Williams Sr. went to the Native Hospital in Tanana with pneumonia and heart problems. I had visited him previously and asked him about his relationship with Jesus Christ. Did he completely trust in Jesus? Grandpa Joe replied, "I do, and I have believed this way for a long time."

We reminisced about the first time we met, in 1960, when I came to Allakaket and stayed with him and his wife. Grandpa Joe remarked, "People asked me, 'Who is that stranger? We don't know him.'" He went on to say that now people are worried we will leave the village. We laughed about the change.

God at Work: Kay

Effie with Kay on the riverbank

This spring, the Lord put Effie Williams on both of our minds, with a sense of urgency to visit her. She had just returned from the hospital in Tanana after a severe bout with pneumonia. She weighed only 79 pounds and was never in good health. Since she was a woman, I went to see her while David prayed at home. Effie was also visually impaired and she lived with Lee and Sarah.

I was surprised when Effie immediately said, "Kay, I want to be a Christian." We talked about John 3:16. She already knew she was a sinner like the rest of us. I told her, "For whoever calls upon the Lord shall be saved" (Romans 10:13). Then we both prayed. I was excited to welcome her into the family of God.

She didn't have a Bible, so I went home to get one. Our son spoke up, "She can have mine." So I gave Dan's Bible to Effie which he had earned by memorizing verses. We continued to meet together and talk more about her spiritual life.

Native Bible Study: David

In the spring, we started a Bible study in our house. I typed out 10 verses for each session. Eleven to 26 adults attended weekly. I read God's Word in Koyukon, and then we discussed the passage. The group served as a translation check for these Bible verses. Summers are busy, so we ended before the river opened up.

Help from Fairbanks

Elliott Canonge flew in from Fairbanks to look at the selected passages from the Gospel of John over six days. Bible translation work is a long, tedious process, requiring checking numerous times to be accurate.

High Water

For two weeks, in June, the Koyukuk River flooded the Allakaket airstrip, which was on lower ground than the village. We planned to fly to Fairbanks but postponed our trip for another week. The water rose to the top of the boat slip beside our house. Natural disasters make one pause and think. We can't control the water level, but we made preparations in case it flooded.

My Father Is Sick: David

On June 18th, I left the village to see my father, who was seriously ill. One of his symptoms was pleurisy, or excessive water in his lungs, hindering his breathing. I stayed for a week, helping my mother with various projects at the house.

I flew as far as Seattle when I broke down at the airport waiting for the next flight to Kentucky. I prayed and asked God to let me see my father alive and not have to go to his funeral at this time. The Lord answered my request. My father returned home and lived for another three months.

Stephen, Beth, Dan, and Debbie showed how thick the ice was at the breakup.

With the Lord: David

My father, Luther Cameron Henry, took his last breath at 70 in a hospital in southeastern Kentucky on September 22nd, 1968. And the very next moment, he was in heaven with his Savior. Martin was with him in the room when he died and watched the life drain from his body. His autopsy listed cancer of the lungs and diaphragm as the cause of death. He wasn't a smoker, but he had worked for years in a chemical factory in Illinois without protective gear.

My father accepted the Lord Jesus when he was 12 and served Him until he died. I'm looking forward to seeing him again. He was buried on the grounds of Camp Nathanael, which is part of the Scripture Memory Mountain Mission in Emmalena, Kentucky.

Bible Study with Effie: Kay

I continued a one-on-one Bible study with Effie Williams. She often read her Bible and studied before I arrived. She wore out the one I had given her. She knew what God's Word said and tried to obey what she read.

Because of Effie, several others wanted to join our Bible study. Jennie Williams and Mary (David) Williams chose to come and learn more about Jesus Christ. Mary's daughters, Rose and Priscilla, who knew English better, would help their mother. If her husband, Henry "Peanuts," came with her, David would go to a different room and read the new Koyukon translation to him.

Fast-forward: Before we moved to Fairbanks, Effie encouraged Mary to enroll in the monthly Bible correspondence courses sent out by Arctic Missions. Mary's daughter Priscilla continued to help her understand and complete the lessons after Rose left for high school.

Even at funerals or potlatches, Effie would bravely stand up and recite, from memory, Bible verses to fit the occasion. Her poor eyesight did not hinder her from being a witness for our Lord.

The Seattle Times: David

Stanton Patty, a reporter for *The Seattle Times,* and Art Patterson with the BIA came to Allakaket. Their project was to visit 12 interior Alaskan villages and write a series of six articles about Native acculturation, the process of changing to the dominant American culture. We were impressed with Stanton as an honest reporter doing research without the political bias of most modern "journalists." The article about Allakaket appeared in *The Seattle Times* on November 11th, 1968, with a nice picture of Eddie Bergman Jr. fishing in the fall. Eddie was jigging for grayling the traditional way, but he was using a modern spinning rod and reel, mixing the old way and the new.

A Bit of Intrigue: David

Several years later, I found out that a friend from Allakaket had lied to the man in Hughes, saying, "David is no good, and he wants to start another church." As a result, the owner of the house in Hughes refused to rent it to us.

The man did this because he wanted us to move to Allakaket. It worked. Does the end justify the means? Sometimes, God works through unusual circumstances. The Allakaket man told me the man in Hughes was upset with him for doing this. The interesting part is that I'm good friends with both men.

Final Packing: Kay

We had a steady stream of visitors as we worked on packing for our move to Fairbanks. As a token of their love, the people of Allakaket and Alatna took a collection to help us with our travel expenses. We sincerely appreciated their thoughtfulness and kindness. We left many good friends and fond memories.

Life in the Big City: Fairbanks (1969~1972)

A Coal Truck Hit Your Mother: David

The first phone call we received in Fairbanks was from the director of Scripture Memory Mountain Mission, where my mother served as a single missionary following my father's homegoing. He simply stated, "A loaded coal truck hit your mother in a head-on collision on a mountain road. She's in the hospital." He wasn't sure if she would pull through or not, but he spared relating that information to me. The Lord's mercy saw fit that she lived.

She had fallen asleep at the wheel and drifted into the coal truck's lane. The steering wheel mangled her nose, fractured her jaw, and lacerated her face. She spent a week in the hospital and finished recuperating at home before resuming her ministry in the public schools.

After that, my mother never looked the same. Her "bad" nose job left her with a flattened, pug nose instead of the more prominent, pointed nose with which she was born. Frankly, my aunt Lillie Bachanz looked more like my mother!

1969
Adjusting to City Life: David

We moved into the Wycliffe Center in Fairbanks on December 10th, 1968, for the best interests of our children. The school system in Allakaket only went up to the eighth grade. We wanted them to live at home with us during their high school years since I had spent three years in boarding schools. We wanted to give our growing family a wider range of activities to do and friends to meet. We planned to continue visiting the 11 Koyukon villages and keeping in touch with the people.

After moving into a house with many appliances, our children were informed, "If you turn off the lights when you leave a room, we'll have more money to buy ice cream." Ice cream was a favorite treat. Every time we walked out of a room, a little hand would instantly flip off the light switch. We purchased half-gallon cartons of ice cream and divided the contents into six servings.

All four children attended Denali Elementary School. Debbie was enrolled in the sixth grade, Dan in the fifth, Beth Ann in the third, and Stephen started kindergarten.

On school days, we got up at seven so our three oldest could catch their bus by eight. They walked half a mile on Wigwam Way to meet the bus at one mile Chena Hot Springs Road. Stephen's kindergarten started at noon. I drove him to school, but he rode the bus home with the big kids.

A Car in Cold Weather: David
One week, we missed Sunday School because we forgot to plug in the head bolt heater for our car. It was minus 40 degrees and it was too cold for the engine to start. In Allakaket, it didn't matter what the temperature was; we walked to church.

When setting in extremely cold temperatures, the car tires would freeze flat on the bottom. I drove the vehicle, but there was a thump, thump until the tires warmed up enough to ride smoothly again. Modern tires don't freeze or thump like that. One time, I drove with a thumping tire down Airport Way and soon realized I actually had a flat tire.

Village Visitors: Kay
Elsie (Williams) Bergman from Allakaket escorted her son, Wilbur, and Gary David into Fairbanks to the orthopedic clinic. After their appointments, we brought them to our place. Elsie's brother and sister attended high school in Fairbanks, so they rode the school bus home with our kids. The little boys were fascinated when they rode on the escalator at JCPenney. They also enjoyed turning our light switches off and on. At this point, the lights in Allakaket had pull-string light switches.

Living in Fairbanks allowed us to visit Natives in their homes, at special meetings, or in the hospital. We were also given access to see inmates in the attorney rooms in prison.

Subbing in Fairbanks: David

The opportunities were different in the big city. I preached through a series on the book of First John at University Park Bible Church since they were between pastors. I also spoke at Bethel Baptist Church, and Kay taught a junior high Sunday School class.

I was a substitute teacher at Lathrop High School and a junior high. It was an excellent opportunity to meet Native Koyukon students in the classrooms.

Printing Projects: David

Three big printing projects were completed in preparation for our coming summer activities: *Hadohzil-eeyah: We Are Reading*, a Koyukon reader; *Dinaahoto' Dinaayił Hanaay: God Talks with Us*, 10 stories from the Gospel of John; and *Dinaaka, Our Language*, a Koyukon dictionary.

Printing projects required much time in the '60s. Thelma Webster drew all the illustrations. The printing plates were paper masters that were easily damaged and could not be reused. After printing, several people helped us collate the pages. Then we punched holes in the pages and fastened them with spiral binders.

New Set of Wheels: David

Our biggest expense in the city was transportation since we lived five miles out of Fairbanks. We bought a new-to-us 1964 red Chevy station wagon with a white roof for 650 dollars. Our growing family required more trips to town.

Cabin Prayer Meeting

The 30-minute radio program *Cabin Prayer Meeting of the Air* was aired weekly on the radio station KJNP in North Pole. We transported Native people to participate.

Back to Allakaket: David

Books were ready for teaching Koyukon, and we were anxious to start our summer on the river, going from village to village. On May 29th, we flew back to Allakaket in a Cherokee Six from Fairbanks. This accommodated the six of us, our dog, and most of our freight. We kept waiting, hoping that our remaining packages would arrive.

All the popcorn we left in Allakaket was popped. We ate rice for several days. Our cupboards were bare. We prayed for food to eat and for our boxes to arrive. Two men ended their river trip at Allakaket and gave us their last jar of chipped beef before they flew home. God answered our prayer for food. The local store was low on food or out of many items as they waited for the annual barge shipment.

I put in a fishnet and caught two whitefish. The fish were a welcome change of diet. A Native friend gave us tasty dried meat.

After two and a half weeks, our packages arrived on the mail plane! Then we had a variety of food, boots, and other supplies for our trip downriver. Our goal was to boat down to Kaltag before turning around, then head up the Yukon River, visiting every village and fish camp along the way as time and energy permitted. It quickly shaped up to be a full summer.

Summer on the River: David

We planned to distribute the three new books to encourage more interest in reading and writing their language. I designed a reading chart to present the Koyukon alphabet. We spent at least five days in each village. The alphabet was introduced during the first two sessions, followed by reading the primer. We would then proceed to look at the dictionary and read Bible passages from the Gospel of John. The river was open, and people needed to travel and fish, so the attendance was unpredictable. The first classes were held in our living room in Allakaket. Together, we read through three Bible passages in the classes.

We headed to Hughes on June 19th. Here is a diary entry by one of our children about our trip:

We started from Allakaket at 3:00 p.m. after saying goodbye to everyone we saw. After going 19 miles, we stopped to pick wild rhubarb. It was plentiful. Later we saw four widgeon ducks. Then we saw a cow moose and her calf from a distance; when we came closer, the cow started running and tried to get up the bank with her calf right behind her. They tried to climb a steep bank but couldn't make it. Then she ran a little farther and climbed up the bank. It kept trying to climb while we came closer and closer in our boat. When we were right across from the calf, it climbed up and stood by its mother, just looking at us. Then they suddenly ran into the woods.

Soon afterward, we stopped at a cabin at Red Mountain to spend the night. We decided not to stay because recently a bear had been there. It tore the plastic off of the windows and clawed a lot of cans and other things. The bear crumpled a five-gallon gas can. (*Editor's Note:* Our dog acted strange and afraid around the fresh bear signs, so we left quickly.)

We traveled two bends from Red Mountain and spent the night. In the morning, we saw a black bear near our campsite. Late at night, when my parents were awake, they saw a porcupine. My father jumped out of the boat, got a stick, and killed it. He skinned it and cleaned it. The following night, we ate it for supper. Then just before reaching Hughes, we saw two Canadian geese. We stayed for six days in Hughes. One day, we saw a duck circle Hughes quite a few times. My parents taught.

We left Hughes at 1:30. The first night we went 40 miles. We saw—

And the diary entry ended there—in suspense! Or the writer had more important things to do.

Hughes: David

In Hughes, we met in the community hall. Indian literacy was a new experience for us. The Natives dubbed our classes "Indian School." In teaching the alphabet, we called one letter "Indian Ł." It was a regular letter L but with a bar through it, which sounded like an L but without vibrating one's vocal cords.

Our outboard motor gave us trouble while traveling to Huslia, and we had to drift for more than a day on the Koyukuk River in a slow-moving section of the river. Drifting meant that we all had to wear our head nets. It was challenging to eat while wearing them. The mosquitoes were bad, but the tiny gnats were worse. We killed as many as we could, but we were outnumbered! There were a bazillion gnats!

After fixing the motor, the gnats still welcomed us every time we stopped! Sometimes we ate our lunch beside piles of driftwood. I told our kids, "I'll make a fire for cooking, but each of you can build your own campfire." They did, too. We made sure they were extinguished before we left.

We stopped at Hog River Landing (Hogatsa), where the barge would tie up to offload supplies for the gold dredge. Our dog got too curious and came too close to a porcupine. My pliers pulled the quills out one by one. Our dog did not appreciate my services, but it had to be done.

Two men from Allakaket and two from Huslia came 26 miles from the dredge to see us. Kay had already boiled our porcupine, so she quickly fried it for the men. They were happy for the meat since liver was served for supper that night.

Huslia

We stayed at Bergman Sam's house in Huslia. The Episcopal Church had Bible School and our children enjoyed that. All the men and boys over 16 were firefighting on a 20-day stint.

It was especially interesting in Huslia, where we held classes in the Episcopal Church. The young adults were very quick to learn to read, but their limited ability to speak Koyukon slowed them down from fully understanding. Middle-aged and older people were slower to read fluently but instantly understood the Bible stories.

The village was waiting for the annual barge to deliver gas and freight. We were grateful to have saved enough gas in Allakaket for our trip. Upon arrival in Huslia, we had 10 gallons of gas left, and the village had limited gas to sell. It was still a long way to Koyukuk.

Koyukuk

In Koyukuk, we gathered in Saunders and Evelyn Cleaver's home. Ella Vernetti had everybody sit in a big circle, Native style. Then they took turns going around the circle, reading the primer.

Nulato and Kaltag

We continued down the Yukon River to Nulato, with classes held in the community hall, then to Kaltag, where we met in the women's clubhouse. The people were blessed with plenty of fresh salmon, which they shared with us. This was our farthest downriver point on the Yukon River, so we turned around.

Sick Leave: Kay

Beth Ann wasn't feeling well when we returned to Koyukuk, so the two of us flew to Fairbanks. Our suspicions were correct; she had pneumonia but recovered quickly.

During our time in Fairbanks, an Eskimo Christian friend, Lillie Lord, saw me in the parking lot at the post office. She prayed on the spot for Beth Ann and our ventures. It was wonderful to have people pray immediately for what was needed. Often, people will say, "I'll remember to pray for you after I go home."

Galena

We were reunited with Kay and Beth Ann in Galena. Classes were smaller due to numerous competing activities: fishing, the US Fish & Wildlife Service, and the Bureau of Land Management's fighting forest fires.

Ruby

In Ruby, our time overlapped with that of a team from *National Geographic Magazine*. They attended a few of our weeklong reading classes in the new community hall. We were impressed that they continuously snapped pictures. Their photographer told us, "We take lots of pictures so we can choose the best ones to put in our publications." Photographers like this kept Kodak in business.

Kokrine Hills Bible Camp

Kokrine Hills Bible Camp was our next stop, 55 miles farther upriver. Earlier this summer, they had two weeks of camp with 85 campers. It was rustic, but it continued to grow and help meet the spiritual needs.

Tanana: David

Several miles below the village, I told everybody, "Time to wash up and change clothes before we get to Tanana." Then we had problems with our outboard motor by a nice beach. We had a long rope, so the kids wanted to pull us upriver by "lining up" as people did long ago. Things were going well. Around the next bend was the village dump, and a caterpillar had pushed dirt out to the edge of the river. We couldn't go through that loose dirt!

I had to swallow my pride and seek help. I climbed the bank and walked past the city dump as I prayed for someone to assist us. It was the middle of the day with pleasant weather. One of the Nicholia brothers towed us the last mile to Tanana.

Buster and Violet Kennedy invited us to stay in their tiny cabin for the week. It was normally the "pan house" for playing cards during the winter. Literacy classes were held in the Episcopal parish hall.

Back to Fairbanks

Returning by plane to Fairbanks gave us time to rest and readjust as our four children prepared for another school year. We trusted that the Gospel of John would encourage people to think about the Lord and seek Him. We connected with many different people in nine villages plus fish camp stops along the way.

Still Adjusting: David

We had to change from a simple lifestyle to a complex way of living. Life in the village took more time to exist, with all the chores and fewer conveniences. Bush living was less rushed, and we lived by "Indian time," where the events in life were more important than going by one's watch. We have spent much of our lives moving back and forth between these two ways of living.

One project in Fairbanks was to raise one side of our home by four inches. This house was built on permafrost that continued to thaw unevenly. It was nine inches lower on one side of our home. We knew if a glass of water was spilled on the higher end of the table, the person on the lower end had to move quickly out of the way. I helped Jerry Lee, the maintenance man, raise the house. We had to dig out and straighten the basement walls as well as realign the plumbing and doors. It was still five inches lower on that side but much more livable.

Our family attended Bethel Baptist Church at 123 Bonnie Avenue in Fairbanks with 40 people. I was elected to the deacon board for a three-year term.

The Second Biggest "Native Village" in Alaska

The number of Indian and Eskimo residents in the Fairbanks-North Pole area kept growing. It was difficult to get an accurate count, but approximately 5,000 Natives lived here. Anchorage was the biggest "Native village," with 20,000.

Athabascans in Canada

In September, Paul Milanowski, a fellow Athabascan translator with the Upper Tanana people, and I drove to Whitehorse, Canada. Then we traveled north, surveying the Tutchone Athabascans to determine the need for translators. It was great to see a related language, but they spoke less of it than the Koyukon people. No translation team was sent there.

Feeling My Doubts: David

I had doubts if we should continue translating the whole New Testament into the Koyukon language. I did not want to produce a museum exhibit! The only fluent speakers were the elders. Many middle-aged people could speak Koyukon, but most used and preferred English. The children in the villages knew a few words in Koyukon but primarily used English. I knew and read about how quickly people groups in other parts of the world have lost their languages.

Christian friends questioned the wisdom of making a Bible translation when people were speaking less and less of their language. We discussed changing to a different Native language group in North America, or even moving to Papua New Guinea. The grass is always greener on the other side of the fence!

The work was progressing slowly. We wanted to do what God desired us to do—no matter what. That was our top priority.

One Bible verse the Lord impressed on both of our minds was, "You did not choose Me, but I chose you and appointed you that you should go and bear fruit, and that your fruit should remain" (John 15:16). There was no lightning bolt from heaven, so we kept plodding along.

Music to Our Ears: Kay

Debbie took flute lessons in school, following in her mother's footsteps. She played a flute solo at church for Youth Night. We also wanted to get the younger kids interested in playing an instrument.

Dan was in the junior high band playing the trumpet, and Stephen was fascinated hearing him practice. Elizabeth was excited to be included in the school chorus and to play the clarinet.

Math Counts

Both Dan and Stephen did very well in math. Dan's teacher was puzzled when he solved a problem on the blackboard. He worked them out in his head, so he didn't need to write out the process to get the correct answer.

Acting

Elizabeth was selected to have a part in a school play and was concerned about how she looked. It comes with maturing.

Typing Projects: Kay

My typing skills were used on several projects for other members. This included the Gospel of Mark for Slavey Athabascans in Canada, First Timothy for the Upper Tanana people, and an Iñupiaq Eskimo dictionary project for the Zibells and Websters.

Good News Club: Kay

Many Indians and Eskimos attended a Good News club I taught at the Birch Park Housing Project. The housing units were close together, so it was easy to advertise for kids to come. I continued teaching clubs every Thursday throughout the winter.

The Happy Hunter: David

Dan was 11 years old when he killed his first moose in October. He went to check his rabbit snares before catching the school bus. Soon he burst into the kitchen, yelling, "Dad, there are three moose in our field!" So we got the 30.06 rifle and ran out. From our living room, Kay watched the hunt. We saw a cow and two calves.

Looking at the young ones, I told Dan, "Shoot the more active one; I'm pretty sure it's a bull." The horns were stubs, three inches long, and hard to see at a distance. Dan didn't have a rack to hang up in his room, but the meat was delicious and tender. He was excused from school as we butchered the moose.

Dan held his moose head.

Native Christmas Cards

We made a Christmas card with the story of the shepherds visiting Jesus lying in a manger, taken from Luke 1:26-38 and 2:1-20. It was good to have something special for our Native friends at Christmastime.

1970
Missionary Conference in Nenana: David

Russ and Freda Arnold were serving at the Nenana Bible Church. They hosted a missionary conference over the weekend, but there was competition from the dog races for attendance. The Arnold children were similar in age to ours, so everyone was happy. After this, we drove farther south to another church in Clear, where I preached at the evening service. Our kids were exhausted from the full weekend.

Grayling: David

In March, I flew 400 miles west of Fairbanks to Grayling. It had a population of 160 on the Yukon River. Arctic Missions sponsored a Bible training session and invited me to participate in the Grayling Bible Church program. The village of Holikachuk (renamed Grayling) relocated to the Yukon River about eight years earlier, where the fishing was much better and it didn't flood. Their Athabascan language is related to Koyukon, and I understood a few of their words.

My subject was: "How to tell people about Jesus in your Athabascan language." I followed the concept of the Four Spiritual Laws: God loves us, we have sinned, Christ died for us, and each one of us must put our faith in Jesus Christ. We translated and sang "Jesus Loves Me" and the chorus of "Only Trust Him." Pastor Ben Neeley, a Native pastor from the Copper Valley region, spoke about Bible prophecy.

I was incredibly impressed with the work of the Holy Spirit when I heard their testimonies and saw their interest in spiritual things. Twenty people were present for the three-hour morning meetings, and 60 each evening. Six adults gave testimonies. A few had heard me on the radio when I read Scripture in Koyukon on *Cabin Prayer Meeting of the Air*. It was an encouraging two weeks.

Minor Surgery: Kay

I spent three nights at the Fairbanks Memorial Hospital. The doctor removed a suspicious lump that was much deeper and more extensive than expected. The lab tests came back negative for breast cancer.

Summertime: David

Jerry Lee oversaw major engine repairs while I did the greasy work on our 1964 Chevy station wagon before our trip to the Lower 48. We left Fairbanks on May 29 and drove down the Alcan Highway. Going up one big mountain in Canada, we went slower and slower while in second gear. I had to shift into first gear, but we made it to the top of the mountain.

Our first stop was Vassar, Michigan, to spend time with Kay's parents and the First Baptist Church, plus other supporting churches in Eastern Michigan. Our four children sang "Jesus Loves Me" in Athabascan. An eight-millimeter movie we produced showed what God was doing in Alaska. While there, we were pleased to trade up to a 1969 red Ford station wagon.

Then we drove to see my mother in Kentucky. Each child who memorized 100 Bible verses earned a free week of camp. Our three older children memorized the verses and attended Camp Nathanael. Stephen was too young to be a camper, so he assisted Grandma Henry in the craft cabin. They had fun at camp with their grandma!

Next, we drove to Bradenton, Florida, to report to Calvary Baptist Church and supporters in Florida. We took summer furloughs to visit as many supporting churches as possible so our children would have a normal school year in Alaska.

Two New Hats: David

I received two new "hats" at the biannual North America Branch meeting in Colorado. One of them was to become the center manager in Fairbanks. The second "hat" was being elected to the executive committee for the North America Branch. I prayed and trusted the Lord to give me wisdom in these responsibilities for His glory.

A Big Encouragement

Sally (Evans) Hudson had a successful time fishing for king salmon in Rampart, Alaska, on the Yukon River. While there, she read Scripture in Indian to her aunt Jenny, who was 88 and spoke very little English. Sally went on to tell us, "If you could have seen the wonderful expression on her face when she heard the Scripture in Koyukon, it would have been a sufficient reward for your labors."

She also read Scripture on *Cabin Prayer Meeting of the Air* during the summer. We were thrilled to hear about Native people reading and using Scripture.

Life in the Big City: Fairbanks (1969~1972)

Sally's Reading Class: David

Sally hosted a Koyukon reading class in her home. That winter, she wanted to transpose the 35 hymns from the old Episcopal Church hymnal of 1907 into the modern alphabet. Of course, this included singing those hymns. We replaced archaic words with current, understandable terms. Most hymns were familiar, such as "Just as I Am," "Look to the Lamb of God," and "Oh Come, All Ye Faithful." We had finished 12 hymns by November. The hymns were sung at the *Cabin Prayer Meeting of the Air*. Luke Titus, Mabel Charlie, and others in Minto were very interested in singing these hymns.

At one session, a relative of Sally's was visiting. She apologized and sheepishly said in Koyukon, "I forgot to spit out my chew before coming here!" We all laughed. She was a big help. Classes continued with strong interest throughout the winter months. Sally was an excellent host, promoter, and encourager.

Dan's 12th Birthday

Dan turned 12 in October. We let him decide what he wanted to eat for his celebration. Dan chose rabbit and an angel food cake. Once a person gets hunting in their blood, it affects their choices in life. Dan claimed that 300 rabbits would equal one moose, but we prefer moose.

Underwater: David

Jerry and Joan Lee, with their two boys, Tom and Mark, lived at the center. Jerry shot a young bull moose near the lake at Camp Li-Wa, which adjoined the Wycliffe property. He asked for permission to use the small John Deere crawler dozer to haul the moose meat home. I told him, "Yes, but stay off the lake; the ice is too thin." There was less than a foot of snow on the ground, so it was easy to see the border of the lake.

Jerry was driving on the marshy plants on the edge of the lake. Suddenly, he broke through the ice and quickly jumped off! The crawler became completely submerged seven feet below the thin layer of ice! It was enveloped by lake weeds in the murky water. We didn't realize the swampy growth extended back over the edge of the lake!

I went to see for myself. A big open hole was in the ice, large enough for the crawler. Red hydraulic fluid was floating on top of the water. My first thought was blood, but I knew it couldn't be since Jerry was safe.

We tried retrieving it by building a tripod over the site. Using a heavy-duty chain hoist, we brought the crawler high enough to see a corner of it, but then the hoist broke.

A Short Move

A housing unit was built over the top of the main building at the center. We helped with painting and the finishing touches before moving in. The second floor had an outside entrance and stairway. The building was on a hillside, so our back door was at ground level. The living room had a great view overlooking Camp Li-Wa. We were also thankful to live in a house with a level floor.

The first floor included offices, a print shop, a general area, two guest rooms, a well for everyone's water supply, and a group washer and dryer. Normally, no one was downstairs in the evenings, so our kids rolling marbles or playing did not disturb anyone below.

New Year's Watch Night Service

Our two older children attended an all-night youth party at Bethel Baptist Church. They joined with the adults. Then, after praying in the New Year at midnight, the youth went to a separate area downstairs to play games for the rest of the night. They had breakfast at 5:00 a.m. and were home by 7:00 a.m.

Tape Recording: David

John Pine and his family served with Arctic Missions in Kaltag. He asked me to record three passages of Koyukon Scripture for an elder who did not speak much English. Later, John reported that several people listened to the recording and enjoyed it.

1971
January Temperatures

January brought cooler temperatures in the negative 50-degree range for several weeks, which created dense ice fog. It was hard to see vehicles ahead or those oncoming. Everybody drove with their headlights on all the time.

A New Member
Dixie Myers, a single lady, joined our staff and lived on the lower level of our building. Her main job was preparing printed literature for the Native languages and secretarial work.

Proofreading Native Languages: Kay
Dixie and I proofread Bible translations and other literature. One of us would read the text letter by letter, while the other person checked what was being read against the original. When we got tired, we would switch roles. We were able to read Eskimo or Indian translations without pronouncing the words or knowing what they meant. The work was tedious but necessary since we always found a few spelling, punctuation, or typing errors. We wanted the translation to be accurate.

Prayer Meeting
Weekly prayer meetings were held on Thursdays for those at the center. There was always a certain need or praise for what God was doing in Alaska. This kept us connected and informed about the Alaskan work.

Snowshoe Racer
Go, Dan! He won the junior snowshoe race at the North American Championship Sled Dog Race in Fairbanks. Dan received a nice trophy and a 25-dollar gift certificate.

Baptism by Choice
Debbie was baptized on Sunday, March 21st, 1971, at Denali Bible Chapel since Bethel Baptist did not have a baptistry. At 14 years old, she made this public profession of her faith in Jesus Christ. We were thankful to see our children growing spiritually and pressing on in their Christian lives.

Out of the Water: David
The recovery stage for the sunken crawler kept dragging on and on, sapping much of our time and energy. We needed assistance and asked the military for help.

The Army sent a 20-ton crane mounted on a truck. They backed down near the site and hooked onto the crawler. Everything worked like clockwork. We thanked the Army and the Lord as it came up out of the water, dripping wet and with weeds hanging on. They brought it to our heated garage. Jerry was so happy. It didn't take him long to restore the crawler to its original condition. Mission accomplished!

Attempting to retrieve the crawler consumed five months of our winter! It was a real emotional drag on Jerry and me. This reminded me of the traditional ending for Koyukon stories: "I have chewed off part of the winter!"

AWANA Club

Our three older children joined the AWANA club at Denali Bible Chapel. AWANA is an acronym for "Approved Workmen Are Not Ashamed," taken from their key verse: "Be diligent to present yourself approved to God, a worker who does not need to be ashamed, rightly dividing the word of truth" (2 Timothy 2:15). AWANA helped children know, love, and serve Jesus Christ. They memorized Bible verses and had fun times together.

Our three older children earned awards with the program. Debbie received a second-place trophy, Dan received the first-place trophy, and Elizabeth received an honor award. They did not include younger children, so Stephen had to wait.

Portland Bound: Kay

The Institute in Basic Youth Conflicts conducted by Bill Gothard was held in Portland, Oregon. It helped give Biblical answers to life's problems. I stayed with our Arctic Missions friends Bob and Carol Moffat and their six children. Flying back to Fairbanks on Alaska Airlines, I was served dinner in the regular economy section, which was the common practice back in the '70s.

New Minto

On our first trip to New Minto, we drove 130 miles on the gravel road system. These people are closely related to the Koyukon people. They had moved to higher ground and a better location after many floods at their previous site. We attended the welcome potlatch and open house celebration for the new site.

A young Native girl with a fractured ankle returned with us. We arrived in Fairbanks at 1:30 a.m. and brought her home from the hospital at 3:00 a.m. She stayed with us for three nights.

More People

The center was filled with people for the summer: Wilfried and Donna Zibell, with their four children, were preparing reading books and the Bible for the Kobuk River Eskimos; Ray and Sally Collins, with their three children, were with the Upper Kuskokwim Athabascans in Nikolai; and Paul and Trudi Milanowski and their three children, in Fairbanks for two weeks for medical needs, were working with the Upper Tanana Athabascans in Tetlin. Our children enjoyed the playmates. Several Native friends spent the night, too. The list can grow when you have extra beds. The Fairbanks Center was being used, and that was why we had it.

Manley

We drove to Manley Hot Springs to visit Jerry and Sally Hudson. It rained all four days and four nights, but they had a large cabin. It wasn't easy to keep a family indoors and occupied during such times.

Additional Jobs: Kay

We were sorry to see Jerry and Joan Lee and their family leave. This meant I needed to take over Joan's job doing the group bookkeeping. One time, a bill to a local plumbing wholesaler was overlooked. They sent us a friendly reminder with this verse, "Render to Caesar the things that are Caesar's" (Mark 11:17). We laughed and paid the bill immediately.

By default, David had several big maintenance jobs at the center. Larger jobs included replacing and burying the fuel oil tank for the main building and replacing the underground plastic water line to the cabin with galvanized pipe. There was the usual stream of missionary correspondence, filling book orders, facilitating printing jobs for our members, and mowing the oversized lawn. We never mowed a lawn in the village! The White man is the only person who plants grass in May so he can cut it in June, July, and August.

Berry Time

All of our children were happy to find a superabundant crop of blueberries nearby. Raspberries were five miles away. We froze the berries to last all winter. The children contributed to the family freezer half of what they picked, and they sold the rest for extra spending money.

Beaver

Beaver, a Native village on the Yukon River, had a population of 101. Debbie assisted Teddy Burhans with Child Evangelism Fellowship for a week of Bible School. All the village kids participated in the programs.

Stevens Village: David

I flew to Stevens Village, with a population of 75 people on the Yukon River in October. Our son Stephen thought this should be his village because of its name. We ran out of time last summer to include Stevens Village and Rampart. I held eight nightly Indian reading classes even though I had to explain changes in the alphabet to fit their Upper Koyukon dialect.

They were very interested in the hymnal project, originally printed in Upper Koyukon. An example of Indian honesty happened when I sang one of the old hymns to a fellow who wanted to learn them. He politely listened. After I finished, he calmly stated, "David, you can't sing." We both laughed. I already knew that before hearing his expert critique.

Memorial Service for Elliott: David

It was difficult to lead a memorial service for Elliott Canonge. He had served as the center manager and translation checker. He was also a personal friend and a strong witness for our Lord.

Fatal Plane Crash: David

Wilfried Zibell was in the process of translating the Bible for the Iñupiaq Eskimos. From Kotzebue, he flew home earlier than expected to be with his family in Noorvik. On Sunday, November 21st, 1971, we stayed up late to verify that Wilfried was on the Cessna 180 plane with four other people that crashed on takeoff. No one had survived. It was hard to believe!

I conducted a memorial service at Denali Bible Chapel in Fairbanks the following Sunday. Many Eskimos gave testimonies about Wilfried's influence on their lives. We can't answer all the "whys" in life, but God has His perfect plan and purpose for each of us and knows the number of our days.

ANCSA

The Alaska Native Claims Settlement Act (ANCSA), was signed into law by President Richard Nixon, a Republican, on December 18th, 1971. It created 13 regional corporations. One corporation was Doyon, Limited, which included all 11 Koyukon villages plus other Athabascan people of the interior. These corporations received one billion dollars and 40 million acres as a result of the settlement.

1972
Minto Ordination: David

I flew with Bishop Gordon to Minto in January to participate in the ordination of Luke Titus to the Episcopal priesthood. Luke was a local Indian and a believer. I stayed with Luke's parents, Robert and Elsie, who were also believers.

In the evening, Mable Charlie led many adults in singing 15 hymns from the old Episcopal hymnal. They sang, "Come Thou Fount of Every Blessing" in Koyukon, which was used during the ordination service. Luke's wife, Alice, was Navajo and sang "How Great Thou Art" in her language.

During the service, Walter Titus read Ephesians 4:7-13 in their dialect. Walter was teaching their language in school. I was encouraged by how much the people were using their language and promoting its use. The ordination service included an abundance of Native food. Charlotte Titus, Susie William Jimmy, and Dorothy Titus worked together preparing Indian "ice cream" made with flaked white fish, which did not taste fishy.

The Minto people had their Sunday evening services in Indian. They translated the Lord's Prayer, hymns, and portions of Scripture. Luke had been going through the Scripture in English, reading a sentence at a time, and then the audience discussed and translated it into their dialect. He also led a Bible study on Wednesdays. Luke wanted his people to understand Scripture and apply it to their lives.

The Minto dialect is related to the Upper Koyukon used in Stevens Village and part of Tanana. I was having trouble making the linguistic adjustments to their Minto dialect. One night, Chief Peter John, an outspoken leader, sat down with me to explain the dialect differences between the upriver and downriver sounds.

Minto Revival 1972: David

The ordination of Luke Titus and Minto's spiritual awareness and aliveness were part of the significant "Minto Revival 1972." God chose to work in a small Athabascan Indian village of 200 people for His glory. Many local Indians were continually praying for Minto and urging others to seek the Lord.

Minto people made a distinction among Christians when asking others, "Are you a Christian or a 'real' Christian?" One is a Christian in name only with no change in their life. But a "real" Christian loves Jesus Christ and seeks to live for His glory. Many people were openly talking about the Lord. The Minto church building was full of joyful people. On Sunday, folks would go to Saint Barnabas Episcopal Church in the morning and the Assembly of God Church in the evening. They were hungry for God's Word and wanted to live and stand as strong Indian warriors in their Christian lives. This revival forever changed and affected Minto and it rippled out to nearby villages for many years, even 50 years later.

Fast-forward: In 2019, my wife, Kay, was living in Hope Haven, an assisted living home in Fairbanks. One of her roommates was Betty Engles. I asked her, "When did you turn to Jesus Christ?" She responded, "1972." I said, "That's when there was a spiritual revival in Minto." She replied, "Yes. At that time, the church was full, and people were praying and talking about Jesus everywhere in Minto."

Reading Classes: David

We continued to have two weekly reading classes in Fairbanks, usually lasting an hour and a half. We were now meeting in the conference room at the Fairbanks Native Center. Kay took over leading the classes when I was gone.

More Paperwork: David

I worked on Wolf and Hildegard Seiler's visas to Alaska. They came from Germany to work with the Kobuk River Iñupiaq people. The Alaska Department of Labor's big concern was ensuring they would not take away someone's job.

Koyukon Bible Study

We wanted to start a Koyukon Bible study in Fairbanks. After praying about it, we decided to wait until someone requested it. Katherine Anderson spoke up and asked us to have the study. We met each Wednesday afternoon with six to nine Indians and read portions of Scripture from the Gospel of John in Koyukon. We talked about what it meant and then applied it to our personal lives. We sang a hymn or two out of the Indian hymnal. A highlight was to have several Indians speak up, give testimonies, and show evidence of the Lord working in their lives.

We began translating short passages from the Gospel of Mark during these studies. One translator was Hilda Stevens from Stevens Village. We would give her a passage to work on at home. The following week, we would read her initial work. She was very gracious and easy to work with as changes were made.

Seminar in Seattle: David

In May, Debbie, Dan, and I flew to the Institute in Basic Youth Conflicts Seminar in Seattle. We learned together. This conference helped our teenagers grow in their faith. We blended in with an audience of 1,300 participants.

Family Visitors: David

My mother and cousin Joan Henry visited for two weeks. We took them to Mount McKinley National Park and they enjoyed the abundance of wildlife.

Fast-forward: Mount McKinley was renamed Mount Denali in 2015. Denali—its name is derived from Koyukon *Deenaalee* (the high one)—is the highest mountain in North America.

The Family Farm

Our children participated in 4-H and submitted a variety of entries at the local Tanana Valley State Fair. A generous 4-H supporter donated animals to any member who would do a project and show them at the fair. Debbie and Dan received a dozen chicks, Elizabeth took three baby turkeys, and Stephen raised four ducklings.

During the winter, they had a pair of gerbils, guinea pigs, a dog, and a colorful oriental bird we found along the road. They wintered chickens to provide eggs. Frankly, it was cheaper to buy eggs. No, we were not starting a zoo!

Dan built a greenhouse, and the children grew different vegetables to display at the fair. Gardening was a good family project, plus it provided potatoes and vegetables for the winter as well as monetary rewards for their effort.

Babysitting Jobs: Kay

Dale and Mary King lived nearby. Mary worked as a nurse and Dale worked for Alaska Airlines. Their daughter Kay was only three, and her big brother, Leslie, was five when we started caring for them. Kay King was known as "Little Kay," and I was "Big Kay." After Little Kay's birthday, she proudly announced, "Me uh uh, 'Little Kay,' me 'Big Kay' now!" I did most of the work while our children were in school, but they took over after coming home.

Camp Li-Wa

The Wycliffe Center bordered Camp Li-Wa, and the only road to the camp was through our property. Debbie was a counselor at Girls' Camp, while Elizabeth helped in the kitchen. Dan was a counselor at Boys' Camp, and Stephen was one of the happy campers.

Bilingual Training (1972~1975)

Preparing for the Bilingual Program

In July, I began a special six-week course with four Koyukon Natives on the University of Alaska campus. The goal was to teach reading and writing in Koyukon so we could develop a program for bilingual education in the village schools. This was a stressful but worthwhile program. Bilingual education reinforced the growing honor of being Indian. We also developed literacy materials with the newly formed Alaska Native Language Center.

Title 1: David

Federal funds from a Title 1 grant were used locally. The basic plan was to recruit local Native teachers from the villages and train them to teach the children in their villages how to read and write Koyukon. Everyone could read English, but this would be their first time reading their previously unwritten language. In January, they would start teaching Koyukon in their village schools of Allakaket, Huslia, Koyukuk, Nulato, and Kaltag. There would also be more potential readers of the Koyukon Scriptures.

Village Trips: David

As the director, I flew to Huslia, Koyukuk, and Nulato to explain the bilingual education program. The village councils selected 11 teachers-to-be to take the training and eventually teach. I also flew to Anchorage to participate in a Title 1 meeting.

Teacher Training Program: David

The Koyukon Bilingual Teacher Training began on October 23rd, 1972, at the Rural School Project Lab on the UAF campus. The participants from Koyukuk were: Arthur Malamute, Eleanor Kriska, and Madeline Solomon; from Nulato: William Ambrose Sr., Paulina

Stickman, Amelia Demoski, Priscilla Sipary, and Victor Alexie; and from Huslia: Edwin Simon, Eleanor Sam, and Shirley Attla.

Others attending from Fairbanks included: Eliza Jones, Sally Hudson, Marie (Brown) Hunter, Miranda (Hildebrand) Wright, Mary Smith, Mary Van Hatten, Nettie Erhart, and Katherine Anderson.

The ones who helped teach were: Winifred Landie with lesson planning and various teaching methods; Dr. Michael Krauss with his linguistic expertise and relations with other languages; Ray Collins with a segment on culture; and I assisted where needed as the leader.

The workshop covered Koyukon literacy and writing, initial material development, curriculum design, classroom management, relations with certified teachers, lesson planning, classroom equipment use, and methods of teaching language and culture.

The lectures were tedious, especially those presenting lesson planning, which was difficult to comprehend and put into practice. At the end of one session, Edwin Simon, stated, "This is like making oatmeal. We pour in the ingredients. We stir them while they're cooking. We end up with mush—and that's the way my brain feels!" He spoke for all of us!

Up until this point, we had not seen any money promised to run the program, but the proposal had been signed in Juneau. People wanted the program to be successful, and they were used to a cashless society. Participants from the bush found housing wherever they could. Enthusiasm kept the program moving ahead, despite all the drawbacks.

Stuck Between Two Languages

In the not-too-distant past, Natives were punished for speaking their language at school. Educators used to think that speaking one's language would hinder the student's ability to learn English and assimilate into the American way of life.

The best way would have been to continue using one's Native language while learning English. This would have given the students vital links with language, culture, and life between their elders and history, as well as with the White man's way. Unfortunately, Natives were often stuck in no man's land, somewhere between English and Indian, thus knowing limited English and being hindered in their ability in both languages.

An Indian Name: David

During a closing potlatch with fish, moose, berries, and Indian ice cream, a beautiful handmade parka with a red fox fur trim and ruff was presented to me. Marie Hunter had sewn the parka. She sized me by observation, and it was a perfect fit!

I was made an honorary member of the Koyukon people and belonged to the *Bedzeyh Te Hut'aane* (Caribou Clan), one of three clans. They gave me the Indian name *Kusge* (Lucky Weasel), an ermine in a white winter coat with a black-tipped tail. The students stated, "We are 'lucky' to have you with us." I was overwhelmed and very grateful.

Huslia Guests: Kay

During the six-week training, Edwin and Lydia Simon and their six-year-old son, Calvin, stayed in the apartment below ours. I enjoyed being with Lydia. She educated me about old-time Native ways and various folk tales. She completed several beadwork and skin sewing projects.

The family set spring pole snares for rabbits. This was a peak year for rabbits, so Lydia caught 70 rabbits in just two weeks. She made a fur blanket for herself to use when riding in the sled behind their snow machine.

School for Everyone

Our children attended four different schools this year, so we were thankful for school buses. Stephen was an active fourth grader at Barnette Elementary School. Elizabeth, in the seventh grade, enjoyed being a preteen. She crocheted afghan squares like a future grandma.

Dan took a taxidermy class in high school and completed several ptarmigan mounts. He skinned a teal duck in four hours so it would mount well. It was time-consuming and tedious work, but he enjoyed it. Dan wrote a paper on creation versus evolution, which will always be a hot issue.

The Lathrop High School Malemute Band raised money for a trip to the annual music festival. Debbie enjoyed traveling with them and playing her flute.

Fortieth Wedding Anniversary: Kay

I flew to Michigan to surprise my parents and celebrate their 40th wedding anniversary. It was a welcome change from all the activities in

Elizabeth sewed this dress for the Golden Days activities in Fairbanks.

Fairbanks. David kept his nose to the grindstone between the Koyukon bilingual program and our four children.

1973
Bilingual Travel: David

I traveled to Allakaket, Huslia, Koyukuk, Nulato, and Kaltag for the bilingual program, spending three to five days in each village. The 12 people who took the six-week training were teaching their language and culture. I was concerned they would run out of Koyukon teaching materials.

Bilingual Team: David

In Fairbanks, I worked with a team of five people to produce reading materials for the classroom. They were Eliza Jones, Sally Hudson, Marie Hunter, Linda (Simon) Demientieff, secretary, and James Grant-Schrock, illustrator.

Sixteen books were printed in Fairbanks for the Koyukon Cultural Enrichment Program in the Central and Lower Koyukon dialects. Everyone was dedicated, and we all worked together. Later, Dr. Michael Krauss and the staff at the ANLC collaborated with us to tweak and improve the Koyukon writing system.

Look at the Big Picture: David

I looked at the Native text, how the book was assembled, and the accompanying illustrations. Out of respect for the hardworking translator, I made sure Sally was the first person to see the book. She looked at it and immediately commented emphatically, "David, you did not look at this picture of the Native family!" Both mother and daughter were wearing short, curvy, tight-fitting dresses. "We don't dress like that to go hunting or do our work at fish camp," Sally stated.

Sally Hudson translated *Paul Hunts for Moose.*

Sally was right. I looked primarily at the text with my tunnel vision, but Sally saw the bigger picture of both text and illustrations.

Baptism for the Boys

Dan and Stephen decided to be more committed in their Christian lives by being baptized at Denali Bible Chapel by Pastor Hokie Moore. It was a tremendous public testimony for young people to take a stand for the Lord before their peers.

Important Advice: Kay

I wanted to encourage our daughters to start praying for their future husbands. The choice of a marriage partner is the second most crucial decision affecting one's whole life. The most important decision is to turn to Jesus Christ.

Debbie was in high school, and it was the right time for her to start praying for her future husband. I entered her bedroom one evening, and we talked about several different topics. Then I challenged her to begin praying for the young man she would marry.

Fast-forward: Debbie stated, "That was some of the most important advice my mother gave me." She accepted the challenge and prayed for several years for this unknown man. She eventually met a wonderful, growing Christian named John M. Rathbun at Bryan College, and they were married in 1981.

Debbie later encouraged her own daughter, Danielle, to pray for her future husband while in high school. As a reminder, Danielle made her computer password "Pray for a godly man." And she was blessed with a God-fearing husband, Isaac Reynolds.

Banquet Time

In mid-April, our church hosted a semi-formal banquet for the youth. Dan looked handsome in a suit coat, tie, and green dress shirt. Debbie looked lovely in a pale blue floor-length dress that she had made. Elizabeth and Stephen were still more interested in making crafts and building models.

Wycliffe Meetings: David

I went to Denver to chair the Executive Committee meetings in early May for the North America Branch. Then, as a representative, I flew to Mexico City for the International Conference for 10 days. This was very profitable and gave me a broader understanding of the worldwide activities.

The travel plan gave me time to squeeze in short visits with my sister, Miriam, and family near Los Angeles, and with my brother, Martin, and family in San Diego.

The End of Title 1

The Title 1 funding ended on June 30th. We worked steadily to complete the dictionary by the deadline. This enlarged Koyukon dictionary, *Dinaaka: Our Language*, had 163 pages. The introduction included an endorsement written by John C. Sackett, President of Doyon, Limited.

Indian friends and eight ladies from Bethel Baptist Church helped us collate the pages. Everyone wore a rubber tip on their thumbs to easily grip a sheet of paper while walking around two long tables, picking up a page from each pile of pages in the correct order, plus the cover, to make one complete book. It was a huge project, but with good help, the dictionary was soon collated and ready.

Wycliffe Day: David

We called the Fourth of July "Wycliffe Day" at the center. Seventy-five people from Bethel Baptist Church came for the picnic, fun, games, and celebration of Independence Day. I explained what the mission was doing in Alaska and answered many questions.

A Hostess Plus: Kay

I did a lot of hosting, which included preparing and serving meals for Native and mission guests staying overnight, making beds, and washing extra laundry. I was not complaining, but the breaks were few and far between.

Another Conference

In July, our family drove to Valdez to see the ocean, the waterfalls, and the glaciers along the way. Then we attended the Victory Missionary Conference. The speaker was Major Ian Thomas. It was refreshing and a fantastic break after all the push for the bilingual program and the busyness of life. Our children enjoyed hiking and camping in the wilderness with the other missionary youth.

Bilingual Workshops: David

Later in July, we organized a one-week bilingual workshop in Huslia. This was for the Central Koyukon teachers and speakers from five villages: Allakaket, Hughes, Huslia, Koyukuk, and Tanana.

Another one-week bilingual workshop was held in Nulato for the lower Koyukon teachers and speakers in Nulato and Kaltag. A curriculum consultant was not available, so I had to fill in. It was good to get together on their home turf.

Prepping for Next Year: David

For 10 days, Eliza Jones and I revised the Koyukon literacy books for next year. The expanded program involved the seven Koyukon schools. Seventeen Koyukon teachers served 350 students in kindergarten through 12th grade.

Village teachers who taught in this year's program included:
Allakaket: Velma Schafer and Jennie (David) Williams.
Hughes: Marilyn (Koyukuk) Evans, and Dorothy Beetus.
Huslia: Edwin Simon, Selina Sam, and Eleanor Sam.
Koyukuk: Madeline Solomon, and Arthur Malamute.
Nulato: Paulina Stickman, Amelia Demoski, and Priscilla Sipary.
Kaltag: Ottie Semaken, and Anna Madros.
Tanana: Mary Dick, Marian Edwin, and Violet Kennedy.

Sixteen books and literacy aids were printed in Fairbanks by the Koyukon Cultural Enrichment Program in the Central and Lower Koyukon dialects. *The River Times* newspaper had a great article about the Koyukon program in their October 25th, 1973, edition. The article with pictures was titled "Old Teach Young Their Language."

A Dream Come True

Velma Schafer wrote the following letter about the bilingual program titled "A Dream Come True" in *The River Times* newspaper in February 1973:

"A Dream Come True"

I didn't speak the English language until I was 8 years old. I remember clearly when we were at our winter camp and our neighbors could speak English and we spoke the Athabascan language. We couldn't understand each other. There were four of us children in our family and two children in the other family. We played every day outside the log cabins. We laughed and played, but we couldn't understand each other. I think that's where we learned our first few English words. Later, our parents brought us to the village and put us in school. Then, when I was in the seventh grade, I tried to write words in Athabascan but I couldn't do it. How I wished I could do it at that time.

Then as the years went by I almost forgot about it and David Henry and his family moved here (Allakaket). I learned a little from them. Then last summer I went to David's class at the U. of A. for four weeks. He was teaching. In November, I went back for another three weeks and now I can write Athabascan.

I am now involved in a bilingual program to teach at the Allakaket State School, which I've been doing for almost a month. So my dreams of writing Athabascan came true after all. I have my own four children who will someday talk, read, and write Athabascan, I hope. Thanks to David Henry, John Sackett, and everyone involved in the bilingual program! Thanks.

Signed: (*Velma Simon*)

Crane for Supper: Kay

Effie Williams and two girls from Allakaket came for a visit. Dan had shot two cranes, so I decided to roast one in a plastic cooking bag in our oven. It was delicious! Cranes have dark breast meat rather than white, like turkeys.

A Ford Station Wagon: David

We were blessed with the gift of a 1972 Ford station wagon. This one had plush cloth seats and a 460 engine with plenty of power. Our monthly support was below the quota recommended by the mission, but God provided for our material needs in this unique way. The air conditioner was appreciated when we drove to Florida the following summer to see my mother and supporters in Bradenton.

Translation Workshop

We hosted a translation workshop at the Wycliffe Center with 14 extra people for three weeks. Six Alaskan translators and two teams from Canada, plus two translation consultants, attended. Most of the translators had a Native co-translator with them. All available spaces and beds were used.

The goal was to translate Galatians. Our schedule included exegesis, extracting the meaning from a passage of Scripture. We discussed certain words that were particularly difficult to translate and listened to lectures on methods for going from English to the Native languages.

Working with Eliza Jones, we applied those ideas to the translation of a portion of Galatians. Post-translation sessions compared what was accomplished, and we shared our solutions for difficult passages or choices of words. A draft copy of the first four chapters of Galatians was completed in October. It was a wonder we hadn't memorized the book! After the workshop, we planned to use Galatians in our Bible study for the final checking before printing.

Eliza was called for jury duty during the training, but she continued working on the translation in the courthouse waiting room. It was encouraging to see her work and dedication. Eliza missed the lectures, but Marie Hunter filled in.

Another Meeting in Anchorage: David

I flew to Anchorage for a meeting with the Alaska Native Education Board. Colleagues attending were Dick Mueller, Paul Milanowski, and Dave Shinen. I spent more time with them when we attended meetings throughout the winter.

Grandma Bertha in Heaven: Kay

My grandmother, Bertha Cecilia (Snover) Temple, departed for heaven on October 11th, 1973. She was 85 and had lived a full life. I was thankful I saw her in Michigan at my parents' 40th wedding anniversary. She had a quiet personality compared to my grandfather, Pete Temple, who was very outgoing. She made the most important decision of her life when she turned to the Lord Jesus Christ in her youth and lived for Him.

Young People's Night at Church

Young People's Night at Bethel Baptist Church put the teens in charge of the evening service. Debbie, with much poise, led the singing. Dan introduced the testimony time by relating how he had worked in the Child Evangelism booth at the Tanana Valley State Fair in Fairbanks the previous summer. He prayed with one young girl who accepted the Lord. Dan publicly cried because he was so thrilled over the girl's decision. He said, "This was the first time I cried because I was so happy!"

Dan saw the need and had compassion. We, as his parents, were awed before our God. In the eyes of many people, this was a relatively "insignificant" child's decision, yet in God's eyes, there was a big celebration and rejoicing over one sinner who had repented.

1974
Progress in Translation

Translations are made, not born. By March, we had a working draft of Galatians. Eliza and Marie worked toward this goal and completed it.

Buried Treasure: David

We continued babysitting Leslie and Kay throughout the winter. One day, Stephen was hiding from Little Kay under our bed. He soon emerged, announcing, "Look what I found!" He held a box of chocolates I had hidden for "safekeeping." So while I was gone, the rest of our family enjoyed Stephen's find.

Sewing a Smock: Kay

Elizabeth purchased material with lavender and green hearts and took her fabric to school to sew a smock in Home Economics class. Then she added green pockets. She also made a pair of matching green pants. Elizabeth babysat one night and was thrilled to make enough money to buy herself a lime green shirt. The new outfit looked lovely on her.

Youth Activities: David

Debbie's band had a spaghetti dinner fundraiser. In March, they planned a band trip to Juneau. That's 625 miles the way the crow flies! We were glad to see them working to raise money, but she missed going to the teenage retreat.

Dan went alone to the retreat at Victory Bible Camp and enjoyed it. He was president of the youth group at church.

Debbie memorized Bible verses with the Bible Memory Association as I did as a teen. Elizabeth and Stephen continued memorizing verses for the AWANA club.

Teen Outing

We sponsored a young people's outing at our house with 25 teens. They had planned to meet at Camp Li-WA, but it was too cold. Then, they enjoyed hot chocolate and donuts, followed by devotions and playing games.

Military Guests

We invited Captain Vernon Stevenson, his wife Dorris, and their three children for a large moose dinner on Sunday. They were delighted to taste moose meat. When they retired from the Army in three years, he wanted to help in missions as a support worker. It's been great to have continued friendships with like-minded believers serving our Lord.

Fast-forward: They followed our recommendation and attended Bryan College in Tennessee, where Vernon graduated in 1979 with a BS degree in Business Administration.

Smelling Smoke: David

In Anchorage for another Alaska Native Education Board meeting, I stayed on the sixth floor at the Gold Rush Hotel. At 2:00 a.m., I was awakened by the smell of smoke. An arsonist had set a fire on the fifth floor. The phones, emergency generator, and exit lights did not work. I lost a night of sleep but made it out safely by the stairs.

Stephen's Friend: Kay

We were glad to have one of Stephen's Native friends over for dinner and spend the night together. I prepared grape juice and crackers for communion at church on Sunday. Our guest had many questions that I tried to answer. He asked, "Who can take communion?"

I responded, "Anyone who has been saved." He replied, "I've been saved four times." So I informed him, "You only need to be saved once, but you need to live for the Lord each day."

Our Young Helpers

Missionaries send prayer letters to churches and friends who support them with their prayers and financial gifts. Relatives and others are also interested in what's happening in Alaska.

Our goal was to write a more formal report or update every three or four months. The complete list included more than 100 people and churches. We printed multiple copies of our letters. Our kids helped by folding them, stuffing the envelopes, sealing them, and putting on the stamps. We all worked together.

Spiritual Activities

Debbie and Dan flew to Tanana and continued to Kokrine Hills Bible Camp by boat. They attended a counselor training session led by Russ Arnold and served as counselors the following week. Elizabeth and Stephen received awards for their Bible memorization in the AWANA club in Fairbanks.

Wedding in Vassar: Kay

In June, we left Fairbanks in our station wagon, heading to Vassar, Michigan, for my youngest sister's wedding. Sheryl Lynn Temple married Harris Gilroy. Debbie was one of her bridesmaids.

Island Lake Bible Camp: David

We drove to Island Lake Bible Camp in Poulsbo, Washington, to participate in our biennial North America Branch meeting. I was re-elected to serve as chairman of the executive committee for another two years.

Final Check of Galatians

In the fall, Jonathan Ekstrom, who translated the New Testament with the Hopi Indians, came to Fairbanks to review our corrections and do a final check of Galatians. Eliza and Marie assisted in the checking. He gave us suggestions and verses that needed tweaking to bring out the best meaning and understandability. We were impressed with his ability and helpful questions. Son Dan said he was a good chess player after work hours.

Galatians was ready to be printed in a trial edition. Very few people will know all the tedious work involved in bringing a book of the Bible to publication in a previously unwritten language. Now we prayed that the Word of God would do the work of God in people's hearts and lives. We mailed copies to various Natives.

Reading Galatians

Frank Titus, who had just moved into an assisted living home in Fairbanks, brought his copy along to read and study. People told us they had read through Galatians several times.

New Bible Study

Dorothy Pitka opened her home to continue the Bible study on Thursday afternoons. We were studying Galatians. One of the new people was Henry Moses, originally from South Fork near Allakaket but now living in Fairbanks. The attendance ran between eight and 10.

Children at Play: Kay

Both Elizabeth and Stephen hosted sledding parties at our place. Elizabeth's party was with the 4-H Junior Leaders. We had a nice hill near our house.

Dan enjoyed his taxidermy class and brought home a stuffed grouse. He mounted a set of caribou antlers from Allakaket for his father.

After Debbie and Dan finished their time in the AWANA program, they taught and mentored the younger boys and girls as assistant leaders at Denali Bible Chapel.

If Jesus Had Been Born in the Arctic

Every Christmas, our family attended the Iñupiaq Christmas Pageant at the First Presbyterian Church in Fairbanks. The Eskimo members performed a unique program of the Christmas story that presented the birth of Jesus Christ and what it would have been like if he had been born in the Arctic.

Joseph, Mary, and baby Jesus, dressed as traditional Eskimos, entered the church. They sat in an open shelter made from willow branches covered with animal skins and a caribou fur hide was on the floor. The children and adults formed a choir of angels with shepherds beside the shelter singing "Hark! The Herald Angels Sing" in Iñupiaq and English. A narrator read the Christmas story from the Bible. The wise men sang "We Three Kings" while they walked up the aisle with beautiful gifts of mukluks, fur mittens, and a parka for baby Jesus. Everyone sang "Joy to the World" to end the program.

Debbie, Dan, and Elizabeth were all in the church youth group. They asked if the youth could put on a similar program for Bethel Baptist

Church during the evening service. So the teens worked together collecting items for this particular Christmas program. Dan's furs were one of the gifts for Jesus.

1975
Cool Start for a New Year: Kay

The cold weather continued its chilly grip on our lives. It was in the minus 50 degrees range and affected our lifestyle. Stephen did not want to go to school, even though we would have driven him to wait in our warm car for his bus as we did for the other kids. Staying home was at the parents' discretion. The school district did not close any schools because of the cold. There was no outdoor recess if it was minus 20 degrees. Maybe that was why Stephen didn't want to go! Church services were discontinued if it was minus 45 degrees.

God's Protection: David

The deep cold chilled the propane's flow to our kitchen stove, resulting in a very low flame. There were different ways to keep the propane flowing to our kitchen stove. One way was to put a small light bulb inside a wooden box close to the propane regulator on top of the tank for a little heat. I put a blanket over the tank, leaving enough airspace for the light bulb.

I left it going for several hours while we went to town to run errands. When we returned home, we saw flames through our kitchen window on the backside of our house. I quickly pulled the blanket away and unplugged the light. I was shaken, but very thankful to the Lord and amazed that the house did not catch on fire. It was only seconds from becoming a devastating house fire!

Pipeline $$$

The pipeline forever changed Alaska! Enormous oil reservoirs were discovered on the North Slope in 1968. The spike in oil prices in 1973 made the 48-inch pipeline economical to build. The pipeline spans the state for 800 miles from the North Slope to the southern port of Valdez, passing near Fairbanks.

The Alaska pipeline was the most prominent private construction venture in history at that time. The impact on Fairbanks was overwhelming, with price increases, more people arriving, and a rapidly changing Alaskan lifestyle. Many of our Native friends worked on the pipeline or in a related job, but big money brought many temptations and problems along the way.

A Native friend working at a local store told us that an unknown man, who had been working up north, came and offered her a large sum of money for one night together. She refused!

More than 28,000 people were employed at the peak of its construction in the fall of 1975. America needed the oil, yet the colossal project aggravated local problems—overcrowding, drugs, drinking, inflation, shortages, prostitution, and the list goes on. Alaska is no longer the quiet frontier state we once knew!

The Pipeline Affected All of Us: David
We felt the effect of the pipeline living in Fairbanks. McDonald's on Airport Way had six cashiers taking orders. Each one had a line of six or more people waiting to place their orders.

One time, I stopped at Samson's Hardware to buy a simple pipe fitting. The clerk told me, "Someone working for the pipeline beat you here and bought every pipe fitting in the store!" It was a large, well-stocked store, but I had to wait until the next shipment came in.

I attempted to buy a piece of land to build our house. The price rose by 1,000 dollars a week! It was not a buyer's market when you were low on cash.

Due to the sudden increase in student enrollment, Lathrop High School had split sessions to maximize the school building space. Debbie, Dan, and Elizabeth attended in the afternoons from 1:00 to 6:00 p.m. Stephen attended a regular full day at Barnette Elementary since he was in the sixth grade.

Debbie was a high school senior saving money for higher education. She had a part-time job as an office worker in the mornings. Her office gave her a 100-dollar pipeline bonus each quarter. Every bit helps.

Dan, a junior, did some trapping in the mornings for extra income. Elizabeth was a freshman and prepared the noon meal before running off to school. Her favorite food choice was ramen noodles. She's an excellent cook today.

Jim Temple: Kay

My brother, Jim, spent time with us during the winter. He was waiting for the union to call him to work up north. To keep himself busy, he worked at Earthmovers six days a week. It was beneficial to us to have him around and entertaining. He had plenty of stories to tell about our earlier days in Vassar.

Jim reminded me a lot of my dad. He was handy with odd jobs and willing to do whatever we asked him to do. When David went on a bilingual trip, the heater in the cabin was not working correctly. Ice was starting to form in the toilet bowl, and one pipe had already burst. When Jim came home from work, he helped me get the space heater. He also got the water pump going, which had shut off automatically. It was midnight when we went to bed.

Cleaning the Office

Dan had been working at Earthmovers before Jim arrived while still attending high school. He was the cleanup man and did odd jobs around the shop and grounds. One time, he decided to do a thorough job of cleaning the shop. *Playboy* magazines were lying around, so he gathered them up and threw them in the trash. Uncle Jim jokingly told him, "That's stealing!" Dan didn't want a smutty office.

Fast-forward: Grandsons Milo Cogan and Josh Rathbun worked for Earthmovers during the summers of 2013 and 2014.

More Meetings in Anchorage: David

I flew to Anchorage for meetings to assist in developing more Koyukon literacy materials through the Alaska Native Education Board. Overall, Marjorie (Williams) Attla developed and worked on 11 easy-reading Koyukon books for students in the classroom. She worked with the Anchorage staff to develop oral language units.

Bilingual Education Conference: David

Another bilingual education conference was held in Anchorage. The state and federal governments were becoming more active in schools across Alaska. The Koyukon area included seven schools, so I served them as a part-time consultant. More proficient Koyukon readers were being trained. We were incredibly thankful to have Scripture on the list of reading materials.

Snow Camp

During spring vacation, our three older children went to Victory Bible Camp for Snow Camp. The regular church bus broke down, so another bus had to be chartered for the 700-mile round trip from Fairbanks. They returned tired but happy. Stephen stayed home since he was still in grade school.

Bowling Night

We took the family bowling at the Arctic Bowl. They gave the kids a free game for each A on their report cards. We did not bowl enough to figure out the scoring system, but we had fun spending time together.

Private Tutor: Kay

I served as a private tutor, teaching Laura Alfred how to read and write her language each Monday morning. Laura was 76 and sharp as a tack. I took her to the Comprehensive Alcoholic Program to visit her son, Tim. He was committed to a six-week program and wanted help. She said to him, "Tim, you need to turn your life over to the Lord."

Rescued

During the weekly Koyukon Bible study, Henry Moses told us that on Mother's Day, he and Mabel had ridden to Nenana with friends. Usually, Henry would not have gone, but he decided to go this time. When they got to Nenana, they stopped at their son's house, but no one was home except two young grandchildren. Henry and Mabel brought them back to Fairbanks. That night, at 3:00 a.m., the house was fully engulfed in flames and burned to the ground. Henry was praising the Lord for sending them to save the lives of their grandchildren.

Another Band Trip

Debbie flew to Nome and Kotzebue with the Lathrop High School Malemute Band. It was an excellent opportunity for her to see and experience western Alaska. While in Nome, she visited Dave and Mitzi Shinen, Wycliffe translators with the Siberian Yupik people. She saw the official finish line of the annual Iditarod sled dog race.

Birthday Waffles: Kay

There were so many activities and visitors that we did not get to celebrate Debbie's birthday with a special dinner and cake. So I made waffles for breakfast with strawberries and lots of whipped cream, with a candle on each person's serving. Stephen asked, "Am I going to get a birthday cake for my birthday?" Soon we had a cake for Stephen when his day came.

Spring Banquet

Elizabeth sewed a cute floor-length pink dress for the spring banquet. Debbie made a pretty yellow one. Both girls were accompanied by young men from church. Dan washed and waxed our car so he could take a young lady to the event.

Finished His Book

Steve finished his AWANA book after memorizing 130 Bible verses. He took the initiative to learn the verses and earned the only trophy in his group. It was great to see God's Word going into his head so it could be transferred into his daily life. He was also growing taller. We had a hard time keeping him in pants that were long enough!

A Wedding Coming Up: Kay

Dixie received an engagement ring from Ron Swanson. She asked Stephen to light the candles during the service. Stephen wore a pastel-blue jacket that Debbie had sewn. She won a grand championship ribbon with it at the Tanana Valley State Fair. Debbie played her flute during the reception. I took care of their gifts.

Mexico City: David

Mexico City presented a different and exciting change from life in Alaska. I served as a delegate at the International Wycliffe Conference for two weeks. Twenty-six countries worldwide were represented by 112 delegates. People from developing and emerging countries were becoming more involved in linguistic and translation training projects.

Henry Family Reunion: David

On the way to the conference, I stopped in Ramona, California, to see Martin and Miriam, their families, and my mother. This was the first time in 20 years that the three of us siblings were together at the same time.

David, Miriam, and Martin

My mother's cancer doctors were just over the border in Tijuana, Mexico. She took vitamin B-17, laetrile, and cobalt treatments for breast cancer that was already well-established in her body. These treatments were not allowed in the US. She felt better, but the final results were in the Lord's hands.

Graduation Day: David

I hurried home from Mexico in time for Debbie's graduation from West Lathrop High School. Kay's parents came from Michigan for two weeks and took in the big event. The whole family was happy to honor our first high school graduate! She continued working at Samson's Hardware to earn money for college.

Debbie was considering the possibility of training as a missionary nurse. A local radio station interviewed her along with nine other seniors. We were pleased with her glowing Christian testimony.

Camping Activities: Kay

Dan went to Kokrine Hills Bible Camp on the Yukon River to train and serve as a counselor for three weeks. The camp was primarily for local Native children. During camp, Dan had the privilege of leading three campers to trust in the Lord Jesus.

We drove to Nenana to pick up Dan. He was excited to have helped at camp, and he talked the whole way home.

Elizabeth worked with the girls at Camp Li-Wa, and then she attended a week of Teen Camp. Stephen attended Boy's Camp.

Indian Leaders

Walter and Virginia Maillelle, from Anchorage, stopped for a visit. They were from Grayling, a village on the Yukon River. Virginia had been saved earlier and prayed for many years for her husband to turn his life over to the Lord.

Conservative Baptist Conference

We drove to Kenai Lake Baptist Church at Cooper Landing, Alaska, to participate in the annual Conservative Baptist State Conference. The scenery was beautiful. We rode in the Wycliffe pickup and camper, which sleeps five. Debbie did not come since she had to be at work on Monday. We met Christians from other parts of the state. The trip was 460 miles each way. The only road went through Anchorage, so we shopped on the way home.

Digging Potatoes or Trapping: David

Dan was given a temporary job harvesting potatoes with Henry Gettinger, a well-known potato farmer. He had 400 tons of potatoes that needed to be bagged. Dan decided it was not a very prestigious job when they ran into some rotten potatoes. Spoiled potatoes affected those around them, and they stunk. His partly underground storage area was kept at 33 degrees for safekeeping all winter.

Dan kept his job helping bag potatoes until he felt overloaded and then resigned. He preferred to trap near our house, catching red foxes, lynxes, and rabbits.

Debbie Goes to College: David

Debbie prepared to attend Alaska Bible College (ABC) in Glennallen, Alaska, 250 miles from Fairbanks. The Lord provided most of her needed money through her job in Fairbanks. The college began on Monday, September 8th, so we drove on the Parks Highway to Cantwell and then to Paxton on the Denali Highway. The fall shades of red and yellow were spectacular. We saw many hunters but no moose.

We borrowed the mission's pickup with a camper, so we had a place to sleep. Most could ride in the camper with room to move around. On the drive home, we stopped to buy gas, and Stephen got out of the camper to go to the bathroom. We started on our way. A little farther down the road, I stopped for something, and we realized Stephen wasn't with us. So immediately, we turned around. It was a lovely day, and he knew we would return.

Back in Fairbanks

The local schools continued to have split sessions due to overcrowding. Dan was a senior. One of his subjects was Fortran Compiler, a computer language he later used at the University of Alaska. On the side, he was helping Carl Horst build a house for Rachael Zook, whose home had burned down. He enjoyed construction work and gained valuable experience. Elizabeth was a sophomore, and Stephen was in the seventh grade.

Translating the Gospel of Mark (1976~1979)

New Translation Helper: David

All winter, I worked with Henry Moses to translate the Gospel of Mark into Central Koyukon. In appreciation for his help, I hauled firewood for him.

New Translation Checker: Kay

I worked with Agnes (Matthew) Mayo Moore to get a literal, "back translation" of the Scripture. She translated the Koyukon back into English. It enabled us to see both languages. Eventually, we needed both translations to show a translation consultant.

Finding her house was difficult. To make matters worse, she had a "guard dog" tied near her porch. Every time I took a step, the dog barked and barked. The dog knew I was afraid.

So I prayed and yelled, "Agnes!" She came to the door saying, "Praise the Lord. I needed somebody to talk to today."

She started crying, so we talked about what was bothering her first. I was grateful that I had gone to see her—dog or no dog. Agnes told me she would help every Thursday afternoon, so that's what we did.

Young People's Night

The young people had the evening service at Bethel Baptist. Elizabeth, Dan, and three others wore masks, showing how Satan comes in various disguises. Elizabeth kept going over her role at home and said her part with great expression. She was getting over her shyness. Practice makes perfect, but Dan did not practice his lines at home. Both did great anyway.

The Farm Inside

Elizabeth brought home a pet mouse a friend had given her. She enjoyed playing with it letting it run over her hands and arms. It was tame but not her parents' first choice. One little mouse was not going to multiply. Stephen had a pet gerbil that made a little bit of noise. He had fun building mazes for them.

Indoor Train

Stephen had seen Cousin Gerry Pedersen's huge basement train layout and decided to make a smaller version in his room. He set up a large sheet of plywood in the corner of his bedroom. He laid out the track, sprinkled green sawdust around for grass, and colored areas for flowerbeds. It looked quite realistic. He could lie in bed and think about riding the train.

Accidents Do Happen: David

In October, Dan took a girl from church on a date to a football game. Their family lived farther out on Chena Hot Springs Road. As they were returning on the first big curve, the car started to fishtail with two vehicles coming toward them, hitting the front of the first one. The other vehicle hit the side of our car, which pushed in the post between the side doors and dented the fender, spoiling the back tire. The girl's head hit the windshield. We thanked the Lord; no one was seriously injured. Cars can be replaced.

Thanksgiving Vacation: David

Debbie came home for Thanksgiving and a short vacation. She shared that everybody was friendly and the food was good. Pancakes with berries, salmon, roast moose, and moose hot dogs were on the menu. The college was on a list to recover moose kills in their area, and their freezer was full.

Dan and I drove her back to Glennallen. Dan, a senior considering his options, sat in a few classes before we returned home.

President Ford in Alaska

President Gerald Ford, Henry Kissinger, and their entourage stopped at Eielson Air Force Base on their way to China. It was good to hear him speak up close. Both Dan and Stephen shook the president's hand.

A New Car

We purchased a new six-cylinder blue Plymouth Valiant with a stick shift and overdrive. It was economical and big enough for our family. The insurance company settled with us after we totaled our 1972 Ford wagon. Fairbanks had little selection to offer during the pipeline boom.

Christmas Party for Junior High Kids

The Christmas party for the junior high kids from Bethel Baptist was at our house. Stephen helped in planning the event. Elizabeth made Rice Krispies treats in the shape of wreaths and holly with cinnamon red hots for the party.

Just a Phone Call Away: David

Paul Starr, an Indian believer, worked at a pipeline camp with several hundred employees. He called on New Year's Eve to wish us a happy new year. He wanted to talk with us Christians and be encouraged in his walk with the Lord. Paul did not know of any other believers in his camp. Paul commented, "I'm only a phone call away!"

1976
Phoning Debbie: Kay

We started using our telephone more, especially to talk with Debbie. Now we understand how our parents felt having their children away at college. One time when I suggested calling her, Dan said, "When we all leave home, you'll be just like Grandma Temple, wanting to talk on the phone constantly."

A Hospital Job Plus

Debbie worked two days a week at Faith Hospital in Glennallen. She filled out insurance forms and liked her job, but we hoped she wouldn't get behind in her studies. Debbie was also giving flute lessons to a fifth-grade girl. She got four A's and one B for her first semester grades, so we knew she was doing well.

More Sewing Projects: Kay

Elizabeth sewed a set of pillows that looked like a giant hamburger when they were put together. Two pillows were tan, representing the bun; a darker brown one was for the burger; a green one for the pickle;

a red one for the tomato; and a yellow one for the cheese. She probably didn't like onions, since there was no white pillow. It made a giant hamburger, in your preferred order, spread out on our couch. We were glad to see that she stayed with her projects to completion.

A Golden Opportunity

Dan had a golden opportunity to lead an Eskimo lady to the Lord while serving at the Fairbanks Rescue Mission. She showed him the scars on her arms from failed suicide attempts! He was in charge of the Monday morning meeting, answered the phone on Thursdays, or did whatever needed to be done.

Hunting in Our Yard

Dan thought he saw something walking behind a tree but wasn't sure. Then Stephen spoke up, "There's a lynx!" A big lynx was walking between our house and the garage. Dan shot it, and it was a beauty! He sold the hide for 205 dollars. He gave Stephen 30 dollars for spotting the lynx. Stephen used his money to buy a pair of cross-country skis.

Bible Memory Association: Kay

Dan and I were enrolled in the Bible Memory Association program. He had 13 verses a week, and I had seven. They expect more from younger minds! Debbie was in the program through the Glennallen Chapel. Stephen memorized verses with AWANA. Elizabeth continued attending the church youth group.

More Good News: David

We received a letter from Calvary Baptist Church in Florida. They were increasing our monthly support. It was the exact amount I had received for working with the discontinued bilingual program. We were missing that income, but the Lord supplied it. It was great serving a God like that!

Primary Sunday School: Kay

I taught a primary Sunday School class at Bethel Baptist. Others had taught our children in classes, and we wanted to be involved.

Stevens Village Again: David

In February, I flew to Stevens Village, staying for one week with a Native family. I worked mainly with Hilda Stevens, the bilingual teacher. She was 65, but still young at heart and eager to teach. Working with Hilda in her classroom and reinforcing her teaching of the Koyukon language was encouraging.

In the evenings, I had time to visit everyone in Stevens Village. Different Natives asked me many spiritual questions. For example: "Where did the different races come from, since we have a common ancestor like Noah and Adam?"; "Can I lose my salvation?"; "Will you pray for me?"

Several people were involved in the charismatic movement, speaking in tongues and prophesying. One 90-year-old blind man, David Adams Sr., told me a person prophesied he would receive his sight. It was difficult to answer, but I felt I had to tell him the truth. "David, I don't think you will receive your sight before you die. But because you are trusting in Jesus Christ to forgive your sins, you will see Jesus with your new eyes when you get to heaven." He responded, "But I want to see now."

Fast-forward: David Adams did not receive his sight before he died several years later. It was hard for me to be brutally honest with him, but he had false hope based on wishful thinking.

I also had a good time talking with a teacher about faith in Jesus Christ and how it all related to our daily lives. He was an agnostic. I enjoyed the open discussion with someone who believed the opposite of me while still being friends.

Snow Camp

Fifteen teenagers from Bethel Baptist, including Dan and Elizabeth, went to Victory Bible's Conference grounds for snow camp. Debbie returned with them to Fairbanks for a few days. Someone asked if it was her spring vacation, but she responded, "It's for recuperation from snow camp." ABC students were in charge of the camp. Debbie oversaw the indoor games and was a counselor in one of the girls' cabins where Elizabeth was a camper. They played in the snow and learned more about God.

Back to Stevens Village: David

I flew to Stevens Village for another bilingual program visit. This time, I traveled with an education specialist. I assisted her in preparing several aural/oral language games and simple dialogues in Koyukon. Unfortunately, liquor preceded our visit and hampered our time with people outside of school. Hilda Stevens greatly appreciated our visit and said it helped distract her from thinking about the adverse effects of alcohol on her family and the village.

Youth Rally

In April, Bethel Baptist Church hosted a weekend youth rally that included the ABC choir. Debbie played her flute for one of the numbers. They sang at the Sunday morning service and went back to Glennallen right after dinner. We were grateful to hear the choir and have Debbie home for a few days. We didn't expect her to get home this often when she went off to college.

God's Word for Today: David

It was an honor to read the Koyukon Scriptures to Sally Hudson's Aunt Jennie. She was 96 but very alert and appreciated hearing the Bible in her language. She was a dear Christian lady who had recently moved here. We had visited her several months previously in Anchorage.

Basic Youth Conflicts Again: Kay

Elizabeth and I flew to Eugene, Oregon, to attend an Institute in Basic Youth Conflicts seminar at the end of April. We enjoyed each other's company for a week as well as the conference.

Spring Banquet

Elizabeth purchased a lovely, light-green long dress for the spring banquet on May 8. She used her babysitting earnings for half, and we paid the rest.

Four Teenagers: Kay

On Monday, we had a surprise birthday party for Steve. Dorris Stevenson drove her son, Greg, to Steve's school, where she picked up Jon Wheelock and took them to Pay & Save. In the meantime, David met the other boys at their junior high school and then drove to Pay & Save to rendezvous with Greg and Jon.

They beat Steve's school bus and were sitting in our living room when he arrived. We all called out, "Surprise!" as Steve entered. He was amazed! They played games and had a hot dog roast. I made a birthday cake featuring an engine and several train cars. That evening, he said, "Thank you, Mom, for everything."

Steve turning 13 launched us into a new chapter in our lives. Now we had four teenagers under one roof for the summer. Life moved faster for all of us during those energetic teenage years.

Another High School Graduate

Dan graduated from West Lathrop High School on June 3rd, 1976. The commencement celebration was held at the concert hall in the Fine Arts Complex at UAF. Their graduation theme was: "Born Free: The Spirit of '76."

Summer Visitors

Gary and Mary Ann Eastty, Wycliffe support workers with the Paiute Indians in Nevada, served as guest helpers for the summer. They did maintenance jobs like painting, repairing leaky faucets, and putting in a sewer line. It was interesting to hear about the Paiutes' simple lifestyle like that of the Koyukon people, even though they were not part of the Athabascan family.

Two Workers: Kay

We left Debbie and Dan in Fairbanks to continue their jobs as we took our river trip. Debbie worked as a receptionist at Tanana Valley Clinic. I called her to make Dan's physical appointment. Of course, she recognized me. She sounded professional and businesslike on the phone. Was I a prejudiced mother?

Dan worked on constructing the new sanctuary for Bethel Baptist Church at 1310 Farmers Loop Road. He volunteered his time and expertise.

Cruising Down the River: David

Elizabeth, almost 16, and Steve, age 13, plus Kay and I, spent three weeks visiting Indian villages along the Tanana and Yukon Rivers.

The primary purpose was to continue relationships and be a witness for our Lord.

A generous friend loaned us his 24-foot aluminum riverboat and a 50-horsepower outboard motor. I mounted a cabin I had built with several helpers onto the boat. It was as large as it could be to fit on the boat. We had a top bunk for a smaller person and a lower bed for three of us. When Steve slept with us, I slept in the middle, and when Elizabeth slept with us, Kay slept in the middle. There was only room for Kay or me to sleep on one side or the other, but not on our backs or stomachs.

The boat was launched in the Tanana River near Fairbanks on June 7th, and we headed toward the Yukon River. We stopped at all the fish camps along the way, plus Nenana, Manley Hot Springs, Tanana, Kokrine Hills Bible Camp, Kokrines, Ruby, Galena, Bishop Mountain, Koyukuk, Nulato, and Kaltag. It was 500 miles one way from Fairbanks to Kaltag.

Before leaving Fairbanks, we had prepared some Scripture to distribute along the way. This trip allowed me to read those Scripture portions and use the "wordless book" with the black, red, and gold pages—representing sin, the blood of Jesus, and heaven—in explaining the way of salvation. People were open and welcoming everywhere we went. This prepared the way for distributing and using future Scripture translations.

Three days were spent in Tanana at the biennial gathering for the *Nuchalawoyya* (Noo-cha-la-WOI-ya) celebration. A variety of races and games, along with Native foods, were there. We reconnected with many Native visitors.

After turning around at Kaltag, we headed upriver toward home. Just below Galena, the lower unit on our outboard motor gave out. We made arrangements to send the boat and motor to Nenana on Yutana Barge Lines, and we flew home.

God's Fish Wheel

Dan was selected to pilot *God's Fish Wheel*, a cabin cruiser with a canvas top. They launched in Nenana and traveled 600 miles one way to Holy Cross and then back to Nenana. The trip was sponsored by Jim Barefoot Sr., who served with the *Cabin Prayer Meeting of the Air*.

One week was spent in each of the six villages where they ministered. Dan preached most of the time, one member had a guitar, and all helped sing gospel songs. They stopped in Koyukuk, Nulato, Kaltag, Grayling, Anvik, and Holy Cross.

Our son got "stoned" in Nulato! Let us explain. Their team met in the community hall. Several village kids had fun throwing stones on the metal roof, which was noisy. The extra commotion did not stop their meeting, but what disturbed the meeting were several firecrackers going off—inside the hall. It was loud and sounded like a gunshot. This took place over the Fourth of July weekend.

My Civic Duty: David

I was called to jury duty in Fairbanks for 30 days. Debbie was on a different jury during this time. My jury case lasted seven days, dealing with an armed robbery at a local liquor store. During the trial, Kay brought Steve to sit in the courtroom and listen to the case. The suspect acknowledged, "The man you saw was a person about my height and build with a car similar to mine, but it wasn't me." Steve looked at Kay and said, "He's lying!" The jury found the man guilty.

Bill Glass Crusade: David

Bill Glass had a July gospel crusade in Fairbanks at Herring Auditorium. Debbie, Dan, Kay, and I took their counselor training sessions and helped during the crusade. We did follow-up counseling with several people. I counseled one elder who accepted the Lord. He was the only one over 80.

Another Conference: David

We drove as a family to California to see Kay's sister, Sheryl, and her husband in Davis. Then we saw Miriam, Martin, and their families in southern California. It was always nice to include a short visit with family along with doing business.

The next stop was the biennial conference on August 24th near Estes Park, Colorado. As the executive committee chairman, I was responsible for the conference and was reelected to serve as chairman.

Sweet 16!

Elizabeth turned 16 in September. We celebrated with six of us in a pickup with a camper. Her special day started with breakfast at the Pancake House. The next stop was for dinner at Kentucky Fried

Chicken. The plan was to go to a Farrell's Ice Cream Parlour, but we couldn't find one in any cities near our route.

The next best thing was to purchase an angel food cake smothered with whipped cream. Driving along to our overnight camping site, we hit a few bumps, and her cake slid. It was the funniest-looking, yet delicious, birthday cake. Then we gave her presents, went to bed, and called it another great day.

Two at Alaska Bible College

We drove straight to ABC to enroll Dan as a freshman and Debbie as a sophomore. Now half of our family is gone. Dan liked his classes and even took fundamentals of music and joined the choir. Debbie's classes included beginning piano and first-year Greek to understand the Bible better.

Fort Yukon: David

Dick Mueller invited me to Fort Yukon for a five-day moose-hunting trip. I had a great time but left the moose for somebody else. Dick's outboard motor broke down while traveling, so we floated, put up a makeshift sail, which helped, were towed, and finally borrowed a smaller motor to get back to Fort Yukon. Such is life—no moose and everybody has outboard motor problems.

Happy B-day, Dan: Kay

We drove to Glennallen to attend a "Walk Thru the New Testament" on Saturday, sponsored by ABC. We celebrated Dan's 18th birthday with blueberry cake, whipped cream, and candles. The children gave me my birthday gifts a few days early. It was fun to be together as a family.

Rollover

We thanked the Lord for His protection when the van from ABC went over an embankment and rolled over twice, landing on its roof. Debbie and Dan, plus three other students, were in the van. They crawled out and flagged down a passing vehicle that took them to Faith Hospital. Dan's roommate was injured and required surgery to remove his spleen. All were shaken, slightly bruised, and sore.

Back to Tanana: David

I flew back to Tanana in November for a few days to check the translation texts. First, we looked at the Christmas story. Mary worked with Florence Albert to discuss and agree on the translation. I checked a few pages with Joe John and David Henry, a Native with my name, in Tanana. This encouraged me since it was tough to find qualified helpers.

Cross-Country Skiing: Kay

I bought a pair of secondhand cross-country skis and boots. Elizabeth used Debbie's, and Steve had his own. Now I could ski with them.

New Cleaning Lady

The construction of Bethel Baptist Church was progressing on time and looking good. Their first service was on Wednesday, the day before Thanksgiving—a joint service with Denali Bible Chapel.

The trustees asked Elizabeth if she would be the church janitor until the end of 1976. She was glad to earn spending money. Pastor Hokie Moore praised her when she cleaned during the construction stage. He remarked, "The church was the cleanest when Elizabeth cleaned it."

The Miracle of Christmas

In December, the choir at church presented the musical, "The Miracle of Christmas." The young people did the pageantry during the singing. Elizabeth was a lovely Mary, and Steve was a shepherd. We all enjoyed the benefits of the larger building.

Was It Worth All the Work?: David

Much effort, time, and work with different Indian helpers were necessary to get the Christmas story ready. Like anything in life, one gets weary after working for a long time on a project. But the Christmas account was finally prepared.

Paul Starr stopped by with a request. He needed a ride to Nenana, 60 miles from Fairbanks. I liked to see his family, but I was not excited to take him. It was late Sunday night, and it would be much later when I returned home. There was no other way for him to get to Nenana. So I told myself, "I'll make the most of it and have a good time with Paul."

Elizabeth Starr, Paul's mother, was home, so I read the Christmas account from the Gospel of Luke to her in Athabascan along with John 3:16. She responded, "That's the first time I realized Jesus came for me, to be my Savior." Then we kneeled down, and I led her in a prayer to receive Jesus Christ as her personal Savior. What a thrill!

I never told Paul that I had not wanted to take him home. I almost missed a tremendous blessing. Elizabeth's salvation was my payday for all the time and work to produce the booklet. Yes, it was well worth it!

Most Indian contacts require a long-term commitment and years of building trust and relationships. Our first contact with Elizabeth was at Arctic Camp in 1958, when we started learning her language and drank many cups of tea together.

We mailed copies about the birth of Jesus in Upper Koyukon to individuals in Tanana and Stevens Village and to Nenana, Minto, Nikolai, and McGrath in a closely related dialect.

More Translation Checking: David

Mary Dick flew into Fairbanks to check more of the translation with Con Naish. It seemed to be a never-ending process. Piece by piece, we were getting the Gospel of Mark ready.

Chess Moves and More

Steve was in the eighth grade when he joined the chess club. He was the top player at his school. At one chess meet, he heard another player say, "That boy in the blue shirt (meaning Steve) is very good." He made himself a chessboard with inlaid pieces of colored wood.

Elizabeth was a high school junior looking ahead to graduation. Her immediate goals were to get her driver's license and a job.

Christmas Fellowship

We enjoyed wonderful fellowship over Christmas dinner at Roger and Bev Wheelock's home. Steve and Dorris Stevenson also came for the potluck with more food than we could eat.

Jail Time

One of Dan's jobs to earn money was as a jail guard at the holding cells in Glennallen. When someone was apprehended by the state troopers, they stayed temporarily in a holding cell until they could be transferred to Anchorage or Palmer for incarceration.

As Dan said, "I make my living off of sin—someone else's sin!" He could do his homework while guarding. Earlier in the year, he had the privilege of pointing two prisoners to Christ Jesus, with God forgiving their sins.

Urbana Missionary Conference: David

When I took Debbie and Dan back to college, I saw posters saying, "Send Debbie to Urbana." She already had 190 dollars at Thanksgiving time. Through generous donations from students and staff at Alaska Bible College and Bethel Baptist Church, she was able to attend the conference.

Urbana was a prominent Christian student mission conference sponsored by InterVarsity Christian Fellowship. The five-day event was intended to inform Christian students about cross-cultural missions and to encourage them to serve. The theme of the Urbana 76 conference was "Declare His Glory Among the Nations." Billy Graham opened the event with "Our Response to God's Glory."

It was held at the campus of the University of Illinois, 150 miles south of Chicago. A large facility was needed over Christmas break for more than 15,000 college students and staff. Debbie left Fairbanks at 1:15 a.m. on December 26th. The conference ended on December 31st with communion. Debbie commented, "It felt like a taste of what heaven would be like with thousands of people joyfully singing and praising God together."

Then Debbie flew to Florida for a few days to visit Grandma Henry, who was seriously ill with breast cancer. Grandma took her to a Ringling Bros. and Barnum & Bailey three-ring circus event in Sarasota, Florida.

The Birth of Jesus

Mary Dick read the account of Jesus' birth and the wise men's visit during the Christmas service at Saint James Episcopal Church in Tanana. Several women sang Christian hymns in Koyukon. Mary was pleased with the favorable response. One elderly Native was so delighted that she phoned Mary at 2:00 a.m. the following morning to thank her again.

1977
House Guest from Tanana

Mary Dick flew from Tanana and stayed with us for two weeks. She was determined to complete the translation of the Gospel of Mark and ensure it was accurate. We appreciated her attitude. We saw the Spirit of God doing His work.

A New Driver and a Job

Elizabeth passed her driving test. Now, we have three teen drivers. The next thing she wanted was her own car rather than borrowing ours. She found a good job working as a courtesy clerk at Safeway, making three dollars and 15 cents an hour.

Progressive Dinner for Missions

The progressive dinner for missions at Bethel Church started with hors d'oeuvres representing Africa served in the church foyer. The Indonesian salad was downstairs. Then everyone went into the fellowship hall for the main course of shrimp with sweet and sour beef over fried rice served in a Japanese atmosphere. The dessert was in a different room, representing Brazil.

It was an enjoyable evening with a purpose—helping us to think about and pray for missionary work. There were displays from each country and prayer letters from the missionaries our church supported.

Another Snow Camp

Debbie and Dan worked with the staff and students at ABC to plan and lead another snow camp at Victory Bible Conference Center. College students served as camp counselors. Elizabeth took time off from work to attend.

The church borrowed a vehicle that had two flat tires on the way down and didn't have a spare. This forced them to get a rental. The rental van broke down 60 miles out of Fairbanks on their way home. As a result, another vehicle returning from camp brought them home. They had a wonderful time but were exhausted.

Fairbanks Rescue Mission: David

I was in charge of the service at the Fairbanks Rescue Mission on behalf of Bethel Baptist and gave the sermon or recruited someone.

Churches in the Fairbanks area took turns and were responsible for one service per month.

Executive Committee Meeting Plus: David

I flew to Denver to chair the executive committee meeting for the North America Branch. We reviewed past activities and assisted in planning for the future. It was another good meeting, but the trip home was even better.

In Seattle, I stopped to visit Jerry and Sally Hudson. Jerry was not Native but was very supportive of the Native community. He was a former smoker, age 69, and he was undergoing treatment for lung cancer.

Jerry was ready to receive Jesus Christ into his life since Sally had been praying for him. He was so open; it was like picking low-hanging, ripe fruit that was just waiting to be selected. We looked at Scripture together, and then he joined the family of Jesus Christ.

Running Track

Steve went out for track in the spring, which met after school. He ran half a mile each night on Wigwam Way to Chena Hot Springs Road and picked up our newspaper. He made a neat chart to keep track of his time. Steve ran in two junior high track meets in Fairbanks. He ran the half-mile and was quite fast. He had good form and stretched out, so he didn't waste any motions.

April Fools' Day: Kay

Elizabeth was full of tricks and awakened us by yelling, "A moose is in the yard!" We jumped up to look out the window, but there was no moose. She made Steve a peanut butter and catsup sandwich for his lunch.

Debbie called from Glennallen at 11:00 p.m. and said she had news for us, "I'm engaged!" I didn't say much, so she added, "That was fast, wasn't it?" Then she exclaimed, "April Fools!" I heard girls laugh when she said that. I guess her roommate and friends were standing nearby. After I went to bed, I lay awake thinking that she would call one of these days—and it would be true.

Jesus Suffered for Me

We worked on the translation of chapter 15 of the Gospel of Mark with Mary Dick. A little later she told us, "I wish my people could hear about Jesus suffering for us in Koyukon." She related how she had gone into her bedroom at our house, shut the door, and cried. It struck her how much agony Jesus suffered in her place. She understood the meaning of the passage more deeply once she heard it in her own language.

Still Feeling Some Doubts: David

I continued to have doubts about ever becoming a fluent speaker of Koyukon. A good book on successfully learning a foreign language is *Language Acquisition Made Practical* by Thomas and Elizabeth Brewster. We read the book and found it challenging. The number one condition spelled out in the book for being successful was that you live where the new language is spoken. We could not meet that condition because people primarily spoke English.

Koyukon Language Appreciation

We believe one should appreciate their language, culture, and history. We wanted Indians to be thankful God had made them Indians with their own distinct and beautiful, complex language, even if they didn't speak it.

Traveling via Florida: David

I flew to Bradenton, Florida, to visit my mother for a week. She had purchased a small trailer with an addition at Bowlees Creek Mobile Home Park just off Highway 41. Many retired and older adults lived all around. Restaurants and eating establishments were close by, and their prices were low by Alaskan standards. I understood why local people ate out so often and let somebody else cook and wash the dishes.

She was a member of Calvary Baptist, a biblical church, preaching and living out the gospel of Jesus Christ. Many seniors were members, and my mother fit right in. She was well taken care of and visited by the church family.

Mexico City Again: David

I was a delegate representing the North America Branch at the International Corporation Conference in Mexico City. Arriving early, I participated in a seminar on Bible translation.

In southern California, I saw Martin and his family in Ramona on my way home. I also spent time with Miriam's family in Orange.

Kay hosted a party at our place for the women of the Native Bible Study while I was gone. They cooked on an open fire and enjoyed visiting.

Summer Jobs for Everyone

Dan had a short-term job painting the outside of Bob and Joan Biggar's house. He had one more day of work to complete the job. What would he do next?

Rollin Loewenstein, from church, called Dan at 10:00 a.m. on his first day without work and asked about a construction job. Dan jumped at the opportunity. He wanted to work, so he started with Rollin at 1:00 p.m. that afternoon. The Lord's timing was just right. Rollin was building a super-insulated home in College, Alaska.

Later, Steve joined their workforce for a month. By mid-August, they had finished laying the concrete blocks for the basement walls.

Debbie liked her job as a bookkeeper, but it was only part-time this summer. She ended up combining two part-time jobs to earn money for college.

Elizabeth was a senior, and she continued working as a courtesy clerk at Safeway all summer. She weighed only 95 pounds at that time, so the cashiers would call for a fellow to help lift the heavy bags of dog food or rock salt. During the winter, she advanced to cleaning the bakery with higher pay. She was saving money for a vehicle.

Summer Trip to the Yukon River

We took a family end-of-summer trip to Circle, Alaska. It was 160 miles on mostly gravel roads. We were interested since a few Koyukon people lived there.

Then we drove to Central, another small town with miners and sourdoughs. Nearby, the Arctic Circle Hot Springs Resort had an Olympic-size warm pool fed by the springs. We arrived at 9:30 p.m., and they let us swim until 10:00 p.m. for free since it was near closing time.

Four slept in our station wagon while Dan and Steve were in a tent. In the middle of the night, a nearby camper yelled, "Bear!" Dan and Steve jumped up and spent the rest of the night in the station wagon. Six people trying to sleep in a station wagon felt like sardines crammed in a can!

Oil Is Flowing

The oil from the Prudhoe Bay fields started flowing to Valdez on June 20th, 1977. Oil became the primary source of income for the State of Alaska. At first, it seemed the state was looking for ideas and projects at which to throw money. It was like telling a kid in a candy shop to eat anything you want and fill all of your pockets!

Tanana Valley State Fair: David

The Tanana Valley State Fair was a major event in Fairbanks. Our family enjoyed making and entering things, often through the 4-H club. Steve won 79 dollars for all his entries and three cases of canning jars from 4-H. He entered a variety of jams: cranberry-banana, raspberry, blueberry-grape (with grape juice), spiced blueberry, Alaska apple butter (canned applesauce with highbush cranberries), and strawberry-rhubarb.

I bought a set of onyx chess pieces in Mexico for Steve's chessboard. He entered his inlaid wooden chessboard, a salad bowl, a carved wooden car, an airplane, a diamond willow pen holder, candles, an ecology box, a matchstick porcupine, a macramé plant hanger, and a few other things. He received mostly blue ribbons and two purple championship ribbons. Steve won the junior grand championship with a broccoli casserole. He had the ambition and more time on his hands than his siblings.

Elizabeth entered three jams and received blue, red, and white ribbons and made nine dollars. At the same time, Debbie entered peanut brittle and a leather purse, receiving blue and red ribbons and eight dollars. Dan was working on construction and didn't take the time to enter things as he had in previous years.

Multnomah School of the Bible

Debbie transferred to Multnomah School of the Bible in Portland, Oregon. She was excited about attending and living in a big city. She enrolled as a junior and appreciated her year there. The school's name changed to Multnomah University and continues today as a private, non-denominational Christian University.

Longtime friends John and Shirley Bartlett, from Fairbanks, lived at Gold Beach on the coast of Oregon. Shirley was Debbie's shorthand teacher in high school. It was encouraging to have good Christian friends to visit nearby.

Bible College Menu

At ABC, Dan caught a porcupine while running his trapline near the college. He skinned it and offered it to the campus cook, who made it into a stew. The other students also enjoyed it. Not many colleges include porcupines on their menus!

Planning Workshop: David

I flew to Espanola, New Mexico, to participate in a planning workshop. We put all the moving pieces together and pointed them in the right direction for the overall work. One side benefit was visiting Debbie in Portland on the way down.

Koyukon Class at UAF: Kay

Furthering my knowledge and understanding of the Koyukon language, I enrolled in the course taught by Eliza Jones at the University of Alaska Fairbanks. This was the third year Eliza had taught this course. She was outstanding at analyzing her language and explaining the meaningful parts of each word. David and I went over the handouts at home. Several Koyukon Natives were taking the course.

Getting Help

We checked over portions of Mark in Upper Koyukon with Virginia "Virgie" (Starr) Newby from Tanana but now living in Fairbanks. She accepted the Lord five years ago. Virgie knew her Bible well since she was a growing Christian and applied what she read in the Bible to her daily life. She was enthusiastic about the translation work. Her elderly mother listened in on our work.

Another Executive Committee Meeting: David

In mid-November, I flew to Denver to chair our executive committee meetings in Thornton, Colorado. It was good to see things moving ahead. During that time, I gave two lectures on culture shock at a Wycliffe outreach program in Denver.

Christmas: David

Our whole family was together for Christmas. I drove to ABC to bring Dan home. Debbie flew back from Multnomah School of the Bible. We brought out photo albums and had fun sharing memories. We played games, completed puzzles, laughed, and celebrated together as a family until Debbie and Dan returned to their schools.

1978
Interruptions or Opportunities

Many interruptions came—or were they opportunities in disguise? God allows us to be interrupted, reminding us to keep looking up to Him and seeking His help and direction in life. We had mechanical problems with our car, funerals for friends, more visitors than usual, a touch of laziness, and the list goes on.

Native Guests

Mary Dick flew into Fairbanks and lived with us for a month. She continued to care for her six-month-old granddaughter. The baby was a precious little girl and easy to care for. Her daughter Kathryn also stayed a couple of nights.

Mary's other daughter, Shirley LeBeau-Anderson, went to Anchorage for the dog races, and we kept her sons, ages six and nine, over the weekend. They were good kids but needed to be kept busy. Hilda Stevens stayed for three weeks to help Mary with the push to complete the translation.

Mary Met an Old Friend

Mary was staying with us, so this presented an excellent opportunity for Mary and Virgie to work together on the translation of Mark. We called Virgie and asked her about bringing Mary along. She was delighted. But while we were driving to Virgie's place, Mary became very concerned and stated, "We used to drink and do all kinds of things together in Tanana. What if Virgie won't forgive me?" We assured Mary that God had worked in both of their lives.

Mary led the way. When Virgie opened her door to welcome us, the forgiveness and acceptance were apparent and joyous. Two sisters in Christ met each other as part of God's wonderful family. Their drinking history was completely behind them. Virgie told us, "I'm glad I was able to live long enough to be saved."

Youth Jamboree

Elizabeth attended the ABC Youth Jamboree for junior, senior, and post-high school young people. It was a combination of snow camp and orientation to college life to encourage the kids to go to college.

Dan led one of the boys from Bethel Baptist to the Lord during the weekend. His parents were divorced, he had recently moved to Fairbanks, and he had dropped out of high school. Help and hope are available for everyone through Jesus Christ.

Sewing and Knitting: Kay

I sewed more of my granny Afghan squares together on a light brown background. Working on it a little at a time, I made progress. My priority project was knitting a sweater for Elizabeth in two shades of blue. I finished the front and back, but the sleeves and collar were more complex.

Easter 1978

On Easter, we attended the early sunrise service and breakfast at Bethel Baptist Church. At noon, Elizabeth treated us to brunch at the Fairbanks Inn. It was terrific, with a variety of food choices and more time to be together.

Whitehorse Workshop: David

I drove to Whitehorse, Yukon Territory, in Canada, to attend an anthropology and planning workshop for two weeks. Following the workshop, I returned home and chaired the executive committee meetings for the North American work. Six extra men stayed for five days at the center. Kay froze desserts and casseroles for this time. Elizabeth and Steve baked cookies to share.

Summer Jobs

When Debbie and Dan returned, the major thing on their minds, and ours was finding summer jobs. Debbie worked several different part-time jobs. Elizabeth took a one-week training session at Safeway to become a checkout cashier. The new job gave her a big wage jump to seven dollars and 73 cents an hour.

Dan worked at the oil refinery 15 miles from Fairbanks in North Pole. On Saturdays and evenings, Dan also worked on finishing the house that was started last summer. Steve worked with him part of the time.

Girls' Bible Study

Debbie and Glenda (Hanneman) Talcott conducted a Bible study on Thursday evenings for college-age girls. They studied together and alternated teaching.

Teen Fellowship Time

Elizabeth and Steve hosted the high school and post-high school groups at our house for an ice cream social. Elizabeth had purchased three gallons of ice cream. One of the guests let a younger brother drive. Unfortunately, he overshot our driveway and dropped partway over the embankment. So everybody helped jack up the car, put boards under the front wheels, and pushed it back on the road—a little work for an excuse to eat more ice cream.

Another High School Graduate: Kay

Elizabeth graduated from West Valley High School on June 6th, 1978. Her ceremony was held at the University of Alaska's concert hall with 180 graduates. We celebrated her accomplishment in completing this milestone in life.

I finished my Koyukon course at UAF with Eliza Jones and received an A and four credits. No ceremony for me—I was just glad to complete the course.

A Mini Furlough: Kay

We left Fairbanks on June 29th with Steve to take a mini furlough. We drove to Michigan to be with my parents and supporters in the thumb area before heading south.

Stopping at Bryan College, our alma mater, in Dayton, Tennessee, we saw Steve and Dorris Stevenson. We visited my brother Tom and his wife, Mary, in Greensboro, Georgia. It was great to see their family. My brother was the football coach for Green County High School.

In Bradenton, Florida, we visited David's mother. Her breast cancer was advanced, and she often gasped for air when she talked. She had a hard time walking. We pushed her about in a wheelchair to conserve her energy.

Elizabeth flew directly from Fairbanks to see Grandma Henry. David started to wheel his mother to greet Elizabeth, but she stopped him and said, "I'll walk to her—she's my granddaughter!" And she slowly walked to her. We realized this could be the last time we would see her.

Rescued from the Undertow: Kay

Aunt Lillie Bachanz, a nurse, stayed for several months caring for her sister, Nelda, David's mother. She liked to swim in the warm waters of the Gulf of Mexico. One afternoon, she was swept away by the strong undertow.

Elizabeth, on shore with me, could swim but not well enough to rescue someone that far out. She ran along the beach to the nearest people who were sunbathing and cried out, "Help! Help!" The fellow jumped up and swam out to Aunt Lillie, but just as he reached her, she pulled off her wig, exposing her bald head. She didn't want him pulling on her hair and merely getting a handful of her wig. She was bald from her cancer chemo treatments and was too weak to swim to shore.

More Activities: Kay

A generous friend allowed us to use his cottage on Santa Maria Island. Plenty of space was available, and Elizabeth stayed with us for a week. We took a day trip to Busch Gardens in Tampa, seeing the animals while riding high overhead. We toured their brewery but left the free samples for others.

We drove to New Orleans so I could reminisce about my high school trip there with my parents. Then we stopped to tour LeTourneau College in Longview, Texas. At the Wycliffe Center in Dallas, we looked around and spent one night before heading to Colorado.

Next Stop: Estes Park, Colorado

Estes Park was a beautiful site for the biennial conference. Steve enjoyed a three-day backpacking trip as well as the other teenage activities.

Flying Home: Kay

Steve and I flew home immediately after the conference so he would not be late for high school. I needed to spend time with Debbie and Dan before they left for Tennessee. They appreciated having me fix the meals so they were free to do other things.

Off to Bryan College

Both Debbie and Dan transferred to Bryan College. They flew to Chattanooga, where Steve and Dorris met them. It was fun for them to have a brother and sister attending college together. The college continued to put "Christ Above All," as their motto stated when we were students in the '50s.

Accidents Do Happen

Debbie and Dan were riding in a car following another Bryan vehicle that was hit by a train. One student had a broken jaw, while the others had bruises. They were fortunate that God had spared their lives. Dan commented, "It was a strange feeling to drive up and see our classmates lying on the ground."

High School for Steve: David

Steve, a sophomore, was happy to be the only one in our family in high school. He took challenging subjects: chemistry, algebra II, trigonometry, literature, small engine repair, typing, and physical education. He worked on our lawnmower's engine for his class project. His teacher said Steve was doing very well, and he enjoyed that work, whereas I do mechanical work because I have to.

First Car for Elizabeth: David

Elizabeth was anxious for me to return home and help her find her first car. It was a long drive from Colorado to Alaska, but I made it without any problems. She continued working as a cashier at Safeway and had saved enough money for a vehicle.

Her first car was a 1972 blue Chevy Monte Carlo. She was thrilled. It was a good one for a single young lady. It had power and lots of class. Elizabeth found out where her parents' money went. Regular gas was 83 cents per gallon.

Fast-forward: Many years later, Andy and their first daughter, Melissa, took it out of the boneyard, restoring it for her first car.

Cancer-Free at Last: David

At 72, my mother took her last breath on August 13th, 1978. Nelda Louise (Bachanz) Henry was now cancer-free in heaven. She spent her last days in a nursing home in Bradenton, Florida. A steady stream of older visitors from Calvary Baptist Church encouraged her through visiting, praying, and reading Scripture together. We were so thankful we had spent time with her earlier that summer. She was buried next to my father at Camp Nathanael in Kentucky. We did not attend her funeral service, but my sister Miriam picked up Debbie and Dan from Bryan College to represent our family at the funeral.

Improving the Translation

Virgie Newby checked her own work for naturalness, consistency, meaning, and faithfulness to the biblical text. Her mother, Belena Starr, was in a local nursing home, and Virgie would read the Scripture to her. She commented, "That is so easy to understand," which greatly encouraged Virgie. Other short-term helpers included Carrie Joseph and Dorothy Pitka.

A Little Black Dog

Elizabeth acquired a little black puppy. She was charming and trained by placing newspapers on the floor. Elizabeth bought a can of No, which she sprayed at the entrance to the living room and the rug. It worked.

Snow Machine Time: David

Steve enjoyed his snow machine, which had plenty of space to travel. It kept him from getting bored, and he was getting practical mechanical experience.

Ministry Opportunities: David

We spoke at the Nenana Bible Church, sharing about our work. We were amazed to see a full church and how the people were active in teaching Sunday School and reaching out to their community.

Christmas with Grandparents

Debbie and Dan traveled to Vassar, Michigan, to celebrate Christmas with their grandparents, Bob and Vera Temple. They took them to Canada to visit Uncle Louie and Aunt Nancy, plus their three cousins, Robin, Denise, and Rick. They were so happy to see their relatives and play in the snow in Canada, as Tennessee rarely has snow.

Christmas with Parents: Kay

Elizabeth and Steve stayed at home to celebrate Christmas with us. Steve cut down our Christmas tree and decorated it. He placed it in front of our picture window. Elizabeth helped make several different kinds of cookies and pecan pies.

1979
A College Student from Nenana

Sue Eldridge, a student at UAF working on her bachelor's degree, stayed with us for one semester. She had graduated from the Montana Institute of the Bible. Her parents, Bob and Dolores Eldridge, were missionaries serving with the Nenana Bible Church. She stayed with us four nights a week and went home to Nenana on weekends. She fit in well with our family.

College Spring Break

Debbie rode to Connecticut with her friend, Norma Jean Jancewicz. Dan went to Baltimore to train with Open Air Campaigners for their outreach in the bigger cities.

Something in My Eye

We couldn't see it, but Steve could feel something lodged in his eye. The ophthalmologist used a high-powered microscope magnified 32 times to pluck out a small piece of steel with special tweezers.

A Tragic Accident: David

In April, Elizabeth Starr, age 61, was struck and killed by a hit-and-run driver in Fairbanks. We attended her funeral and potlatch in Nenana. Her daughter, Anna, wrote a letter and asked me to read it publicly. The letter eulogized her mother and commented on how she used to drink and had many complex problems that would drive her to the bottle. She also expressed her mother's faith in the Lord Jesus even though her mother couldn't read or write. When I finished, there was a long round of applause, which was most encouraging to Anna.

Two More Weeks: David

Mary phoned to express concern about completing the Gospel of Mark to make it available to her people. There were several recent deaths, and she stated, "We have got to get the gospel to them before they all die off!"

Mary flew in from Tanana in May and stayed with us for two weeks. Our primary focus was to continue translating Mark. We worked on two chapters and further checked with Virgie. After Mary read the Bible in Athabascan to various people, she received comments such as, "It's so real. It's like everything was brought out in the open." And, "It really hits you when you hear it in our language."

Dating and Marriage Panel

The church held a dating and marriage panel consisting of two singles, a married couple without children, and a couple with children. Elizabeth was one of the singles on the panel. She commented later that she was surprised at how low the standards were. She did very well, spoke up, and had good insight.

Coming Home

Debbie took a concentrated three-week course in consumer chemistry following her college semester. She analyzed motor oil and chemicals in food to determine what those additives meant for us consumers.

Dan arrived home on May 12th. He rode from Tennessee to Chicago with John Rathbun, who then flew directly to Seoul, Korea, to be with his parents.

Summer Jobs

Twelve percent was the unemployment rate in Fairbanks. Despite that, Dan landed a job as an insulator at the North Pole Oil Refinery. He started and taught a college and career-age class at Bethel Baptist Church and preached at the Rescue Mission.

Debbie worked at the Bureau of Land Management, plotting Native land allotments on bigger maps. She enjoyed volleyball each noon at work as she got ready for the college team.

Elizabeth rented an apartment in town with a girlfriend. She continued her full-time job as a cashier at Safeway.

Steve had temporary jobs. He was a cashier at the Tanana Valley State Fair. An enormous cabbage that year weighed 56 pounds! Alaska's garden produce grew well with the long hours of sunlight.

Steve also worked at McDonald's on Airport Way three evenings a week and received a free hamburger each time he worked. He wasn't excited about working there, but he did get paid. After returning to high school, he worked on Saturdays to help finish the house he and Dan had started building together.

We prayed for our children to make the choices in life that are pleasing to the Lord. Choices have consequences! We all need more "fear of God" in our lives.

Popcorn and Hot Chocolate

Young people came to our house for popcorn, hot chocolate, and games with our family. Many rounds of Pit were played. Two of the fellows, Ron Hack and Doug Isaacson, were new to our church, studying linguistics at Eielson Air Force Base, and were interested in our work.

The Fairbanks Native Fellowship

We continued to support the monthly Native fellowships to meet the spiritual needs of Native Indians and Eskimos. During the winter, we met at Denali Bible Chapel. For summer fellowships, 30 to 40 people attended at our home. During one meeting, we showed slides of our time in Kokrines. Several from there reminisced about their village life.

How Many Can Fit in a Car?

One night a week, the skating rink offered special rates of seven dollars and 50 cents for the carload or 50 cents per person. Our children piled 16 kids into a station wagon and went roller skating. The car with the most kids got a special prize. Our kids thought they would win until another vehicle showed up with 25!

Developing Educational Materials: David

I flew to Anchorage for two days to assist in developing Koyukon educational materials for beginning levels one and two. It was encouraging to work with others in the overall process.

Then I flew to Galena to support Madeline Solomon in preparing language lessons. We wrote out 30 pages of level-one schoolwork. Madeline was a "born" teacher and it was easy to work with her.

A side benefit of going to Galena was fellowshipping with a group of local believers. I witnessed the baptism of a Koyukon lady in a nearby lake.

Native New Life

Walter and Virginia Maillelle and their three children spent three nights with us. Walter was the leader of the Native New Life movement in Anchorage. It was always good to see them and promote our shared goal of reaching Native people in Alaska for the Lord. We continued our close friendship throughout the years.

Walter, Virginia, and Kay

Traveling Again: Kay

I flew with Elizabeth to Michigan to spend time with my parents. We traveled together to Blind River, Ontario, where we witnessed the wedding of my nephew Robin Horwath to Susan in July. This included time with my sister Nancy. Next, we drove to Speculator, New York, to see my sister Shirley and her family.

Summer Sewing: David

Debbie sewed two shirts for John Rathbun, plus one for Steve, and one for me, which earned her a championship ribbon plus seven dollars at the local fair. She also sewed a shirt for Elizabeth's birthday and one for herself.

Travels with Steve: David

Steve and I flew to Portland, where we stayed at the Multnomah School of the Bible while attending the Institute in Basic Youth Conflicts seminar. More than 30 people from Alaska participated, including several Koyukon Natives from Tanana. All four of our children have benefited from these seminars.

Dale and Mary King invited us to their home. Steve reconnected with Leslie and Kay, whom he had helped babysit. We even took the Kings' boat out for a spin in the lake. They lived five minutes from the Seattle airport, so it was easy to meet Kay and Elizabeth when they arrived from New York.

Flying Back to College: Kay

We celebrated Elizabeth's 19th birthday by eating at The China Garden in Fairbanks. Then we looked at our family slides until it was time to take Debbie to the airport for her trip back to college.

A couple of service members came by the airport just before Debbie's plane left. They had spent time with our family over the summer. It was encouraging for Steve to see these young adults following the Lord. Dan worked a few more days before he flew back to Bryan College.

Debbie met my cousin Diane at the airport in Chicago and went to her home for a few hours. I knitted a sweater for Diane's daughter, Marcy.

Rally Day: Kay

September 9th was Rally Day at Bethel Baptist Church. My third- and fourth-grade Sunday School class joined with the fifth and sixth graders to put on a puppet show. One puppet had a birthday, and three puppets sang "Happy Birthday." Then the rest of the class had various puppets that popped up one at a time, saying, "Happy Birthday."

In the end, the cake spoke and asked one of the students, "Do you have two birthdays?" as we tried to get across the spiritual message of being born again. Your first birthday was when you were born into your earthly family. Your second birthday is when you accept Jesus Christ and are born again into God's family.

How Does Your Garden Grow?: David

I gave priority to planting vegetables, while Kay wanted more flowers. Freezing Chinese peas and rhubarb supplemented our winter meals. Zucchini squash, carrots, potatoes, and cabbage grew well, but it was a hassle to keep the moose away.

Steve grew six purple cabbages. Guests arrived, and he wanted to share one. It was getting dark as he went to the garden to cut one. He ran back, yelling, "Something ate all my cabbages!" The tracks proved a moose had come by and eaten them, leaving only bare, headless stems!

Birthday Celebrations: Kay

I celebrated my birthday with Elizabeth and Joan Biggar at a Chinese restaurant. Then Doug Isaacson took David, Steve, and me out to celebrate my birthday at a different restaurant. Several folks from church were also eating there. Doug asked the organist to play "Happy Birthday," and they all sang to me.

A Volvo: David

We purchased a 1968 Volvo with an automatic transmission. It was cream-colored with a tan top and was small but comfortable. I was pleased to get 25 mpg in city driving.

The DeMers Family: David

We hosted Pierre and Meggie DeMers and their son, Rocky, at the center. We helped them get oriented in Fairbanks and buy supplies, propane, and a new snow machine for their move to Venetie. I took them to meet Dr. Krauss and the Alaska Native Language Center. It helped set the stage for future working relationships.

More Translation Checking: David

Mary and Virgie were anxious to finish the Gospel of Mark. Con Naish checked over the entire book.

Virgie, Con, Mary, and I are reviewing the Gospel of Mark.

The Upper Koyukon Athabascan translation was finally ready at a first-draft level. It had been a tedious process, yet working with various Koyukon people amounted to great Bible studies in Mark.

1980
Spiritual Ministries: Kay

I worked with a weekly Native Ladies' Bible Study. At Bethel Baptist Church, I continued teaching the primary Sunday school class and a ladies' Bible study involving non-Natives. David preached when called upon and taught a Sunday school class as a substitute teacher.

We spent numerous hours together visiting Natives in the hospital and their homes. It was wisest to go as a team, so if only the lady was home, there would be no questions about why David was there.

Other Activities: Kay

I continued keeping the books for Wycliffe along with hosting members and others. David served as center manager, kept the vehicles going, answered mission correspondence, and filled out book orders.

Translation Checker: David

Henry "Hank" Hildebrandt, who worked with the Babine Carrier people in Burns Lake, British Columbia, came to Fairbanks to check over the Upper Koyukon Gospel of Mark. We appreciated having a helper who worked in another Athabascan language. Agnes Moore, Henry Moses, and Dorothy Pitka worked with us.

Hank offered many practical suggestions. He was scholarly as well as biblical. It took months to check and work through Hank's suggestions and answer his questions concerning the translation.

Helpful Henry: David

Henry Titus from Ruby came to Fairbanks, and we worked overtime on translation for six hours each day. Then I spent four to six hours typing up the translation and preparing for the next day. After two full weeks, Upper Koyukon Mark was transposed into the Central Koyukon dialect. This gave us a tremendous boost toward getting the Gospel of Mark published.

Henry's Older Brother: David

Henry's older brother, Frank Titus, and I were reviewing a few minor details about the translation. He reminded me, "Don't rush, young man; you've got until you're 100 to get it all translated!" That was Frank's humor.

Building Our Log Home (1980~1983)

Land or House—Still Looking: David

We kept looking for land on which to build. I made some offers based on getting a construction loan, but others had cash and acquired the property. We had a low income and very little cash on hand, so we continued praying and waiting.

H & R Block: Kay

H & R Block hired me during the tax season. I wanted the training and experience working on other people's taxes to help with our paperwork. It was a pleasure to run into several Native people from the villages needing tax assistance. I worked on taxes in the mornings, and then on the translation project in the afternoons.

A Grease Monkey

Steve was gaining mechanical experience as a grease monkey. In his high school class, he replaced the pistons in his snow machine engine. He also worked on our 1968 Buick, which had engine trouble.

Growing Pains: David

I talked with Steve about being sure the girls he took out were Christians. When he went to bed, he read his devotional book, which focused on dating only Christians. He showed it to us. He was a deep thinker, and we had good discussions. If we had that discussion today, I would emphasize the need to date a "growing" Christian, but not to choose one who is a Christian "in name only."

Voice Your Values: David

I participated as a delegate at the Republican State Convention in Fairbanks, which involved two full days and one long night. This allowed me to voice my values and include them on the Republican platform. I believe strongly in the beauty and sanctity of human life. We are the result of God's unique creation. King David wrote about the first nine months of human life starting at conception in Psalm 139:13-15. All lives matter to God!

We are pro-life and respect all human life. We have voted in every election since we turned 18 and we plan to continue in all future elections. Research each candidate, pray, and vote for your values!

Qualifying for Boston

Dan was anxious to run in the 1980 Boston Marathon. He went to Birmingham, Alabama, and qualified for the big race. Out of more than 300 runners, he placed 17th and received a trophy. He was the only participant from Bryan College to run in that year's marathon.

Running the Boston Marathon: David

Dan traveled to Boston and was excited to enter the 84th running of the Boston Marathon. The starting gun was a loud cannon boom. He quickly mixed in with thousands of other runners, wearing number 2,343. They ran the winding course of 26 miles and 385 yards to the finish line in downtown Boston.

Dan ran with 5,417 contestants and completed the grueling competition in three hours and seven minutes. He had perseverance and endurance. The Bible talks about that kind of running in our Christian lives:

"Therefore we also, since we are surrounded by so great a cloud of witnesses, let us lay aside every weight, and the sin which so easily ensnares us, and let us run with endurance the race that is set before us, looking unto Jesus, the author, and finisher of our faith, who for the joy that was set before Him endured the cross, despising the shame, and has sat down at the right hand of the throne of God" (Hebrews 12:1-2).

Bryan College Graduations: Kay

Elizabeth, David, and I witnessed two of our children graduate from Bryan College on May 5th in Dayton, Tennessee. Debbie received a BS degree *magna cum laude*. Dan received a BA degree *cum laude*. Their service brought back memories of when we graduated in the '50s. We even knew a few of the staff. John Rathbun, our future son-in-law, graduated in the same class with a BS degree. My parents, brother Tom, and family joined the celebration.

Steve Is 17: Kay

Steve's 17th birthday was celebrated late since Debbie and I spent extra time in Michigan with my parents after the big graduation event. Steve relished his choice of pizza fondue, salad, and blueberry pie

Dan and Debbie

Two in Fairbanks: Kay

Elizabeth now worked her night job at Safeway, stocking non-food items from midnight to 8:00 a.m. We watched her play in a softball game. She was a good pitcher and player in the outfield.

Steve cashiered at a self-service gas station. He was also a camper at Camp Li-Wa. He stopped home briefly to pick up the car keys and a cookie or two as he drove off to work. There were advantages to having camp so close to home!

In the fall, he took another job at The Jade Place for John Haley, where he cut the green gemstone for clocks, coffee tables, and special orders. He liked the work but often came home covered in dust. His lungs must have been tinted green. He gave us a jade clock, eight inches by 10 inches, with small pieces of ivory tusk glued on marking each hour. Their clocks weren't selling well, so Steve was laid off. The boss hated to let him go.

Open Air Campaigners

Dan traveled to Moody Bible Institute in Chicago for a concentrated three-week seminar and practical training in preparation for witnessing to people in the big cities. In Atlantic City, New Jersey, he began a summer ministry with Open Air Campaigners (OAC) on the famous

Boardwalk by the beach. They preached the gospel to lost people and mobilized the body of Christ through effective open-air outreaches.

His sermons were 10 minutes long on one side of a wide sidewalk to attract people to stop and listen while others walked by. Dan started with a simple rope "trick" showing one long cord versus a short one to demonstrate that we all have committed different lists of sins. Then he opened his hand and revealed two even lengths of the rope since we are all sinners in God's sight. He used a sketch board with an illustration and block letters for a keyword which he filled in during his message. In the end, he and other team members would distribute gospel tracts and talk to those interested in knowing more.

When I Finish

Dan enjoyed the Street Ministry and Donna Prettyman often worked with him. She was gifted at music and singing, which added to the messages. One time in the middle of preaching, Dan said, "When I finish, I'm going to ask my girlfriend to marry me." She knew what was coming and said, "Yes!"

Haiti, Here I Come

Debbie joined Teen Missions International for their summer outreach to Haiti. God provided all the funds through generous people. She flew to their home office on Merritt Island, Florida. Orientation and boot camp lasted one week. She was trained to be an assistant leader for 25 teens. Their summer job was to build a brick school building in Haiti. Debbie helped cook for the team. After finishing their work early, they walked around sharing the gospel during the last week.

Debbie saw one man sitting on the asphalt pavement to keep warm. When he stood up, the holes in the back of his pants revealed more than a string bikini! It was a stretching experience for her to see people living in extreme poverty, much worse than anywhere in the US, with little, if any, hope for a better future.

Daily Vacation Bible School: Kay

I was in charge of refreshments for Daily Vacation Bible School (DVBS) all week at Bethel Baptist Church. I would much rather teach, but serving refreshments was necessary for the overall program. I would make the Kool-Aid and count out the exact number of cups and cookies for each class. I had to be ready at the proper time for their breaks.

End-of-Summer Ministries

Debbie and Dan fulfilled their summer ministry commitments with many fond memories. They both found winter jobs at the North Pole oil refinery and worked four 10-hour days, which gave them long weekends. Debbie purchased a Datsun B210 to solve her transportation needs.

"You've Won!": Kay

"You've won!" was my phone message. The manager at Safeway called with the good news. I asked, "Can I put it in the trunk of my car?" He responded, "You'll need a pickup truck. It's pretty big."

I had completely forgotten I had entered my name for a big charcoal grill. We used it for our family and the monthly Native Fellowships.

More Training: David

A refugee family from Vietnam was sponsored by Bethel Baptist Church and brought to Alaska. Moi Pham knew very little English, and his wife, Bah, knew none. Both of us enrolled at the Fairbanks Literacy Center in a seminar called English for Speakers of Other Languages.

I took Moi to English literacy classes while Kay stayed with Bah and their children. Working with this couple reminded us of when we first came to the Koyukon area and couldn't comprehend or pronounce their language. Bah was pregnant, so we taught her to say, "Help me. I'm going to have a baby. Take me to the hospital." We gave them a literacy machine with key phrases on cards, which they could play and listen to on their own. Our goal was to improve Moi's speaking ability in English so he could get a job, feel comfortable in Alaska, and give them an opportunity to hear the gospel. They brought a Vietnamese Bible but did not believe it.

Bah delivered a healthy baby at Fairbanks Memorial Hospital in mid-October. A local Vietnamese lady served as their translator. Then we set up realistic dialogues for them to practice for everyday living.

Philadelphia Bound

Just before he left for Philadelphia, Dan snared two red foxes. Now, the Alaskan trapper could go to the big city and preach.

Dan flew there in November to serve with Open Air Campaigners. He rented an apartment in a complex of row houses. His two-story home was 18 feet wide. The backyard was 18 feet by 50 feet. It gave him sunshine, a little grass, and fresh air. A pumpkin he planted grew to 40 pounds in the hot weather.

Call from the Police Station

Steve called from the Fairbanks police station at midnight and asked us to come pick him up. He said, "I'll tell you what happened when you get here," and hung up. He was driving Elizabeth's car, and the engine sounded strange, so he stopped by the police station and used their phone. As the kids grew older, nothing surprised us anymore!

December Native Fellowship

At the last monthly Native fellowship for the year, Mary Dick read the account of the birth of Jesus in Athabascan. Then several Indians sang "Oh Come, All Ye Faithful." We who know the Lord Jesus thrill at singing carols and hearing the account of His birth again and again in English. Now the same good news is available in the Indian language.

Craving Pickles

Over the Christmas holidays, Dave Czech was our house guest. He was looking for a better job since his bills were growing faster than his income. He temporarily left his wife, Lilly, and their two young children in Minnesota with her parents. Dave craved dill pickles and wondered what that meant.

A little later Dave reported, "You remember when I craved dill pickles? Well, I found out my wife is pregnant!"

A Big Sparkler

Elizabeth received a ring on New Year's Eve from Andy Cogan. It had a single diamond surrounded by gold nuggets. She was very ambitious and started to sew her wedding gown with help from Andy's sister.

1981
The Red Camaro: David

Steve purchased a red 1974 Camaro muscle car and rebuilt the 350 V-8 engine. I advised him that it is best to partner with Christians for business deals, and I ended up loaning him enough money combined with his to buy it. He could do the repair work with good tools and advisors during shop class. He was excited to plunge into this project.

A young man's first car!

Alyeska Resort: David

We took Steve and his buddy, Jon Wheelock, on a trip to the Alyeska Resort in Girdwood, Alaska, near Anchorage. They skied on the biggest ski trails in Alaska's Chugach Mountain Range and thoroughly enjoyed their day.

Traveling Abroad

Debbie and Glenda traveled to England and Scotland for two weeks. It was cheaper to fly from Alaska to London than to Seattle! They met up with Karen Esckelson from Vassar, and the three had a wonderful time together.

Recording Studio

We were excited to have Bill and Julie Odom, with Bible Translation on Tape, record for us. We chose a quiet room, think "recording studio," with Mary Dick reading the Gospel of Mark. Everybody at home was asked to be quiet. If a jet flew over, we stopped and waited until it had passed. In the beginning, Mary read haltingly, so we had to redo those parts. But in the end, she read at a normal speed with good expression.

After each recording session, Bill would listen on a seven-inch reel-to-reel tape recorder so he could quickly move the tape to cut out any extraneous noise such as clearing one's throat. He could add space or even a breath of air that he had copied. Much work goes into improving a recording. Mary liked the final product, and it was top-quality to share with anyone.

Was Satan Trying to Hinder Us?: Kay

In the middle of our recording sessions, Mary came down with a sore throat for several days. Then both Bill and Julie got sick. Next, Mary got some frightening news that two men tried to abduct her daughter in town. A policeman saw her struggling and rescued her.

One of my jobs was to keep Mary's three-year-old granddaughter occupied along with my other activities. Additionally, we accommodated extra company for a month. It was quite a challenge!

We Are Human: Kay

Yes, we are humans. One day I was getting uptight and tired of so many extra people. While preparing the meal, I realized I was becoming resentful of having to cook all the time and keeping up with so many other things as well. To help me, I kept repeating, "Be hospitable to one another without grumbling" (1 Peter 4:9).

The Lord Provides

The Lord provided through this hectic time. People shared meat, fish, and other food items. Several even gave us money to help with our extra expenses. It was amazing to see how the Lord laid it on people's hearts to give without saying anything. We knew the Lord was supplying our needs.

The Vietnamese Family Leaves Alaska

Moi and Bah moved to Los Angeles. The wife, Bah, cried as she left her loving friends in Fairbanks. A dozen of us from Bethel Baptist Church saw them off.

Imagine riding a bus on the long Alcan Highway twice with three small children! US Customs in Washington State refused to let them cross the American border and sent them back. Consequently, the church had to fly them to LA.

Exciting News from LA

Moi and Bah connected with the large Vietnamese community in LA. They liked the warmer California area. The most exciting news was a phone call from the Vietnamese pastor of a Christian Missionary Alliance Church saying, "Moi accepted Jesus Christ as his personal Savior."

Back to H & R Block: Kay

I received an urgent call from the manager at H & R Block. A new field manager in Fairbanks had been fired, and several employees resigned, so they needed qualified help. We needed extra money to send David to New Jersey for Dan and Donna's wedding. So I decided to work for them until the end of the tax season.

Mentally Challenging

When we were in our forties, we committed ourselves to memorizing Bible verses. We learned eight or nine verses each week from the book of 1 John with 105 verses. After several weeks it became tiresome, but we pushed ourselves until we memorized the whole book.

Dan Heard Wedding Bells: David

Flying to Philadelphia, I spent two weeks with Dan before his wedding. Dan's presentation encouraged people to listen to the gospel.

Their ministry van is on the street in front of a housing development.

Dan and Donna were married in Egg Harbor Township, New Jersey, in May. It was a beautiful church wedding ceremony with tuxedoes. Kay's parents, Bob and Vera Temple, were also at the celebration.

Last High School Graduate: Kay

Steve graduated from West Valley High School in Fairbanks on May 29th. I took in all the ceremonies. David was gone for Dan and Donna's wedding. Now all four of our children have finished the high school milestone.

Debbie Flies Again

Debbie flew to South Korea on a five-week trip to visit her boyfriend, John Rathbun. She wanted to meet the rest of his family on their turf. His parents, John and Joyce Rathbun, served as missionaries with TEAM, The Evangelical Alliance Mission, with a lifelong commitment to Korea.

Twenty-five Years for Us: Kay

We jointly celebrated our 25th silver wedding anniversary with Roger and Bev Wheelock at Bethel Baptist Church. A great potluck dinner, a chance to feed each other a piece of anniversary cake, slides from our wedding, and lots of good memories were shared with family and friends.

Building Our Log Home (1980~1983)

The following night, Dr. and Mrs. Ted Mercer, president of Bryan College, hosted a Wheaton Tour Group in Fairbanks. They invited us for dinner and a 25th-anniversary cake. Then David told about Wycliffe's work in Alaska.

I wrote the following note to my parents: "I'd like to thank you for the fine Christian upbringing you gave me, which has contributed much to our happy 25 years together. Thanks so much for all your time, effort, frustrations, and consistency when you gave of yourselves to raise six children. You'll never know how much I appreciate both of you."

Adding Spice to Our Lives: Kay

My niece, Becky Spear, who was Elizabeth's age, stayed with us part of the time. I took her to an interview for a job in Barrow, Alaska. We were surprised at how quickly she got the job and was on her way up north. She added spice to our lives with her energetic personality.

Dog Sitting a Dobby: Kay

Steve was dog-sitting a Doberman pinscher (or a Dobby) for 60 dollars a month. It was a watchdog type but very friendly. She followed me around and often sat at my feet.

Jim came to visit us during this time. When he opened our front door and saw the Dobby coming down the hall, my brave brother slammed the door and yelled, "Doberman!" and ran.

Elizabeth Heard Wedding Bells: Kay

Elizabeth and Andy Cogan were married in July. The ceremony took place at Andy's parents' homestead on Haystack Mountain, 25 miles out of Fairbanks. It was a beautiful day for their outdoor ceremony. Her dress had a light-blue sash that went through two applique-lace roses down the back.

The Cogans roasted a large pig for the potluck reception. Out-of-town guests on our side included my parents, Bob and Vera Temple, my sister Nancy

Elizabeth looked lovely in her satin wedding gown, which she had sewn.

and Louie Horwath, and my brother Jim. Andy's relatives included most of his 10 siblings and their families.

Word of Life

Steve headed to Word of Life Bible Institute in Pottersville, New York. The institute is a collegiate-level program offering intensive Bible study and ministry training to help students deepen their faith and discover God's vision for their lives. He stopped to visit Dan and Donna in Philadelphia on his way there.

Steve read the Institute's catalog and agreed to it. We did not point out that the male students wore a tie to classes and a suit for dinner. This was a big stretch for him. Studying there was much more challenging than in high school.

We liked their modular approach to learning, such as a gifted Bible teacher who came and taught Revelation for two weeks in concentrated form. We highly recommend a one-year Bible school to help youth become grounded in God's Word.

Upper Koyukon Mark on Tape

The sets of four edited audio cassette tapes with Mary Dick reading the Gospel of Mark in Upper Koyukon arrived. We quickly packaged the tapes in plastic albums and started to distribute them.

The first set of tapes went to Mary in Tanana. She wrote back and told how she went to Philip Kennedy's house so he could listen, then to Walter Nicholia, Gladys and Joe Johns, and Helen Peters. Rocky Riley wanted a set for his father. News spreads quickly in a small village.

Final Typing and Printing of Mark: David

The complete Upper Koyukon Gospel of Mark was in print! It is closely related to Central Koyukon, but there are enough differences to warrant separate printings. It took extra work, but we wanted God's Word to be readily available to everyone. Now, we could promote and distribute the audio cassettes and the gospel together in Tanana, Stevens Village, Minto, Nenana, and Fairbanks. The title on the cover was *Good News from Mark* with Koyukon beadwork.

We encouraged people to listen to the recording while looking at and reading the text. In the end, most put the text down and only listened to the oral reading. God had brought all the pieces together so the Athabascans could hear what He wanted to tell them from the Gospel of Mark.

One More Off My Bucket List: David
Roger Wheelock made numerous trips north to Prudhoe Bay in a 10,000-gallon tanker delivering gas. He invited me on a two-and-a-half-day 1,000-mile round trip. Roger knew every curve on this gravel highway. There was oodles of good food in the company cafeteria.

The Henry Side
Steve spent 10 days in New York City to work with the Open Air Campaigners as part of his Word of Life assignment. He saw the Thanksgiving Day Parade before flying to Fairbanks. He was home for most of December and Debbie's wedding. It was a joy to see what the Lord was doing in his life.

Dan and Donna came to Fairbanks for two weeks, so our family was able to meet her. She was well-organized and cooked our Christmas dinner. They both shared their work and testimonies at Bethel. The church surprised them with an offering that covered all their travel expenses.

Debbie's Wedding Bells
Debbie married John Rathbun at Bethel Baptist Church in Fairbanks in December. His parents, John and Joyce Rathbun, came from Korea, along with his two brothers, Joel and Jim, and sister, Joanna. They had a Christ-honoring service, and we gained another son-in-law. Three of our children were married in 1981!

Their first home was an efficiency apartment in Fairbanks with a Murphy bed (wall bed) in their living room. It gave them more space when the bed was put back against the wall during the day.

Money Down
We completed the paperwork and put money down on a piece of bare land in a new subdivision just off Farmers Loop Road next to the Dog Musher's Hall and race track. It became our future home five miles from Fairbanks at 915 Union Drive. We were on higher ground than Fairbanks, with a view of the majestic Alaska Range, but not up in the hills.

1982
The Empty Nest Syndrome!

Since Steve returned to Word of Life in New York, we were free to travel more. The "empty nest syndrome" was a new feeling, but we knew it was a normal part of aging. Both newlywed couples, Elizabeth and Andy, plus Debbie and John, lived in Fairbanks. Dan and Donna continued with the Open Air Campaigners in Philadelphia.

Recording with Madeline

Flying to Galena in January we stayed with Madeline Solomon to record her reading the Gospel of Mark in Central Koyukon. We completed the taping phase in a short time. Madeline had moved from Koyukuk and taught in the Galena School's bilingual program for eight years. At 76, she was still going strong with a sharp mind as a well-known Koyukon elder.

Coming from a Catholic background, she was happy that both the Catholic and Protestant Gospels of Mark were the same. She was also pleased that our goal was to make the Gospel of Mark available to all churches.

We worked all weekend, as well as each night after her teaching sessions at school. Our recording equipment was of average quality. The microphone broke in the middle of the project, so we spent several hours phoning and scrambling around town until we found one to borrow. Numerous visitors stopped by, but breaks were necessary, and the local people needed to know what we were doing. Planes flew overhead, and snow machines drove by, adding to the village flavor of the recordings. We didn't erase anything but rerecorded parts to save time. We were interviewed by the school and videotaped for their bilingual program.

Visiting in Galena

We spent many hours visiting in Galena, a city of 700 people and the center of the Koyukon region. It was interesting to see how many people had moved here from Koyukuk. A young lady told us, "I bet Dave and Kay thought no one from Koyukuk would ever get saved."

Building Our Log Home (1980~1983)

Seven-Inch Reel-to-Reel

A friend lent us a seven-inch reel-to-reel tape recorder to edit the audiotapes of Madeline. The bigger reels are much easier to turn back and forth to find the exact spot in the tape needing to be fixed or cut out. Mechanical problems and more delays slowed us down but we continued looking to the Lord for His help.

House Logs: David

We received permission to cut house logs near Two Rivers off Chena Hot Springs Road. First, I had to mark the trees on state land by tying a bright orange plastic ribbon around each one. Then the state charged 25 cents per linear foot. We spent a week and a half cutting logs.

David cutting down a large spruce tree

The very first tree, a huge white spruce, fell on my chainsaw. I had set it down in a low spot before the tree fell. To free it, I had to dig out the snow and moss under my saw. It had a noticeable dent but it continued working. A friend with a caterpillar dragged the logs to our loading site. Then a logging truck made three trips and brought them to our building site. One day later, the access road was officially closed. We met their deadline, just before the state also raised the linear price to one dollar and 25 cents for personal use!

A state inspector checked out our building site. Somebody thought we were selling our logs and reported us to the authorities. He also talked to the truck driver in Fox who hauled our logs.

Hospital Visits: David
We often visited people at the Fairbanks Memorial Hospital. One trip in March was to see Clyde Hunter, the husband of Marie Hunter, a helper with the translation of the book of Galatians. Clyde was dying from cancer. I shared the gospel with him and he turned to the Lord. We wish people would be open to receiving the Lord before they're on their deathbeds. But as the saying goes, "Better late than never!"

Sunday School Commitment: David
For three months, I taught the adult class on the book of Nehemiah at church. He was a man of vision and prayer who oversaw the building of the wall around Jerusalem in record time despite many obstacles.

On My Way to Israel
Steve called from the New York airport. He was excited about going to Israel with a group of classmates from Word of Life. We were glad he had this opportunity. Seeing the sights in Israel helped his understanding of the Bible.

Printing the Gospel: Kay
I typed the Central Koyukon Gospel of Mark in preparation for printing. Then David proofread the book and pasted in the biblical illustrations.

Due to technical problems, the accompanying audio cassette tapes were delayed. If everything fell into place too easily, we would forget to depend on the Lord and miss seeing Him at work.

Closing the Wycliffe Center: David
Since the work in Alaska was closing, I began packing and mailing boxes of books and other basics to the office in Colorado. The move took much of our time. The property was sold to Camp Li-Wa to expand and continue their ministry. This had been a great place to raise our family.

Our First Granddaughter: Kay

Holding and welcoming our first granddaughter, Melissa Marie Cogan was a delight. She was born in April to Elizabeth and Andy in Fairbanks. I knit her a baby sweater as we started a new phase in our lives—grandparenting.

Peeling Party

Log peeling parties were organized at our building site. One could sit, straddling a log while pulling a drawknife toward themselves, to remove a section of bark. Sap could get on your seat unless you sat on a cushion. We stacked the peeled logs on top of other logs to dry in the sun, wind, and fresh air.

Chuck Mowery loaned us a farm tractor to roll the logs, one at a time, on two slanting logs to build our house. With minimal help, I prepared, notched, and cut out two big logs a day.

Elder's Conference

Denakkanaaga (Our people speak) is a Koyukon word for the elders' organization representing 42 Native villages in the interior of Alaska. We drove to Minto for this conference with 300 people.

Prepping to Build

Permafrost is a problem in the Fairbanks area, so I hired a driller to dig a test hole in the center of our building site. He went down 45 feet before he hit frozen muck. I made the house footing 30 inches wide so there would be less reason for our house to sink or shift. In June, we dug and prepared to pour the concrete foundation.

Great Guest Helpers: Kay

My parents came for part of the summer. My brother Jim generously paid for their plane tickets. Mother helped with cooking, washing dishes, and keeping up with our family and guests. Another good friend, farmer Don Foster from Fairgrove, Michigan, came with practical knowledge. Dad and Don helped lay the joists and the plywood for our floor. We were very thankful for other helpers who showed up—even for just a day.

Needing More Money: Kay

David ordered the cement for our footings from Fairbanks Sand and Gravel on credit with no money down. They gave us 30 days to pay for it, but we didn't have that much money. We were building on a shoestring budget and had faith that it was God's time for us to build. We prayed and waited.

Unexpected help showed up as we prepared to lay the concrete blocks. Steve Fluth, a professional block layer, quickly laid the corner blocks and left the others for us to fill in.

While the men were laying the concrete blocks, I went to the post office, then hurried to our building site with a big smile and announced, "Here are our dividend checks. We'll live off of these the rest of our lives!" I held out two checks for 1,000 dollars each. That was the first year the Alaska Permanent Fund Dividend was paid. We thanked the Lord for providing at just the right time.

Church Activities at Bethel: Kay

As I served on the pulpit committee, we continued to look for and evaluate who we should present to the church as a potential pastor. We extended a call to a pastoral candidate at the end of October.

David served on the deacon board, requiring more time since the church was without a pastor. He continued to lead a weekly Bible study and prayer meeting.

A Nephew Visits: David

Kay's nephew Rick Horwath from Ontario, Canada, visited us in the fall. I was surprised when he popped his head over our log wall, about six feet high, peered at me on the inside, and said, "Hello, Uncle David!"

Steve and Rick had driven up the Alcan Highway. They brought a six-pack of Vernors Ginger Ale, an iconic drink in Michigan that Kay liked. Steve had spent the summer counseling at Word of Life Ranch Camp in New York. This completed all his coursework and practical service work.

Remembering the First Snow: David

As a kid, I was always excited to see the first snow in northern Illinois. I would run out in it barefooted. It was easy to remember the first snow this winter in Fairbanks since we had just finished the ceiling with two-inch tongue-and-groove lumber. Then we laid heavy plastic over the unfinished roof and the windows. We closed our home for the winter by installing a door. The building would resume after we returned next spring.

Our Second Granddaughter: Kay

Our second granddaughter, Kara Joy Henry, was born in September to Dan and Donna at Fairbanks Memorial Hospital. We liked their choice of names since "Kara" means (Joy) in the Greek language. "Kara Joy" used together strengthened the meaning. I knitted her a baby sweater with Winnie-the-Pooh on it.

Sharing the Good News

Mary Dick shared the Good News with people in the village of Hughes. As a result, Alice (Williams) Ambrose wanted a copy of the Gospel of Mark and the audio cassettes. Alice wrote about the hunting and food-gathering news and her family. "Thanks for the book and the tapes. I enjoyed listening to it. My mom Susie Williams came over and listened to it with me. I cooked and ate and listened to it until it was time to go to bed; the tape was good."

Alaska Christian Education Conference: David

An Alaska Christian Education Conference was held in Fairbanks. I gave a presentation on "Tips for Cross-cultural Communication." We learned these things while living in another culture.

Half a Century of Love: Kay

Bob and Vera celebrated their golden wedding anniversary with a dinner at the Bavarian Inn Restaurant in Frankenmuth, Michigan. Their celebration was also a Temple family reunion with everyone attending. After 50 years, my parents had six children plus grandchildren and great-grandchildren.

Recalling family memories and hearing stories from the past 50 years was part of the fun. Debbie and Beth organized a memory quilt, sent out squares to all the relatives, and sewed them together.

Kay Henry, Shirley Spear, Nancy Horwath, Bob and Vera Temple, Tom Temple, Jim Temple, and Sheryl Cavazza

1983
Show and Tell

Whatever you call it—show and tell, deputation or special assignment—the primary goal is to report what you have been doing to advance God's kingdom. So we followed the Apostle Paul's example: "Now when they had come and gathered the church together, they reported all that God had done with them, and that He had opened the door of faith to the Gentiles" (Acts 14:27).

We reported on the work in Alaska by preaching in churches, showing slides, giving reports, going to missionary conferences, setting up displays, being interviewed for newspaper articles, having one-on-one conversations, or eating with friends. Over the winter months, we traveled to Michigan, Kentucky, Florida, and Ontario, Canada.

Back to Alaska: Kay

While driving back to Alaska, we visited my cousin Diane and her husband, Gerry Pedersen, and their children, DeLynn, Scott, and Marcy, in Illinois. Then we drove to Madison, Wisconsin, to see David's aunt Lillie Bachanz. Seeing family and friends along the way helped to break up the long, tedious hours of driving.

Our next stop was with Dick and Shirley Walker in Fort Saint James, British Columbia. We also attended the Sunday service with them on a Carrier Indian Reserve. Another stop was with Hank and Barb Hildebrandt in Burns Lake with the Babine Athabascans. It was exciting to see what God was doing in other areas. We visited Indian museums and drove through their villages.

Cindy Is in Heaven: Kay
My niece Cindy Lois Spear, passed away from cancer on April 20th, just before her 21st birthday. We were glad we had seen her recently. But we were most thankful she had recommitted her young life to the Lord Jesus.

Housesitting
Upon returning to Alaska, we house-sat for Harvey and Margaret Martin near our building site, which was very convenient. Harvey had helped install our electric meter for our power.

Trees and Shrubs: David
Our plot of ground used to be in a huge potato field that had returned to nature. Consequently, we had many willow and cottonwood trees, which I cut down. I planted birch and spruce trees.

Fast-forward: Those birch and spruce trees have grown to more than 12 inches in diameter. I wish I had planted more birch for future firewood.

Domestic raspberries were planted in a long row on the backside of our lot. They continue to grow and produce many berries. They brought back good memories from Kay's childhood when she went to her great-grandma's house in Michigan and picked raspberries every summer.

My Old Hospital Roommate: Kay
Planning to travel to the Koyukon villages this winter, I was concerned about a place to stay. We were greeting a line of women at an Athabascan conference when we came to Olga Solomon from Kaltag. I expressed that we planned to visit her village. Immediately she responded, "You two can stay with me and James." Olga and I met in

1960 at the Native Hospital in Tanana waiting to have babies. I had not seen her since then!

Steve Flew to Texas

Steve enrolled at LeTourneau College in Longview, Texas, to pursue a degree in engineering. His cousin, Rick Horwath, was also attending. One of their extracurricular activities was to parachute. They both jumped six times and enjoyed it. Certain things in life are better left to the younger generation.

Moose Meat

Dan and Joan Beckdahl shot two moose. They invited us to their house to help process them. When we finished, they insisted on giving us enough meat to fill our freezer. The Lord provided for us since we were too busy to go hunting that fall. Generosity is a wonderful Christian trait for showing love.

Housebuilding Continued: David

We let the grass grow on our path to the outhouse as soon as we had an indoor toilet. I installed the wiring and wall outlets in our outer log walls. The first one fit perfectly. I went into the crawlspace and drilled a hole for the wire at an angle. The cutout for the electrical box was ready with an angled hole. The two angles met, so I pulled the wires through and connected everything. Unfortunately, all the other angled drill holes did not align exactly and required more fiddling and time to get the wires pulled through.

More Visitors, Comments, Questions: David

Visible from Farmers Loop Road, many people noted our building progress with a steady stream of visitors, both Native and non-Native. Our house was 40 feet by 40 feet with no partitions, and our bedroom was behind the curtains in one corner. When people came inside, their values in life showed in their comments such as: "This would make a good church," or "This would make a good dance hall." Many people asked questions about building a log house.

Axel Carlson wrote a book called *Building a Log Home in Alaska*. I went to see him with my most pressing questions. His first comment was, "You know, more couples have gotten divorced by building log houses!"

Fairbanks Native Fellowship

Walter and Virginia Maillelle and their two grandchildren spent the weekend with us. Walter played his guitar, sang, and shared about the Lord at the monthly summer fellowship.

The Native Fellowship in front of our unfinished house

A Korean Birthday

Our missionary friends in Tennessee, Francis and Hazel Neddo, couldn't come to Alaska to see their new grandson and asked us to fill in for them. Their son, Joel, had married a Korean lady, Suk, and now lived at Eielson Air Force Base. We were amazed to see all the great ethnic food. We thought others were coming, but we were their special guests as their son's "grandparents."

It was the celebration of the boy's first birthday, following Korean tradition to bless him with a prosperous future and a healthy life. He wore a colorful Korean outfit. We were honored to play a role in his life. Traditional Koreans count a baby's age as one year old at the time of birth. Unborn babies are real humans.

Reel-to-Reel Editing: David

The plastic sheeting over our windows was difficult to see through, and we needed to replace it before getting colder. After the first glass window was installed, we drove to Palmer to edit the recording of Madeline Solomon reading the Gospel of Mark.

We stayed with Barney and Ruth Furman at the Arctic Missions property on Lazy Mountain in Palmer. Barney, who ran the Multi-Media Center, set up a large tape recorder. I spent many hours listening and comparing the recording to the written text. I cut out extra spaces, blips, or unwanted noise. I spliced in myself, announcing each chapter like, "Chapter One." Barney helped make the master tapes, dubbing others and labeling them. It was very tedious and grueling work.

David with the seven-inch reel-to-reel tape recorder

Often a Central Koyukon sentence would start with "Go Jesus" (meaning "This Jesus"). It didn't take Barney long to begin saying, "Go, Jesus. Go, Jesus." He understood that much in Koyukon with an American twist.

I was concentrating and focusing on editing hour after hour. On October 12th, Ruth made a cake for Kay's big 50th birthday. This was the only time I completely forgot about my wife's birthday. I never forgot her special day after that!

Our Underground Vault: Kay

I painted a wall partition in our house with water-based paint. Some paint got on my wedding ring, so I put it in a small container to soak for easy cleaning. David saw the little container with white, cloudy-colored water and threw it in the toilet. This was before we had a bathroom sink installed.

I asked David, "Where's my ring?" We checked all the little containers and even looked in the toilet bowl, but there was no ring. It had been sent on a one-way trip to our septic tank, which became our underground vault!

When our septic tank was pumped out, David told the driver about our problem. He could not assure me that his suction hose would pick up the ring. It would also be a major problem to find it when he dumped his truckload of sludge.

Fast-forward: Several years later, while picking cranberries, I saw something flash in the sun. It was a ring, and the sparkling diamond caught my attention! I put an advertisement in the newspaper, but no one claimed it. So that's how the Lord gave me a replacement wedding ring.

Growing Pains: Kay
Our well was drilled before the first snow. Now, David didn't have to haul water from town. We purchased a used Maytag heavy-duty washer and dryer set for 150 dollars. Someone had bought several truckloads of used appliances from a pipeline camp. They served us well for many years.

Steve's RA
Steve's Resident Advisor (RA) at LeTourneau College sent us several letters expressing his appreciation for Steve's Christian life and fellowship. We were thrilled to hear this kind of news. "I (we) have no greater joy than to hear that my (our) children walk in truth" (1 John: 4).

Tanana Trip
We flew to Tanana and stayed with Esias and Mary Dick. Mary continued to promote Mark's gospel. We gave her 100 copies of the Lord's Prayer in Koyukon and a portable Sony recorder to play the tapes.

They lived in the new housing circle, two miles upriver from the old part of town where most people still lived. To visit the 400 people in Tanana, we walked many miles each day. Several commented about the tape recording and expressed their appreciation. In contrast, one man told us that Jesus was just a lesser shaman (medicine man), yet we were able to share the truth with him.

Allakaket and Hughes Trip

In Allakaket, we stayed with Lindberg and Lydia Bergman. On Sunday, we attended Saint John's-in-the-Wilderness Episcopal Church. Joe Williams Jr. was the local ordained Native priest in charge. After we sat down, he leaned over and asked me to preach. It was the first Sunday of Advent, so several Natives read relevant Scripture portions. I read the parallel passage in Mark's gospel and spoke about John the Baptist. One of the lay readers was Jennie Williams, who read the passage in Koyukon.

We visited every home. Two Indian ladies prayed in Athabascan as we prayed together for them and their families. At that time, most homes were in a compact row near the riverbank, which made visiting much easier. We were glad to see the interest in the Gospel of Mark in their language.

Next, we flew to Hughes and stayed with Joe and Cecelia Beetus. We visited every home in Hughes. Their son, Bob, held services in his home since there was no church building. The Lord had convicted Bob about his past life and his need to turn to Jesus Christ. Effie Williams from Allakaket, now living in Hughes, encouraged Bob to start church services in his home.

Goodbye 1983: David

We closed the year 1983 by attending the Fairbanks Native New Life meeting. Their final meeting was held at the Fairbanks Rescue Mission, which doubled as an evening service for the people there. James Johnson, Jr., led the Native New Life service and asked me to speak. Following the service, we visited several Native elders at the local rest home. And thus ended another year!

1984
Back to the Yukon

The first trip of the New Year was to Koyukuk, and we stayed with Josie Dayton, Benedict and Eliza's oldest daughter. After the major flood of 1963, most people built new houses on three ridges a short distance from the river. Everyone knew about the cassette tapes with Madeline Solomon.

Then, we flew to Kaltag and stayed with James and Olga Solomon. It was enjoyable to stay with a couple who preferred to speak Koyukon and knew English.

We had a pleasant visit with the resident Catholic sister and the Jesuit priest, Ted Kestler. The village was 100 percent Catholic. Each week, he alternated holding services between Kaltag and Nulato. He had a set of Gospel of Mark cassettes and encouraged local people to listen to the Bible in their own language. The gospel is for everyone.

Meeting with the Bishop: David

Back in Fairbanks, I visited the Episcopal Bishop, George Harris. I informed him of our activities in the Koyukon villages with Episcopal Churches: Allakaket, Hughes, Huslia, Tanana, Rampart, and Stevens Village. He was in favor of our work and expressed appreciation for our mission's work in Alaska and the Philippines, where he previously served. He offered to help in any way he could.

Nulato Next

We flew to Nulato, population 350, and stayed with Andrew and Hilda Johnson. They were older than we were but looked healthy as Andrew reported, "I don't hurt anywhere!" The people appreciated the Gospel of Mark in Central Koyukon. Nulato and Kaltag spoke Lower Koyukon, which had minor dialect differences from Central Koyukon.

Great Response

Mary Dick reported that people came to her and asked about the tapes. They wanted her to pray for them and followed along when she quoted "Our Father," also known as the Lord's Prayer, in Athabascan. She was so thankful to have played a part in sharing the Good News with her people.

We visited people from Tanana at the Fairbanks Memorial Hospital. One elder, Jason Edwin, had cancer. It was good to talk with him about the Lord and especially to hear his testimony. A week later, he went to meet the Lord.

Another cancer patient in the hospital was Violet Erhart from Tanana. She told us, "I have many visitors, but I tell the last visitor of the day to turn on the tape recorder with the Gospel of Mark when they leave."

Native New Life: David

The Fairbanks Native New Life was now meeting in the lobby of Bertha Moses Patient Hostel. This was a convenient location, with Bertha serving as the facility manager. She believed in Jesus Christ and encouraged residents to attend the meetings for spiritual help. Natives from the "bush" would reside here when they came into Fairbanks for medical assistance. Jimmy Johnson Jr. led the monthly meetings. He preferred to lead the singing. He often asked me to share a devotional.

More Time in the Hospital: David

We spent more time in the Fairbanks Memorial Hospital—visiting. One friend was Richie Ketzler from Nenana. When I visited him, he was taking the maximum dosage of morphine for his pain from cancer, which did not help. We met him during our Arctic Camp days. He had great hunting stories and he even let us try out his dog team under strict supervision. My visit was an excellent opportunity to remind him of God's love and his need to turn to Jesus Christ for forgiveness.

Several months later, he passed away while we were gone. A friend told us, "Richie made peace with God before he died." No church can do that for you. They can only point you in the right direction. Before we die, each of us must decide whether to accept or reject Jesus Christ.

Music, Music, Music

Bob Beetus from Hughes accompanied us on the drive to Anchorage for the annual Native Leadership Conference and *musicale*. We discussed spiritual things along the way.

For three nights, it was music, music, music, with 1,000 people attending each evening in the auditorium of West Anchorage High School. Hearing numerous Native groups singing praises to God and giving their testimonies was a taste of heaven.

The *musicale* was started during the mid-'60s at Victory High School, which was operated by Arctic Missions. They encouraged the students to share their testimonies in the *Native Musicale* during the annual Fur Rendezvous. This was continued by the Anchorage Native New Life Fellowship.

Fast-forward: The weekly Native New Life Fellowship and the annual *musicale* are continuing to honor the Lord Jesus. This is the longest-running Native Christian ministry in Alaska to see God glorified among their people.

"Life" at LeTourneau

Steve told us about the new "life" at LeTourneau College. He wrote, "I tell you, God is real, and I feel the presence of God today. I've seen the power of God and how God prepares the hearts of those who need to hear the gospel and the paths of others to tell them. I'm getting really excited about how all of this is working here on campus."

He and another student, Kevin, started praying for a revival several months earlier. Two guys from the dorm came up with the idea of going to the parking lot of Showbiz Pizza Place to witness to the youth who hang out on Friday and Saturday nights. Six other students joined them. He used ideas like the rope "trick" he had learned at Open Air Campaigners to share the gospel.

Ruby Bound: David

In Ruby, a village with 200 people, we stayed with George and Judy Richardson. The Ruby Bible Church was growing. We visited all the people who moved there from Kokrines.

Harold and Florence Esmailka, an ambitious Koyukon couple, owned Ruby Trading Company, the former Northern Commercial Company store, and started Harold's Air Service. Harold was a younger Koyukon speaker in his 40s, so we gave him a set of audiotapes.

Bad Knees: David

Walter Maillelle and Fred Mamaloff stopped in Tanana and encouraged Esias to turn to the Lord Jesus during a ministry trip with Native New Life. Mary Dick's husband's knees were severely crippled with arthritis.

Later while visiting him in the Fairbanks Hospital, I said in Athabascan, "I believe in Jesus Christ with my heart and depend on him." His eyes lit up, and he responded, "I also believe in my heart." We thanked the Lord for this affirmation of his faith. This was the first time we heard it directly from Esias.

He had seen a difference in his wife after she started translating the Gospel of Mark. A changed life is the strongest testimony of God working in our lives.

Huslia

We flew to Huslia and stayed with Wilson and Eleanor Sam and their four children. Wilson led the work at the Episcopal Church. Eleanor offered to read from the translation of Mark when any reading came up on the church calendar. We had two Bible studies at their home. Huslia is a growing Koyukon village of 200 people.

Where Are the Easter Eggs?: David

The main church in Huslia was the Episcopal Church. There was also a smaller Catholic congregation. This Easter, there was no resident Episcopal priest, so the itinerant Catholic priest from Galena took charge of the joint service held in the larger Episcopal Church building.

The church was packed on Easter morning, with people standing along the back wall of the church and outside the door. The priest asked me to read the Easter account in Koyukon, which I did from Mark 16:1-8. The priest went on to explain that the women coming to the tomb seeking Jesus was like asking, "Where are the Easter eggs?" This was his main point. What a missed opportunity! We were shocked that there was no explanation or excitement about the resurrection of Jesus Christ and what this means for us today.

After the service, Wilson phoned the Episcopal bishop's assistant in Fairbanks, expressing his disappointment. Most of the day, we visited and talked about Jesus in people's homes. We knocked on one door at the end of the day—and got the occupants out of bed. Then we knew it was time to quit. That was one Easter we will never forget!

Sanding Job

Dave Czech sanded the entire interior of our log house, bringing out the natural grain and beauty of the spruce logs. Another fellow put clear non-glossy Varathane polyurethane on the inside walls. We liked the lighter and brighter walls.

Ferry Trip

We drove to Estes Park, Colorado, for another biennial Wycliffe Conference and Ministry Workshop. The first night was spent in Tok with fellow members Paul and Trudi Milanowski. We continued to Haines to catch the Alaska Marine Highway System ferry and toured the Inside Passage through southeastern Alaska. Our car was parked on the lower deck with all the other vehicles. We had never seen so many bald eagles in one place! It made us wonder, "Why are they considered an endangered species?"

We saw our state capital, Juneau, from the ferry. We wished they had moved it to Wasilla so it would be readily accessible to all of the citizens of Alaska.

When the ferry stopped in Sitka, we visited our friends Mel and Bev Holmgren. Then we continued to Bellingham, Washington. The berths were already filled, so we slept in the solarium on the top lounging deck. In the stern, there was seating under the stars, with a gentle breeze. During poorer weather, we spent two nights inside on the hard deck with many other travelers.

Steve's Girlfriend

Steve had a girlfriend, Cheryl Stanton, from our church. Cheryl's car broke down, and Steve wound up driving her in her parents' car to Alaska Bible College in Glennallen to enroll. He didn't mind taking her!

Becoming More Independent

Steve became more independent as he moved into our "guest house." It was a small, "dry" (no running water) house, eight feet by 16 feet. Steve studied in a quiet place for his final year at the University of Alaska.

Berry Picking: Kay

I went cranberry picking with Elizabeth, granddaughter Melissa, and Dolly Titus-Yaeger. Melissa put more berries in her mouth than in her bucket and wandered back and forth between her mother and me. So much fun to have grandchildren! I found a few berry fingerprints on my jacket.

Rampart

We flew to Rampart with a population of 50. One Native man and his wife listened to the recording of the Gospel of Mark and exclaimed, "Half my relatives are becoming Christians!" We prayed for his acceptance of Jesus Christ.

Around the World (1984)

A Trip to Israel: David

Aunt Lillie M. Bachanz passed away in February 1984 and left me an inheritance of 4,000 dollars. What a wonderful surprise! This would almost pay for a trip to Israel for both of us. I asked Kay, "Would you go to Israel with me?" Of course, she was all for the idea. I had dreamed for many years of taking such a trip.

Around the World: David

After many phone calls, I worked out the details for our trip to Israel, which included going around the world. One of our main reasons for the trip was to check out Southeast Asia, where we considered going with Wycliffe. We wanted to include Papua New Guinea on our route, but that was expensive. There was a particular travel route the major airlines flew, and tickets were much cheaper, so that's what we did.

We bought joint tickets with Northwest Airlines and Malaysia Airlines. At that time, Northwest flew into Fairbanks. They gave us frequent-flyer booklets with 15 coupons. The airline credited every flight segment we flew, and we made lots of stops along the route. When the booklets were completed, we each received two free tickets to anywhere they flew in the US. This was a very generous program before they switched to using miles.

We're Off: Kay

Traveling light, we were on our way with only medium-sized backpacks that we wore on our backs. Elizabeth and Andy loaned us the packs they had used when they went to Europe after their wedding. We put them in the overhead bins, so we avoided waiting for luggage. Our tickets were open to book the next flight according to our schedule and available space. Most flights were not full.

Elizabeth drove us to the airport. I held Melissa, who snuggled up to me. Farmers Loop Road was pretty rough, with one bump too many as Melissa lost her breakfast on me! We turned back, and I quickly changed my clothes.

Our first major stop was in Los Angeles. While there, we shared Alaskan salmon and moose meat with our family. A couple of days were spent with Martin, Cuca, and Angela, then Miriam, David, and their children, Christopher, Abby, and Liz.

Tokyo, Japan: David

We spent three days in Hawaii, a warm vacation spot for many Alaskans, and did the tourist things. The next stop was Narita Airport in Tokyo, Japan. International airports have lots of helpful information in English, so we booked a local hotel and exchanged money before seeing their sights.

The taxi driver drove like a teenager in a hurry! He told us that by law, one could not operate a vehicle in Japan with any dents or rust. We had seen our share of dents and cracked windshields coming from Alaska.

Nothing was written in English at the small restaurant. I pointed to the food a nearby couple was eating and ordered the same. Everything included noodles.

On the street, we met groups of 30 or more schoolchildren out for a walk. It was obvious we were Americans! Every student said hello individually in English as they passed by us, and we responded.

Seoul, South Korea

It was a short flight to Seoul, South Korea. We made reservations at a small Korean hotel and took a taxi. Debbie recommended we see Namsan Tower, which we viewed from our hotel. Debbie had visited John and his family in Seoul before they were married. They walked to the top of the tower, where John proposed to her. Distinctly Korean cultural events showed their dances, music, and dress.

Manila, Philippines: David

We spent several days in Taipei, Taiwan, and then flew to Manila, Philippines. Manila was hot and muggy, so we understood why people want to take a shower daily—to cool off. We stayed at the Summer Institute of Linguistics (SIL) Center in Manila.

We met a Christian nurse who lived in a Muslim region of the islands. Her superiors told her, "If you become a Muslim, you can move up the ladder in our hospital." She chose Christ and continued at the lowest-paid position.

Traveling by jeepney north to the SIL facility in Bagabag took six hours. After World War II, jeepneys were initially made from leftover jeeps, which were well known for their flamboyant decoration and crowded seating. Five sat on the front bench seat with one person even on the driver's left side. The two sides in the back each held seven people on makeshift boards, plus a few more who squeezed in or hung on. They were the standard mode of public transportation.

The rice terraces, which took years of hard work to make, were numerous on the sides of steep mountains. They were impressive, stretching for miles and still producing rice. Farmers spread the rice to dry on one side of the warm concrete road, so the vehicles drove around rather than over those areas.

Crates of live chickens were on top of our jeepney. A large hobbled pig was also thrown on the roof. He quit squealing after a bit. They picked up a man chewing betel nuts who sat next to me. It was gross to see the red juice run down the sides of his mouth, but he didn't seem to mind.

Using the Translated Scripture

At Bagabag, we met Joe and Barbara Grimes, who came for a linguistic workshop. We knew them from our early days at SIL in Oklahoma. They were both linguistic "geeks" yet down-to-earth people. Their questions to us were about the spiritual impact of the Koyukon Scriptures on the people and our involvement in getting the Word of God out.

Many translators and others working in the Pacific area asked us one common question, "What was the North America Branch's policy about working where Native people were losing their language?" It was helpful to discuss common dilemmas.

Sabah, Borneo

At Kota Kinabalu (KK) in northern Borneo, we stayed at the SIL Center. The staff was excellent and asked the usual questions.

We drove to a Native village and stayed with a missionary couple. The wife was a local, but the husband was a foreigner. Consequently, the husband could not teach or do anything in this Muslim country or he would be expelled.

This was the era of big money for Muslim oil producers in the Middle East after the crisis of the 1970s. Malaysians were offered money and material benefits if they turned to Islam. There were many "conversions."

Another Granddaughter

While we were traveling, another granddaughter, Cheryl Michelle Cogan, was born in November. She was the second daughter of Elizabeth and Andy, but we had to wait to meet her until we returned home.

To Southeast Asia: David

We flew from KK to KL, short for Kuala Lumpur, the capital of Malaysia, with good Asian food and service. We stayed in the guest houses of other missions in KL and Bangkok, Thailand.

In Bangkok, we were impressed with all the gold and ornate decorations as we toured the presidential palace. I commented to one of the guards about the gold. He pointed to a piece of gold leaf that had fallen on the ground and said, "Take."

Singapore was a different experience. It is a multi-religious country and a melting pot for the Far East. The government strictly controlled their lives. No chewing gun was sold at the airport.

Via the Middle East

From Singapore, we switched back to flying Northwest to Frankfurt, Germany, with a fuel stop in Abu Dhabi in the United Arab Emirates. It was the middle of the night, but their stores were open to get tourist money. Northwest liked the cheap gas.

Europe

Arriving in Frankfurt, Germany, we took a quick tour of Europe. We had bought Eurail passes before we left Alaska. This gave us three weeks of unlimited travel, so we went to 12 countries.

For a few nights, we slept in an overnight berth on a train to sightsee during the day. These "private" berths accommodated six people. You had no idea who your roommates would be, and they entered in the middle of the night. While traveling to Austria, we met one lady who worked for the UN with deep spiritual needs, so we listened and talked most of that night.

In Italy: Kay

Riding in a gondola, we viewed Venice, Italy. It was eerie to see the sinking buildings with no one in them. The Leaning Tower of Pisa looked too dangerous to climb.

The Colosseum was impressive. It could hold more than 50,000 people. We thought about the first-century Christians who were fed to the lions for the enjoyment of the spectators. Ben-Hur also rode there.

In Rome, while walking down the main sidewalk in the middle of the day, three young boys approached us hollering, but not in English. One boy had a piece of cardboard draped over his arm with something written on it. Then I felt him try to open my purse under the cardboard cover. Thankfully, my purse clasp was hard to open. They realized we were wise to their "game" and ran off.

Elsewhere

In Spain, we were informed at the border that we could not check our backpacks in a locker due to a bomb scare. We didn't walk too far wearing our packs, so we saw very little of Spain.

We went to Nice, a modern city in southern France. The high-speed train took six hours to cover the 427 miles to Paris. It was an exhilarating ride up to 165 mph. It was smoother and faster than other trains.

On Sunday, we visited a large mainline Protestant church in Denmark. There was room for 2,000 people, but only 50 attended the service. We stopped in Catholic churches to look around. They housed a lot of history and relics from the past, but they forgot their spiritual heritage!

Israel at Last

We flew out of Frankfurt on the German airline Lufthansa to Tel Aviv, Israel. Several heavily armed vehicles arrived on the tarmac to watch our plane leave. They met and escorted every plane headed to Israel. We arrived in Tel Aviv on Friday afternoon and caught a bus to Jerusalem. A kind person informed us the buses had stopped running at sundown since it was Saturday, their Sabbath, so we jumped on the last bus. The lady directed us to a hotel. The following night, we checked in at a youth hostel. It was cheaper, and it was right in the Old City section of Jerusalem. They even gave us "married" people a room to ourselves with 14 empty cots.

It was very convenient to walk on the streets, buy long bread from a street vendor, and see the sights. In the Dead Sea, we started to float when the water was just above our waists. We also toured sites around the Sea of Galilee. The 1967 war was a survival miracle for Israel. The odds of winning were enormously stacked against them, but God intervened.

Three Outstanding Sites

The Great Isaiah Scroll and other Dead Sea Scrolls were discovered in 1947. This scroll, the oldest complete book of the Old Testament, is included in the outstanding displays at the Shrine of the Book. We walked 26 feet around the scroll that was open for all to see. A guard was overseeing the glass-enclosed, humidity-controlled display. It was dated about 100 years after Jesus was crucified and it is 1,000 years older than any previously existing fragments of Isaiah. The scribes were super strict in how they hand-copied the Bible, so God's Word has been preserved accurately for us today.

We took a day tour of the Masada Fortress, where we rode a cable car to the mountaintop. It overlooks the Dead Sea, which is 1,300 feet below. Masada is famous for the Zealots' last stand in the Jewish Revolt against the Roman Empire. The Romans built a dirt ramp to attack in 73 A.D., only to find 960 Zealots had committed suicide rather than be tortured to death. We think that dying in battle fighting the Romans would have been better since the Bible does not support suicide even in a "hopeless" situation.

One of the mottoes of the modern Israeli army is "Masada shall not fall again!" and military recruits spend their last night of training trekking through the desert to see dawn break over the great fortress.

Another impressive site was the Garden Tomb. This rock-cut tomb in Jerusalem, just outside the Old City walls, was discovered in 1867. The curators operating the site could not say positively that it was the actual tomb where the body of Jesus had been placed. However, everything here and in the surrounding area fits the description given in the Bible. The angel told the women who came seeking the body of Jesus at dawn: "He is not here; for He is risen, as He said. Come, see the place where the Lord lay." (Matthew 28:6).

Many Other Sites to See: David
On the edge of Bethlehem, we saw Rachel's tomb (Genesis 35:19-20) with a van parked in front. Soon, an Israeli police vehicle pulled up. Everybody stopped and watched but we gave them plenty of room, while one officer put on a heavy flack suit. He walked over and tied a wire to the door's handle. Then the police jerked open the van door. No bombs this time! Three teenage Israeli girls came close to us for "safety." We all scrambled to find a ride back to Jerusalem before dark.

We had trouble flying out of Tel Aviv. The agents said my face did not match my passport photo. I offered to shave, but they said, "No." They took our backpacks and x-rayed everything. They were excited when they found the two steel supports in our packs to make them stand straight. Perhaps they were making an example of our being searched while everybody watched. Next time we recommend going with a tour group.

Forever Changed
We returned to Fairbanks on December 19th. It was minus 35 degrees and 28 inches of snow fell two days later. The trip around the world and to Israel affected our lives just as the pipeline forever changed Alaska. We had a fantastic time and we praised the Lord for making it possible. The trip enlarged our view of world missions, particularly the tremendous spiritual need worldwide. It was a sweet mixture of business and pleasure.

See the People
One of our goals was to see the people. The masses of people were overwhelming. We rubbed shoulders with hundreds of thousands of people. Everywhere, they were worshipping something! The extremes ranged from the teachings of Islam to trusting exclusively in Jesus Christ. A number idolized their technical and material advancements as well as their past. One Buddha in Thailand was made of four and a half tons of gold! We saw a tremendous need everywhere for people to have the freedom Jesus Christ offers.

The Koyukon Konnection

We thoroughly enjoyed Southeast Asia but realized that the Lord wanted us to continue working in Alaska. The Lord gave us many years of trust and connections with the Koyukon people. Altona Brown, an Indian elder from Ruby, firmly stated, "You will never leave Alaska!" We weren't sure at that time, but now we are.

Going Mobile (1985~1987)

1985
Going Mobile

Wycliffe leaders asked us to consider going mobile and helping translators serving other languages in the United States. This would involve a wide variety of activities: typing Indian Scripture, proofreading, layout for printing, being a general handyman, ministry with local Natives, and encouraging the use of Scripture. We would be exercising our "gifts of helps" (1 Corinthians 12:28). This gift is very general to help fulfill the needs of others with strong spiritual overtones.

Normally, the children leave their parents to go to another location, but we left our children to serve temporarily in the Lower 48. Fairbanks continued to be our home base. We kept our Koyukon connection and trusted we could continue to be witnesses to these wonderful Native friends.

Steamboat Mountain: David

Steamboat Mountain, along the Alcan Highway near milepost 392 in British Columbia, included a long, steep climb to the summit at 4,250 feet. Our Oldsmobile ran slower and was losing power as we approached the highest point on the highway. Smelling something burning, I pulled to the edge of the road rather than have our car die in the middle of traffic.

When I got out, I could see smoke and a flashing light. I shouted to Kay, "The transmission's on fire!" I prayed as I quickly grabbed snow and threw it on the flames. Plenty of snow was available and the fire was extinguished. Even a few seconds later, it could have been a major disaster, and we would have lost everything if the car had burned.

We waited and let it cool. Then I started the engine, but the car would not move. The transmission was cooked!

We contacted a garage in Fort Nelson at Mile 300 and waited for help to arrive. Ninety-two miles was a long tow, but the company said that was normal for them. We booked a room at a hotel within walking distance of the garage. In two days our transmission was rebuilt. When I filed for insurance reimbursement, the agent was not happy with the hefty bill, but that's why we had liability.

Fast-forward: the Alcan Highway was rerouted in 1998 with a more gradual climb up Steamboat Mountain. This had been a legendary bad spot on the route.

Crow Indians

Our first assignment was with Fred and Jeanne Miska, who worked with the Crow Indians in Prior, Montana. A little travel trailer in their yard became our home. Approximately 10,000 enrolled tribal members lived on the Crow reservation, and 90 percent spoke their Native language.

Many pheasants were in the fields and ducks were in nearby ponds or drainage ditches close to the road. The people preferred eating their beef cattle and deer.

Our Jobs: Kay

I worked with Jeanne to clean the forced air ducts under their double-wide trailer. We tied a string to one of their children's toy cars and sent it through the vent. Jeanne was at the other end, and then we pulled a cleaning rag through to get rid of most of the dust while Fred was gone. Fred had asthma, so removing the dust helped him.

A team of Crow people worked on translating the Bible in an office building and had completed half of the New Testament. David worked with frozen water pipes and wiring problems at this building.

Miska's baby boy was turning one. They planned a big birthday service for him. I helped Jeanne with carving three turkeys and making the stuffing. We were disappointed to miss the celebration, but we had to move on to our next place.

On our last day in Montana, we ate lunch with the translation team. Then, they prayed for our travels to Nevada. One of the men prayed in Crow. Although we couldn't understand him, God understood, and we appreciated it.

Our First Grandson: Kay

Timothy Daniel Henry was born in February in New Jersey. Dan worked on a pipeline-related job on Alaska's North Slope, while Donna flew back to New Jersey to have their second baby. He was named in honor of two people in the Bible. Timothy worked faithfully with the Apostle Paul to tell people about Jesus Christ (See 1 Corinthians 4:17). His middle name, Daniel, is after his father. I knitted a blue and white sweater and helmet set for Tim.

Our Burden: Kay

We were so burdened for the Koyukon Indians' spiritual needs when we left Alaska. I was feeling down because of our roving assignment and the lack of deep relationships with them. One morning, the Lord awakened me early and showed me I could pray for the people, and He would continue to build His church. I started praying Scripture verses for them.

Teamwork with the Paiutes: David

The next stop was McDermitt, Nevada, to assist John and Joy Anderson, who translated the New Testament with the Northern Paiute Indians. They were overseeing the final stages of their work. Gary and MaryAnn Eastty were developing more primers and teaching people to read. LeRoy and Shirley Frye were typing the New Testament for printing. Kay keyed in First Corinthians in Paiute and the Gospel of Mark in the neighboring Shoshone language on the computer.

I tape-recorded Paiute Natives reading Scripture, thus making the Bible more readily available so those who couldn't read or were poor readers could hear God's Word. I was privileged to go to different Indian homes with the Andersons and record the people reading Scripture. After returning home, the real work began as I edited the tapes. This involved correcting errors by removing repeats and cutting out long pauses. One challenge was that many of their words ended in voiceless whispered syllables, and I had to be careful not to erase any of them.

We participated with the people in their prayer meetings, Bible studies, and church services. We enjoyed being with Native people at the grassroots level. They sang gospel songs accompanied by a skin drum or guitars. Several hymns and choruses had been translated, while others were original hymns sung to an Indian tune. Their hymnal contained 20 songs.

A Fifth Wheel

The Fryes needed to sell their 24-foot fifth-wheel trailer. The price was right, so we agreed. We bought a 1980 one-ton Ford pickup to pull it. Now we had warmer housing, more room, and a place to live without infringing on other people. It was easier to pull and maneuver than a standard trailer.

Going High-Tech

We went high-tech and purchased our first computer, an Osborne Executive portable with a five-inch screen, speed of 4 MHz, and ROM of 8K, for 800 dollars on a special deal. It was the size of a portable sewing machine and used five-and-a-quarter-inch floppy disks. It still works, but we've moved on.

Ten members participated in a computer workshop at the Wycliffe Office in Eastlake, Colorado. Most were beginners like us. We used a word processing program called WordStar. Computers can speed up Bible translation!

Havasupai Indians: David

Next, we drove to Cottonwood, Arizona, to assist translators Scott and Lynanne Palmer. His parents were remodeling the living quarters and adding space for their work. One of the more challenging parts was hanging sheetrock on their super-high ceiling. I assisted with long support poles.

Their principal village was Supai at the bottom of the Grand Canyon, but we did not get there. After completing our work, we drove back to Fairbanks.

New Neighbors

George and Judy Richardson, with their four boys, lived in our log house in Fairbanks for two years while we traveled in the Lower 48. We thought we would be gone the entire time but we came back sooner. So we moved into our little "honeymoon" cabin in the back where Steve and Cheryl would eventually live. We had electricity but no running water and an outhouse. Most of the time we traveled to Koyukon villages.

IRS Calling: David

The IRS notified us we were being audited. Kay had our files organized and ready. The auditor looked over our paperwork and informed us, "Since you are missionaries, you can take a one-time deduction for your expenses and time you spent in Israel." We had a good experience with our only audit.

Tragedy on the River

On the Fourth of July, Steve and his friend Reuben Bawell went kayaking on the cold, glacier-fed Gulkana River not far from Glennallen, Alaska. Suddenly, their kayak flipped over in the swift water. Both of them were wearing life jackets, but hypothermia and the shock of tipping over stunned them. Steve swam to shore, while Reuben hung on to their overturned kayak, drifting around a bend and out of sight. Steve found himself on the wrong side of the river, so he prayed and swam across. He stumbled through the woods singing gospel hymns at the top of his voice as he headed toward the highway to get help.

A small airplane spotted their kayak and Reuben's body by a log jam farther downriver. This was a very trying time for Steve, but especially for Reuben's family in Pennsylvania. Reuben was a solid Christian and had put the Lord first in his young life. God felt Reuben's work was finished, and Steve had more to do.

Wedding Bells: David

Steve married Cheryl Marie Stanton at Bethel Baptist Church in Fairbanks in August. They asked me to perform the special ceremony.

Going Mobile (1985~1987)

The wedding party: Chris Stanton, Cindy (Stanton) Rohl, Cheryl and Steve Henry, Dan Henry, and Jon Wheelock. Our granddaughters, Melissa and Kara, were the flower girls.

Fairbanks Native Fellowship

The Fairbanks Native Fellowship was the first move toward starting a church. In August, the fellowship met at our home for a barbeque, a time of singing, sharing testimonies, and hearing the Word of God. Forty-two adults and 20 kids filled our house. Six Natives played their guitars as Jimmy Johnson Jr. led the worship time. Another Indian gave the Bible message. Our home has an open living room, dining room, and kitchen area that is 20 feet by 40 feet. It was packed!

Fairbanks Native Bible Church

We met with Arctic Missions and SEND International missionaries about a joint effort to formally establish the Fairbanks Native Bible Church. It felt like now was God's timing with more than 6,000 Native people living in Fairbanks.

The church started meeting weekly in our log home. Natives had tried various churches in the city but did not feel comfortable. The goal was to meet the spiritual needs of Alaskan Natives but not be limited to Natives while providing a loving, safe place to share the gospel.

New Life: David

Andy Demoski from Nulato was in the hospital. We had been friends since we moved to Koyukuk in 1960. His wife, Amelia, worked with us in the bilingual program. Andy was not doing well with cancer. I had the honor of praying with him as he put his trust in Jesus Christ. Several weeks later, he met Jesus in person. At this time, I was under a lot of pressure from another problem affecting my life. Andy's decision was a tremendous encouragement and boost from the Lord.

A Change of Direction

We had favorable talks with Wycliffe and Arctic Missions leaders about going "on loan" and continuing with the Koyukon people. We could see the Lord's guidance in this and were excited about the possibilities. We wanted to be where the Lord wanted us to be.

Help from the Bible

Different people are impressed by various verses in the Bible. We like "Trust in the LORD with all your heart,
And lean not on your own understanding;
In all your ways acknowledge Him,
And He shall direct your paths" (Proverbs 3:5-6).

Another helpful verse was: "And let the peace of God rule in your hearts" (Colossians 3:15a). If we don't experience His peace, we are making the wrong decision in life. We've done that when we moved ahead on decisions on our own.

Another Concert: Kay

Dan and Donna invited us to a Christian concert highlighting a folksinger. When we arrived, Kara looked up at her mom and said, "Grandma's here!" Then she hugged and hugged me. Of course, I gave her hugs back too. What a thrill to be a grandma!

A Young Lady from Allakaket

A young lady from Allakaket, whom we have known since she was a little girl, came to Fairbanks and spent a few days with us. She was waiting to enter an Alcohol Rehab Center because she wanted to change her life. We had an opportunity to talk with her about the best Helper, Jesus Christ.

Disaster on the Koyukuk River

Six young adults from the village of Hughes drowned in the Koyukuk River on September 2nd, 1985. They were on their way home following an event in Allakaket. Their 24-foot boat collided with a moose swimming in the river.

To the Koyukuk River: Kay

We flew to Hughes to offer assistance wherever possible. I spent many hours preparing food. Henry and Sophie Beatus lost three of their children. We sat with them and looked at their magazines without speaking for half an hour. There was nothing to say. They knew we were there and that we cared about them and their children.

One weekend, a gospel team of four came from Minto to minister. These Native Christians offered real hope and comfort. We were greatly encouraged to see strong Indian believers serving the Lord and reaching out to others in need.

One night, Catherine Attla from Huslia came to visit until after 1:00 a.m. She was raised by her grandparents and had a wealth of Indian knowledge. It was interesting hearing the Indian beliefs about creation and the great flood because they were similar to the accounts in the Bible.

1986
Decisions, Decisions, Decisions

Life is full of decisions, and God gives us the freedom to choose. We are not robots! In the end, we desired to please the Lord and hear Him say, "Well done." So we made a two-year decision leading to a long-term change.

We started the New Year being "on loan" for two years to Arctic Missions after serving with Wycliffe for 27 years. We were under Barry Arnold's field supervision. Barry and his wife, Vicki, were serving with the Galena Bible Church in Galena.

What Was Driving Our Decision?

Our decision to change missions was driven by the desire to fulfill our ministry with the Koyukon people. We wanted to see more Koyukon people turn to the Lord and grow in their relationship with Jesus Christ.

Our Ministry

Over the next two years, our ministry focused on discipling one-on-one, visiting Native people, teaching Bible studies, encouraging people to read and obey the Bible, and ministering in the 11 Koyukon villages. We lived in Fairbanks, where more than 500 Koyukon people resided.

When we flew to Nulato and Koyukuk, we shared VCR tapes from Anchorage Native New Life and the *musicale*. People especially wanted to see and hear what Walter Maillelle had to say since they knew him.

The *Jesus* Film

The *Jesus* film is a two-hour motion picture about the life of Jesus based on the Gospel of Luke. There was room in the plane, so we flew with Tom Lamphere with Campus Crusade to show it.

The movie was shown 10 times over the next two weeks in nine Koyukon villages and Anaktuvuk Pass, an Iñupiaq village. A third of the residents came to view the film in most villages. We knew of two Indians who chose to follow the Lord after viewing it.

Spring Break

Steve and Cheryl organized a two-night retreat during spring break at the Beckdahls' cabin 30 miles north of Fairbanks. Snow machines provided the only way to travel there. Seven couples and 17 singles attended, so there were wall-to-wall people.

Medical Need in Michigan: Kay

I flew to Michigan for a month to help my father care for my mother after she suffered a stroke. She was making good progress, but the aftereffects of the stroke continued.

Going to a Wake: David

Nome Stickavan and I went to Agnes Oslund's house in Fairbanks for a Catholic wake for Vivian Peters from Nulato. Nome was a great witness as he played his guitar, sang gospel songs, and shared his faith. Returning home after 1:00 a.m., we thanked the Lord for an opportunity to give people hope in the face of death.

Fairbanks Native Bible Church Meeting

The people continued to meet weekly at our house during the winter, and the attendance averaged 50. We looked for a larger, more central meeting place since our house was five miles out of town.

On June 1st, the Fairbanks Native Bible Church moved their meetings to Denali Bible Chapel on 1201 Lathrop Street. We appreciated their support and kindness in making their facility available. The afternoon service allowed people to sleep in or relax before church. The building met all our physical needs and was more accessible to everyone. The church had outgrown our log home.

Rafting to Nenana: David

Steve prepared his raft for the Great Tanana Raft Classic Race from Fairbanks to Nenana on the Tanana River. Four 55-gallon gas drums were required for good floatation, and they could use any man-powered way for faster travel. He had connected seven bicycles with a shaft and chain to power a paddle wheel. Oars were available for backup power. The "ride" included lots of exercise and fun for the crew: Steve and Cheryl Henry, John and Debbie Rathbun, Dan Henry, and others.

They launched their raft with many contestants at 6:00 a.m. on Saturday, June 7th, and arrived in Nenana at 2:30 that afternoon. Steve's raft came in third in both years he participated. I drove our truck with a fifth-wheel flatbed to haul the raft back to Fairbanks and parked it in our backyard.

Guest from Kaltag: Kay

Olga Solomon came to Fairbanks for medical needs and stayed with us for a week. We drove to Nenana to see the rafts arriving and transport crew members home.

Alfred's Final Days: David

Paul Starr's 87-year-old father, Alfred Starr Sr., was in the Fairbanks Memorial Hospital. I read Scripture in Upper Koyukon, talked with him about the Lord, and prayed for him. We left the tape recording of the Gospel of Mark playing when we departed.

His memorial potlatch involved many people. Paul asked me to bless the food in the Koyukon language. The new Alfred Starr Cultural Center and Museum in Nenana was named in his honor.

Our Fourth Granddaughter

Our fourth granddaughter, Andrea Beth Cogan, was born in July to Elizabeth and Andy. She was named after her parents. Now, we have four granddaughters and one grandson.

Visiting Nulato

On which side of the road should we walk? Facing traffic was our first instinct. But in Nulato, everybody walked on the side of the road that the wind was coming from, so the cloud of dust from the pickups and four-wheelers would blow away from the walkers. That was a practical solution.

Marie (Mountain) Dayton excitedly asked us, "What do you think we named our baby girl?" We had no idea. They named her "Kay Dee" since we were the first ones to visit Marie in the hospital after her baby was born. We felt so honored!

Our goal in Nulato was to visit people and be a witness. Summer was a busy time, and the salmon were running. Many people were at their fish camps.

Doing Time: David

I had a clergy authorization for visiting the Fairbanks Correctional Center, so I was able to see inmates in a semiprivate attorney room. A prison employee ushered in an inmate and handcuffed his hand to a connection on the table as standard procedure. The door was left open. This was a wonderful opportunity to talk freely with a client for a limited time. Security cameras were everywhere. The only way I could distribute any literature was to give it to the person at the front desk to pass on to the individual.

Sometimes, I had to wait for an available room during roll call twice a day as they accounted for every inmate. Other times I visited the standard way through a special glass window while both of us held a telephone to communicate. Native friends often gave us the names of inmates requesting a visit and spiritual help.

Filling In

We filled in at Bettles Bible Church for a month, 35 miles north of the Arctic Circle. The combined population of Bettles and Evansville was 60. It was a great opportunity to relate to Indians, Eskimos, and non-Natives.

More Time in Allakaket: Kay

In Allakaket, we rented a 20-foot by 24-foot place and had Bible studies. Several asked if we could have the study twice a week, so we gladly accommodated their request. Our normal schedule was to study in the mornings, eat lunch, visit homes in the afternoons and evenings, or hold a Bible study. Two ladies helped with the children's studies. Anchorage Native New Life meetings on VHS tapes, SonRise Indian Gospel Band, *The Cross and the Switchblade* movie, and kids' Bible stories were shared with the people who had VCRs.

One young mother told us she thought about Bible study every time we had one but couldn't come because her husband worked at night. So I went over to her house with the study. She was anxious to learn and had good questions. Later she told us, "I believe, but I haven't said anything to anyone." She studied her Bible, didn't drink, loved her husband, and took good care of her children as she quietly lived for the Lord.

One day, a young fellow stopped by and gave us a sled load of firewood. That was very thoughtful of him. We would usually buy a small sled load of firewood for 20 dollars as people were glad to earn some money.

Reminiscing: David

After the ice on the river was safe to walk on, we visited the Eskimo people living in Alatna. Oscar Nictune Sr. reminisced about our sheep-hunting trip up the Alatna River in 1967. That was a memorable hunting trip!

Opposition: David

Our testing and challenges came in the form of strong opposition. I was talking to a young man about the Lord when his brother entered the cabin. He swore at me for witnessing to his brother. Later in the day, he climbed on an unfinished building and insulted me while I was walking to the outhouse. I did not answer him. At times like this, we stood firm in our trust in the Lord and were sure of God's timing for us to be in the village. A month later, I was able to talk to this person briefly about Jesus.

Thanksgiving in Allakaket: Kay

We prayed for a turkey for Thanksgiving, and the Lord sent two. One flew here with Barry Arnold from the Galena Bible Church. The other turkey arrived on the mail plane from the Fairbanks Native Bible Church and included everything for the celebration. What a blessing!

I cooked one of the turkeys and a couple of pies for the village Thanksgiving celebration in the hall. The feast included many cranberry sauces, blueberries, and Indian ice cream (made with cooked, boned, and flaked sheefish, but without the fishy taste). Someone brought a swan, which was bigger than our turkey. It was all dark meat and tasty. That was a first for us.

Failed Attempt

A young lady failed in her attempted suicide soon after Thanksgiving. God granted her more time to live and make better choices in life. We were grateful to have had a good talk with her. God only gives us one life to live on this earth.

Protection from Unseen Spirits

People in the village believed they needed protection from unseen spirits. We agreed and also wanted protection. Several sprinkled wood stove ashes outside in their yards or hung a dried fishtail or fin inside above their doors and windows to "feed" the spirits. Others wore a cross necklace.

We asked the Lord to put a fence of protection around our cabin. God is much stronger than Satan or his evil spirits, so we trust Him for our safety.

Birthday Party for Jesus

All the children were invited to a birthday party for Jesus in our cabin. We retold the story of the birth of Jesus and served refreshments to 25. One big candle was on the cake. The children said Jesus was 1,986 years old, but since Jesus is God, He doesn't have a beginning or birthdays as we do.

Experienced It All

We experienced everything during our two and a half months in Allakaket: short hours of daylight, dancing northern lights, beautiful sunsets, 56 degrees below zero, lots of snow, temporarily plus 33 degrees and rain, many Bible studies, wonderful friends, open opposition, open houses, birthday parties, a funeral, a lot of walking, and no vehicle to maintain. Young and older people expressed an interest in spiritual things and wanted to know more, but they were not ready to commit their lives to the Lord.

The second turkey was cooked for a farewell dinner with moose soup, moose meat, and baked fish for 35 adults. We watched a video of Anchorage Native New Life singing, giving testimonies, and preaching. It was a great time of fellowship together.

More Conveniences

Flying home for Christmas, we looked forward to being with family, catching up on our mail, and everything else in Fairbanks. Debbie and John's downstairs apartment on Third Avenue was easier to live in than in our dry home without running water or sewer. It was convenient living in town, and we could walk to many places if necessary. We stayed here until we flew to Kaltag.

1987
High-Risk Mother Gives Birth: Kay

Rhea Williams from Allakaket asked me to be in the delivery room when she gave birth. Rhea had high blood pressure and only one lung after surgery for TB. She was 44 years old, not the ideal age to have a baby either. This was a time of testing Rhea's faith in the Lord.

God blessed Rhea with a healthy baby girl named Rebecca who had no complications. Right after her birth, the doctor said that something strange was written on her chart: "The patient can only go shopping <u>after</u> a safe delivery." Rhea flashed a big smile and responded, "Yes, that's why I wanted to have my baby in Fairbanks!" I took her shopping the following day.

On the Iditarod Trail: David

Kaltag was 400 air miles from Fairbanks. We came for a month of ministry, going from house to house. We were like "trail markers" pointing people to Jesus Christ. The mighty Yukon River was frozen solid, but the Indian residents had warm hearts. One lady commented, "I know God sent you here for a reason." She was open to hearing the gospel.

We watched the annual Iditarod Sled Dog Race as they passed through this official checkpoint and headed out to the Alaskan coast. This was "the last great race," covering 1,150 miles from Anchorage to Nome. A few times, a dog team was tied outside our door, resting on fresh straw and too tired to care about us walking close by them. All 57 dog teams were required to check in as they continued toward the finish line.

We admired how each musher was motivated to continue. It reminded us of running the "Christian Life Race" with endurance. It takes total commitment. We thanked the Lord for the people who, for many years, had faithfully prayed for us and given financially so we could keep pressing on.

James and Olga Solomon invited us for a supper of moose soup and sweet bread cooked in beaver fat. The next day, I went with James to check his 60-foot gillnet set under the ice of the Yukon River. He caught 16 whitefish. James told me he was very impressed with the change in a mutual friend's life after he had turned to Jesus. Our lives speak louder than our words.

Boating Disaster on the Yukon River: David

We were enjoying family time in Fairbanks when we learned about a major boating disaster on the Yukon River on May 31st. Seven young adults in the prime of their lives were traveling in a riverboat from Koyukuk home to Galena. All seven were lost when the boat capsized. The husband of one of the victims had stayed with us in Koyukuk when he was a boy. He was left with five children to raise.

All the planes to Galena were booked, but I got on a charter flight. I went to offer comfort and help for a week. It was a tough time for everyone. People were soul-searching as several said, "I think the Lord is trying to tell us something." People were more open to talking about the Lord Jesus during a crisis.

Bible School in Allakaket

In June, a team of summer missionaries led by Lynne Lingaas flew to Allakaket for Daily Vacation Bible School. Eight adults worked together to teach all the lively children. We combined fun and spiritual training. Our indoor activities were held in the church.

One teenage girl accepted the Lord, and several others took extra Bible lessons. Two young men were asking questions and seeking the Lord. However, it was difficult to grow spiritually in a village with no missionaries or regular church services. Allakaket is like many villages in Alaska with a few struggling believers and little spiritual help.

To Valdez and Beyond: Kay

We were thrilled to have Aunt Leola and Uncle Ed Erb visit us in Alaska. On a tourist boat in Valdez, we saw the Columbia Glacier up close. Uncle Ed and David reached out and touched the glacier.

Driving back to Fairbanks through western Canada via Whitehorse, Dawson City, and Chicken on the Taylor Highway, we saw more of Alaska. Above the tree line, we could see beautiful scenery along winding gravel roads for miles.

Stopping at the "Welcome to Alaska" sign with Ed and Leola

Heading North

After driving north on the Dalton Highway along the pipeline to Jim River near Bettles, a friend flew us to Allakaket. It was encouraging to visit several Christians who wanted to talk about the Lord and the Bible and apply scripture to their daily lives. Numerous hunters were in the area since the beginning of moose, caribou, and sheep seasons. We came home with a donated moose hind quarter.

Tanana Next: David

A shortage of missionaries in remote villages was a pressing need. We flew to Tanana for a five-week ministry trip to assist a group of believers who wanted church services and Bible studies. Kay worked with Ruth (David) Grant on starting a women's Bible study using Philippians.

Paul Starr married Mary Miller, and they were living in Tanana. Several months previously, he had visited us in Fairbanks and told us about the spiritual needs in Tanana. We worked with him to start a Bible study in the Gospel of Mark through the Tanana Bible Church. I asked him to lead it for one week, which he did. On Sundays, I asked Paul to pick out the songs and help lead the service, and then I would preach.

The Apostle Paul told the younger Timothy: "And the things that you have heard from me among many witnesses, commit these to faithful men (and women) who will be able to teach others also" (2 Timothy 2:2). God wanted us to keep sharing and passing on leadership roles.

Late one night, an Indian man knocked on our door. He realized his life was going downhill with alcohol and sin. He repented and put his trust in the Lord. We prayed that the local Christians in Tanana would help him grow spiritually.

October Birthdays: Kay

Dorris Stevenson, Bev Wheelock, and I celebrated our October birthdays at a Chinese restaurant. I have enjoyed many years of close friendship and Christian fellowship with these wonderful ladies. The waiter brought our bill with three fortune cookies. As we opened and read our fortunes, we realized all three of ours were identical. We laughed at what happened. We didn't believe in their predictions, but it was fun to read and laugh about them.

Another Grandson: David

We were excited to welcome our second grandson, Jonathan Stephen Henry, who was born in December to Steve and Cheryl in Fairbanks. He was named after two great men in the Bible and his father. Jonathan was a man of character, faith, integrity, and a close friend of King David (see 1 Samuel 23:16-18). Now, we had two grandsons to continue the Henry name.

Return to Koyukon Country (1988~1990)

1988
More Paperwork

The career applications for Arctic Missions, Inc. were completed. Their goal was to make Christ known throughout the northern region and establish churches in Alaska and western Canada. Switching missions after 29 years was difficult.

A Lot of Water: David

A lot of water had flowed down the Yukon River during our years with Wycliffe. They closed the Koyukon work. There were approximately 500 Koyukon Natives who used their language, but over 2,000 spoke English.

Eliza Jones taught two levels of the Koyukon language at the University of Alaska Fairbanks. Dr. Michael E. Krauss, with the Alaska Native Language Center, told me that with the current declining rate of spoken Koyukon, there will be few speakers after 2020. We were thankful those speakers were able to read about Jesus Christ in their language.

The Lord directed us to join Arctic Missions as career missionaries. We continued ministering to Natives in 11 Koyukon villages and in Fairbanks at the hospital, the jail, and assisting at the Fairbanks Native Bible Church.

Name Change

Shortly after joining Arctic Missions, they changed their name to InterAct Ministries. The original name was great for us, but it did not fit the expanding ministry in Canada. The new name reminds us that we must interact with people about the Lord Jesus daily.

Brenda: Kay

A young Native, Brenda (Roberts) Meeker Taylor from Minto, was saved in 1976. She was concerned about reaching out to her people for the Lord. Several were ones I had been working with, so we visited them together in the hospital. Brenda was only 28 and legally blind.

Radio KIAM Nenana: David

I prepared a month's worth of recorded studies from the Gospel of John to be aired on Nenana radio KIAM. They were geared toward Native people in the villages who were the majority of the listening audience.

Sharing with Churches and Christians: David

We drove to the home office of InterAct Ministries in Boring, Oregon, and met the personnel serving there. The facilities were small yet adequate and efficient to meet the needs of the missionaries serving in Alaska and western Canada. We also visited my brother and sister and their families in California.

Then we drove to Vassar, Michigan, to see Kay's parents and five supporting churches in and near the thumb area of Michigan. Calvary Baptist Church in Bradenton, Florida, was another stop.

God has been faithful to us, so we wanted to tell others. The Bible says, "With my mouth will I make known Your faithfulness to all generations" (Psalm 89:1b). We were encouraged to hear from many people who were praying for us and the work in Alaska. It is a spiritual battle requiring prayer. All our supporting churches and individuals continued to back our service with the Alaskan Natives.

He Made It: David

Steve graduated on May 8th from the University of Alaska Fairbanks with a BS in Civil Engineering. He planned to use his degree and experience for mission

Steve and Cheryl with their son Jon

work in Africa. They had their first child during his final year of college like we did.

Family Camp: David

Speaking at Family Camp in June at Kokrine Hills Bible Camp was a great opportunity with wonderful fellowship. People came by boat from Tanana, Ruby, and Galena.

Off to Allakaket: David

We flew to Allakaket with a team of five summer missionaries for two weeks to conduct a Bible school. All 35 children attended. After the classes, we walked the younger children to their homes. Our team led two adult Bible studies and Sunday church services each week.

One young man sought to fill the emptiness in his life. I told him we planned to stay two weeks, to which he replied, "You have two weeks to work on me." So I did. We prayed that he would accept the Lord.

One evening, I visited another man. After three hours of talking and several cups of coffee, he was ready to accept Jesus Christ into his life. Our biggest concern was how to help him and his wife grow spiritually without stronger Christians to encourage and help them.

Bible School in Fairbanks

When we returned, we taught 18 excited primary children in the Bible school at the Fairbanks Native Bible Church. An answer to prayer was that church attendance was up that summer, averaging 50 people.

God's Trail

We finished writing *God's Trail*, a 31-day devotional in English similar to the *Daily Bread* but culturally relevant to the people of the north. We distributed 250 copies in Fairbanks and outlying "bush" villages. We wanted to help Natives apply God's Word to their daily lives. We all need that!

Memorial Potlatch

In Huslia, we rented a vacant house for a month. The relatives of Jimmy Huntington put on a huge memorial potlatch during our first weekend there. He was a well-known Indian leader who had turned to the Lord in his later years. The school in Huslia was named after him.

The potlatch included a wide variety of Native foods: caribou, ducks, geese, moose nose soup, and Indian ice cream. They gave away blankets, calico material for sewing, personal effects, handmade pillows, beaded slippers, gloves, hats, wolf and wolverine fur ruffs for parkas, other skins, and the list goes on. This was their way of thanking people who helped at his funeral a year ago and distributing presents to his extended family to honor Jimmy. Many wonderful speeches were given. Celebrating, Indian dancing, and having fun with old friends continued late into the night.

Kay, Pray for Me: David
A small charter plane brought guests of the Baha'i religion during these events. We were asked if they could stay with us overnight. Since we had room, we agreed but were not aware of their religious connection.

One man in their group was a practicing Eskimo shaman. He had a unique skin pouch about two feet long and 15 inches in diameter beside his sleeping bag. The shaman lay there, whispering as he fingered his bag. We took the opportunity to tell them about Jesus Christ. Their response was, "You're 2,000 years behind the times since Jesus lived long ago."

Kay and I prayed together for our witness and that they would consider Jesus as we fell asleep. In the middle of the night, I was awakened with severe, painful stomach bloating that I'd never had before! I whispered, "Kay, pray for me, pray loud enough so I can hear you." Kay prayed. Immediately the pain and swelling subsided. Satan is strong, but Jesus Christ is stronger. We thanked the Lord for His protection and loving care. We were also thankful we had each other's support.

End of the Trail for Madeline: David
It was sad to hear about Madeline Solomon's passing. From Huslia we chartered a plane with five other people to Koyukuk. We stayed two nights for her funeral. For two hours, everyone sang gospel music with Wilson Sam leading while others played their guitars in the community hall. I shared how Madeline had read the entire book of Mark in Koyukon on cassette tapes. Then in her later years, she would turn on her tape recorder and listen to the gospel every night as she fell asleep. Consequently, several people asked us spiritual questions.

SonRise Indian Gospel Band

The SonRise Indian Gospel Band came to the "bush" of Alaska. They were Cree Indians from Canada who came to Huslia to play their music and minister in the community hall for two nights. I sat next to Richard Derendoff, who had significant hearing loss. After the band played, he told me, "I heard them real good." Then we flew with the band to Allakaket and Tanana.

Plane Crash

Roger Huntington crashed his small plane while circling over a bear den site. He was severely burned over 50 percent of his body. Miraculously, he and his friend walked seven miles to Huslia in zero-degree temperatures. Roger required many surgeries, skin grafts, and months of healing. The fiery crash helped him become fully committed to Jesus Christ.

Fast-forward: Roger and Carol Huntington minister full-time at Kokrine Hills Bible Camp on the Yukon River. They lead the camping ministry and disciple people.

1989
Even Alaskans Were Complaining

It was so cold in late January that even Alaskans complained about the weather. The two-week cold snap featured frigid temperatures of minus 76 degrees in Tanana. That was just four degrees off Alaska's all-time low of negative 80 degrees. In Fairbanks, our temperature was minus 60.

Grandson #3: Kay

Cheryl was anxious for their baby to be born. The doctor prescribed rest, so she was confined to bed for two months to prevent any problems or premature birth. This was a challenging time since Jon was a toddler. Steve and Cheryl welcomed Wesley Reuben Henry into their family in February 1989. Reuben in the Bible was the forefather of one of the 12 tribes of Israel. The name was also given in honor of Steve's close friend. I knitted a little blue and white sweater and cap for him.

Tanana for Easter

We flew back to Tanana for the Good Friday and Easter services as we celebrated the resurrection of Jesus. Since the sun rose at 6:45 a.m., we planned to meet at 6:30 a.m. It felt colder with the wind chill than the thermometer reading of zero degrees. We walked up the hill behind Tanana, overlooking the Yukon River for our service.

The Easter celebration at 11:00 a.m. was packed with 50 people at church. This was a great opportunity to share about Jesus' dying for them so they could experience the power of His resurrection in their lives.

Tanana Again: Kay

We flew to Tanana for the third time this year to assist at the church. I helped start a Sunday school class for kindergarten through sixth grade. The kids listened well and were anxious to learn. Mary Starr taught the following week.

The *Denakkanaaga* Conference lasted for three days. We connected with many "old" friends. Elders are highly respected by Alaska Natives.

Side Job in Valdez: David

When we returned to Fairbanks, Dan invited me to work 363 miles south with him on a project after the big Exxon Valdez oil spill on March 24th. This was the second-largest oil spill in US waters. There was a rumor going around that Captain Joseph Hazelwood would make a good governor since everybody knew him and he created thousands of jobs.

Dan's contract was to insulate self-contained units to generate steam for cleaning the oil off the rocky beach. The work site was crowded with activity with people working everywhere. The mess hall was huge but served good food. We insulated the units quickly, and Dan signed off on them. Then we visited Kay's brother, Jim Temple, and his wife, Dale, in Valdez.

A Cree Indian Pastor

Roy Cardinal, a Cree Indian pastor, and his wife, Leona, stayed with us for several days. Roy's story was typical. He had prayed the sinner's prayer but his lifestyle did not change. God gave him time to recognize his sins and ask for forgiveness. Then Roy totally committed his life to God. Now he is pastoring a church in Canada.

Grandson #4

Our fourth grandson, Milo Alexander Cogan, was born in August to Elizabeth and Andy in Fairbanks. Three of his older sisters cared for and played with him. Kay knitted Milo a baby sweater.

God Had a Different Plan: Kay

We were preparing for a ministry trip to Tanana in October. The Bible tells us, "A man's heart plans his way, but the LORD directs his steps" (Proverbs 16:9). The Lord changed our plans, and I ended up in the hospital with a major problem.

Tiny red dots on my lower legs looked like a rash. I felt fine, but this was strange, so I saw a doctor. It was easy to diagnose but not cure. I had immune thrombocytopenia (ITP), a disorder that can lead to excessive bleeding from unusually low levels of platelets, which help the blood clot.

Hospital Time: Kay

An Indian friend had asked me to pray with her seriously ill mother in the hospital. When the nurses wheeled me into my hospital room, I was pleased to have her mother as my roommate for several days. This gave us plenty of time to pray and read the Bible together.

At first, the doctor tried easy fixes for my blood disorder, but none worked. This gave me another week in the hospital. The doctors sought answers while we looked to God, the Great Physician, for His help.

My children and grandchildren frequently visited the hospital. Once, Elizabeth brought her four children. Each one decided they needed to use my bathroom. Someone pulled the "Help" chain, which set off the beeper. The nurse ran in, but she was understanding.

Many Indian friends visited and prayed for me. Johnson Moses Sr. reminisced about how our children would visit their kids when we lived in Allakaket. He felt bad that he had been a poor example since he used to drink. Johnson commented, "It's been 12 years since I drank!" We thought he had been a good influence on all of his children. His wife, Bertha, came in the evenings and we played Scrabble.

Time to Vote: David

During election season, Kay was still in the hospital, waiting for her blood condition to change. So I said to Kay, "Let's go vote." She put on her street clothes, and I took her to our election site. A nurse saw Kay returning to her room and read us the riot act. "If you bumped yourself or fell, you could bleed to death." Anyway, our votes were counted in that election.

Surgery: Kay

The traditional methods to bring my blood platelet count up to normal did not work. Consequently, the surgeon removed my spleen. This procedure was not always successful, but mine was. I was sternly warned if we traveled to a "bush" village and experienced any problems to return to the hospital immediately. The Lord greatly blessed me, and I never had any further issues.

Granddaughter #5: Kay

Suzanna Grace Henry was born in Fairbanks in December to Dan and Donna. She was named after a woman in the Bible (Luke 8:3) who provided for Jesus out of her resources. I knitted a sweater for her. Now we had nine grandchildren under nine as we closed out this decade of the '80s.

New Year's Eve Party

A New Year's Eve party at the Fairbanks Native Bible Church included games, food, and fellowship. This was a real encouragement to have an alcohol-free event when so many used to celebrate by drinking with friends.

1990
More Travels: David

Starting the new decade, we traveled to Tanana to assist Paul Starr with the work. We enjoyed interacting with the church people and those around town.

Then we flew to Huslia to work with the believers there. It was easier to work in Tanana because they frequently met together. We wished that villages like Huslia would meet in someone's home to sing, pray, read the Bible, and share what they were learning. In the early church in the book of Acts, Christians met in private homes before there were any church buildings.

Return to Koyukon Country (1988~1990)

We cleaned the rental home for the new InterAct missionaries, Don and Brenda Ernest, who arrived in April. They started immediately teaching kids' Bible clubs and having adult Bible studies. God answered our prayers meeting the needs of the people in Huslia.

Babysitting Opportunities: Kay

While babysitting Elizabeth's children, we had a tea party for snack time. Melissa helped fix lunch by stirring up Jiffy blueberry muffins. Annie set the table and carried the food. Cheryl was responsible for the dessert and she made a cute little container full of gummy bears. Everybody was helpful. Milo was too young to have a hand in the activity but he enjoyed eating it.

AWANA Awards

Melissa completed all three books in AWANA, so she was honored at the closing program. She received a Sparky plaque with her name on it, a baseball hat, and a cute little ceramic hand with a girl leaning on it for memorizing 37 additional Bible verses in a month-long Verse-a-Thon.

Tanana for Easter

Tanana Bible Church celebrated Easter with a sunrise service on the hill overlooking the river. A campfire was built to heat water and keep us warm. Somebody forgot to bring the tea! Paul Starr suggested, "Let's make spruce needle tea." We hung our coffee can full of snow over the fire to melt and then we boiled the needles. This held us over until we went to Pat White's house for hot drinks.

Life Cut Short: Kay

During our time in Tanana, we were saddened to hear that Paul Starr's sister took her own life. I had been working with her in Fairbanks as she was seeking spiritual help. Everybody has their share of tragedies in life.

When we are afraid in life, we need to follow the example in this Psalm. "Whenever I am afraid, I will trust in You. In God (I will praise His word), In God I have put my trust; I will not fear. What can flesh (people) do to me?" (Psalm 56:3-4)

Yukon River: Galena (1990~1991)

The Move to Galena: David

Our Alaska Director, George Schultz, asked us to fill in at the Galena Bible Church. Barry and Vicki Arnold were leaving after eight years. The new church building was 40 feet by 60 feet with four Sunday school rooms, a pastor's study, and a nursery. It was located back from the Yukon River on Louden Loop, and the attendance was around 40. The church was ready to be independent of the mission when they could find a pastor who would work with this cross-cultural mix of Native and non-Native people.

After filling the gas tank in our 1974 Chevy sedan, I drove to Nenana, and Yutana Barge Lines shipped it to Galena. There were roads, although one could only go a few miles around town. We mailed boxes of personal effects.

On July 8th, we flew to Galena and lived in the mission house across from the "old" community hall. We knew a few of the residents.

Galena was on the north bank of the Yukon River, 300 air miles west of Fairbanks. There were 800 residents plus 300 personnel at the adjoining SAC Galena Air Force Station. This was the largest Koyukon village and the major airport hub that coordinated freight and flights to the surrounding villages.

Air Force Chaplain: David

The Air Force chaplain, Charles Davidson, was motivated spiritually. Their station was small, so he served as the Protestant and Catholic chaplain, preaching the same message in each service. I asked him to preach occasionally at the Galena Bible Church.

Annual Mission Conference

In Palmer, the special speaker for the annual InterAct Missionary Conference was Pastor Price, originally from Wales, so he spoke with an accent. He had not finished high school, but he knew the Scriptures! He just preached from the Bible without using notes. Pastor Price stated that when he was younger, he asked A.W. Tozer, his spiritual mentor, about going back to school. Knowing they were undereducated, Tozer replied that the Holy Spirit taught them. They both studied the Bible and depended on God to teach and work through them.

Side Trip to Anchorage

After the conference, we continued to Anchorage and connected with Donna and our grandchildren, Kara, Tim, and Suzanna. We went shopping together, took the children ice skating, and then we all bowled. We even went to the Anchorage Zoo. It was a fun time!

Back in Galena: Kay

Carol Huntington, Roger's wife, was slaughtering chickens. I held the chicken's wings while she cut off their heads. Then she plunged them into a big pot of boiling water and quickly plucked them. It was fun working together.

Roger took David fishing in his Super Cub. Flying west over the hills near the coast, they landed on a clear river, casting their lures for silver salmon. Every cast hooked a salmon! They returned with fresh fish to freeze for the winter.

Visitors: Kay

Uncle John and Aunt Jean Temple from Vassar, Michigan, flew to Galena to visit us in the "bush" of Alaska. They joined us for a potlatch in the community hall across from our house. They did like we did, eating a bit of everything at the event and taking the rest home for later. We enjoyed introducing them to our friends and experiencing life on the Yukon River.

Later two of our children, Dan and Elizabeth, visited us in Galena. People from Koyukuk remembered them. They came during a memorial potlatch for Madeline Solomon when they could see everybody in town and share lots of good food.

Prince George, BC: David

Koyukon Natives Jenny Pelkola and Mae (Grant) Gunderson, along with missionaries Russ and Freda Arnold, Alan and Linda Ross, and ourselves, flew to Prince George, British Columbia, Canada, in a nine-passenger airplane. It was beautiful flying over the snow-covered range of mountains. I was seated in the back of the plane, enjoying the scenery until the door popped open beside me. The plane's speed kept the door from flying wide open, but it was enough to suck out the warm air and chill me.

The mission-sponsored conference was held at Ness Lake Bible Camp to improve indigenous outreach and discipleship among Athabascan Indians. It also included people serving in other missions: SEND, the Northern Canada Evangelical Mission, the United Indian Mission, and Native people from western Canada. Jenny reported on our activities to the Galena Bible Church after we returned.

Regional Mission Meeting

We hosted a regional InterAct Mission meeting in Galena. Kenny Hughes Jr., the missionary pilot, brought Brian Anderson from Nenana, Don and Brenda Ernest and their three children from Huslia, and Ken and Janet Gibbs Jr. from Grayling for our get-together. The Galena Bible Church provided a potluck dinner.

Funeral in Tanana

A funeral service was held in Tanana for Richard Grant. One of his 10 children was Mae Gunderson, who said: "My father gave me the gift of human life. We even share the same birthdate. In October, I helped him receive the gift of eternal life through Jesus Christ." Richard was now in heaven.

A Koyukon Missionary Serving Outside of Alaska

Mae Gunderson shared at the Galena Bible Church how she started following Jesus Christ and her desire to become a missionary. In 1991, she went to the "foreign land" of Minnesota and worked for the Lord among the Chippewa Indians on the Red Lake Indian reservation.

Weddings in Galena: David

I had the honor of performing two weddings. The first was for Curtiss Carlo and Harriet Silas at the Galena Bible Church. The second church wedding was for Wilma Ambrose and Kenneth McEnulty.

Galena Bible Church Activities

The Sunday church services in Galena averaged 40 people with Sunday school classes for the children. During the week, there were Good News clubs and a ladies' Bible study.

1991
A Cold Start: David

On Sunday, January 13th, the temperature read negative 50 degrees. I was reluctant to plug in our car and drive the mile to church because of the extreme cold. I felt like the man in the story where the husband doesn't want to go to church. But his wife reminds him, "You're the preacher; you have to go." The car started right up after being plugged in for two hours. Dense ice fog slowed our drive. Sixteen people attended. I didn't have the nerve to tell them I had wanted to stay home. We were thankful for the loving fellowship together and the warm church building.

Another Henry Grandson

In January, Curtis James Henry was born to Steve and Cheryl at Fairbanks Memorial Hospital. He was named after James, who wrote a book in the Bible with the same name. Among our grandchildren, we had 10 under 10 years old.

We Need to Talk: Kay

Margaret (Dayton) Huntington, who had been a young girl in Koyukuk when we lived there, phoned from the Galena Clinic where she worked. She urgently stated, "I need to talk with you as soon as possible." So I replied, "I'll be glad to talk right now. Where do you want to meet?"

I walked over and listened to her. Margaret related how she and Gilbert felt so overwhelmed about the sin in their lives and thought, "All we can do is fall on our knees and cry out to God for forgiveness." The Holy Spirit prompted them to do just that. Gilbert had returned home from a hunting trip and looked gloomy like someone had died. Margaret asked him, "What happened?" They both realized they were sinners and felt the same way about things in their lives.

Curtis in his first family photo with his brothers Jon and Wes

David phoned Gilbert and drove to their home. After they talked, the men prayed together for Gilbert to have peace with God about their feelings of guilt and shame.

The following Sunday, David told them. "I want you to have the privilege of telling others this wonderful news." And they did so in front of an excited congregation. The Bible tells us: "I say to you that likewise there will be more joy in heaven over one sinner who repents than over ninety-nine just persons who need no repentance" (Luke 15:7). This time there were two—husband and wife. It was a thrill for us to witness this incredible work of God.

Margaret was anxious to share their great news with her good buddy Mae Gunderson from Tanana. Mae had shared her faith with Margaret, who had seen what God was doing in her life. People are reading and watching our lives.

Other Interested Parties

Several other local people crossed our path with an interest in spiritual things. They wanted God to forgive their sins, but they didn't want to give their lives to the Lord and make a total commitment to follow Jesus Christ. It has to be their choice. We can't decide for anyone else, not even for our children or grandchildren.

The New Preacher

The pastoral candidate and his wife, Dave and Chris Rush, flew from California. The Galena Bible Church board met and encouraged them to move to Galena by the end of June. We found other housing and made the mission house available.

Baptisms in the Yukon River: David

The Yukon River was cold. But Margaret said, "We want to be baptized in the river, just like Jesus was baptized in a river."

On a Sunday afternoon in August, Gilbert, Margaret, and two other believers were baptized in the Yukon River. The water was shallow, so we walked far out to baptize them.

**David is on the left with Margaret,
while Pastor Rush is on the right with Gilbert.**

They wanted to be baptized together since they had accepted Jesus Christ into their lives together.

Then Gilbert and Margaret drove their boat to Bishop Mountain Fish Camp to share their good news and answer any questions about baptism and their faith in Jesus Christ. Everyone at the fish camp was happy for them, including Margaret's parents, Roger and Annie Dayton.

Back to the Big City: Fairbanks (1991~1993)

Moving Back to Fairbanks

Since Galena had a pastor, we moved back to our almost-finished log house in September. We did not want to leave Galena and the Yukon River, but it was the right time to go. We were thankful to be near our family and grandchildren in Fairbanks.

Gospel Crusade in Fairbanks: David

I was asked to lead a follow-up committee for the Tom Claus Indian Gospel Crusade in Fairbanks in early October. Tom, a Mohawk Indian evangelist, held three days of meetings. Several Native people served as counselors for those who wanted to receive the Lord Jesus and follow Him. Nine people accepted Jesus Christ as the new "Chief" of their lives, while others recommitted their lives to the Lord.

Huslia: David

Indians from Huslia, Galena, and Tanana took charge of special gospel meetings in Huslia. Many believers, including Gilbert and Margaret Huntington, shared their testimonies, prayed, and sang gospel songs. Margaret stated, "My life is just a little flashlight, shining for Jesus!" The Saturday night service lasted four hours!

Big Changes in Alaska

In 30 years, we have seen many changes in the villages of Alaska:

From cutting ice or melting snow for water ~ to village wells and running water.

From washboards or gas-powered washing machines ~ to community laundromats.

From filling and heating washtubs for baths ~ to village showers.

From dog teams for travel ~ to iron dogs (snow machines) and four-wheelers.

From loose-fitting barrel wood stoves ~ to airtight wood stoves.

From dim kerosene lamps ~ to bright electric lights.

From making coffee on the woodstove ~ to instant electric coffee pots.

From one or two mail planes a week ~ to one or two planes a day.

From school through eighth grade ~ to kindergarten through high school.

From playing basketball outside ~ to indoor gyms.

From messages sent via radio ~ to cell phones and internet access.

From no TVs ~ to satellite dishes and HDTVs.

We have changed, hopefully for the better—with God's help. Fortunately, Almighty God does not change. He is consistent and faithful. "Jesus Christ is the same yesterday, today, and forever" (Hebrews 13:8).

Big Changes in the Soviet Union

Major changes were occurring next door in the Soviet Union with the fall of communism in 1991. A complete breakdown in their way of life had occurred, and their currency, the ruble, kept declining in value. They experienced difficult, lean years of existence.

Special Assignment: Kay

Our mission director encouraged us to go on a "special assignment" to share what God was doing among the Native people of Alaska. In November, we headed down the Alcan Highway. Our first stop was the InterAct Home Office in Boring, Oregon.

1992
Continuing our Special Assignment

Vassar was our home base for trips to nearby supporting churches in Michigan and Blind River, Ontario, Canada. We spoke with numerous individuals and at many different churches in New Jersey and Florida gaining more support.

Time in Northern Illinois: Kay

On our way back to Alaska, we visited Cousin Diane and her husband, Gerry, in Joliet, Illinois. At the Adventist Medical Center in Hinsdale, Illinois, we saw Cousin Joanne Temple and her mother, Alma. Joanne was so surprised to see us! At Moody Bible Institute in Chicago, David spoke in two of Dr. Leonard Rascher's classes for cross-cultural ministry.

"We Are Sending You to Siberia!"

Those dreaded words, "We are sending you to Siberia!" used to mean a one-way ticket and brought fear into the lives of believers and their families in the former Soviet Union. Many never returned home. During these friendlier times, Gale Van Diest, our general director, asked us to consider going to Siberia to reach out to the tribal people.

At Northern Illinois University, we sought out their Russian department. They sold us an old textbook and accompanying audio cassettes. The very first lesson included the formal way to say "Hello" in Russian Здравствуйте pronounced ZDRASTvooytye, a real tongue twister. We wished they had started with the informal greeting of "Hi" or Привет, pronounced preeVYET.

Grandchild #11

Danielle Mae Rathbun was born in Fairbanks to Debbie and John in April. They had to wait 10 years before God gave them a beautiful daughter. Debbie claimed Psalm 113:9, "He grants the barren woman a home, Like a joyful mother of children. Praise the Lord!" God answered her prayer in His time.

The Russians Are Coming!

The arrival of 18 Russian missionaries was delayed due to government red tape and weather. They waited for flight clearance from Provideniya on Russia's Chukotka Peninsula to fly to Nome, Alaska.

Provideniya, or Providence in English, was a former military seaport with 5,000 people. While waiting, the group shared their faith, visited homes, and even baptized a new believer. God's delays present new opportunities. Then they flew the 222 nautical miles to Nome, Alaska.

For six weeks they trained in cross-cultural principles, biblical studies, family living, and Christian educational aids at the InterAct Office in Palmer. This was the beginning of a partnership with *Svet Evangelia* (Light of the Gospel), the largest evangelical mission in Russia. Many hungry hearts were searching for spiritual answers. God had thrown open the doors for the gospel!

Russians in Allakaket

We attended the training during the first week, then accompanied four of their group to Allakaket: Alex, a Sakha missionary; Nadia, a youth worker; Sergei, the Russian mission director; and Nikolai, their translator.

They sang several songs during a memorial potlatch in Allakaket. One was composed by a Christian serving time in a Russian prison. He looked through a small window, six inches by six inches, and could see a crane flying freely like his family was enjoying freedom. Even though he was locked in prison, he sang like Paul and Silas when they were in jail. His joy was in the Lord Jesus. Prison songs remained fresh in their memories.

Local family members sang memorial songs about Joe Williams Jr., who had died recently. Effie, Joe's sister, sang one she had composed for him to the tune of "How Great Thou Art."

Russians in Huslia

In Huslia, Sergei preached to a packed church on Sunday through Nikolai, his interpreter. People had their photos taken with the guests. Russian Christians brought hope and Good News to Alaska, whereas the Native people had been warned in the past about Russian spies.

Going to Siberia

Our mission encouraged us to fly to Siberia and determine if the Lord was calling us to serve there. We had many questions; a personal visit would answer a few of them.

Sister Cities

Fairbanks, Alaska, and Yakutsk, Siberia, are "sister cities." They promoted exchanging information, travel, and ideas. This was a great connection for us, providing orientation to Siberia. They handled our visas, which were something new to us, and it was cheaper to fly on their charters.

Survival Russian

We talked with Melissa Chapin, an active member of the Sister City program, about learning Russian. Given our limited time before our charter flight, her crash course was called "Survival Russian." It got us off on the right foot for the trip.

Our Charter Flight

A chartered Soviet Aeroflot plane brought Russian and Sakha people to Fairbanks for two weeks. Sixty people, including ourselves, left Fairbanks on the return flight on July 31st. Our plane stopped in Magadan, Russia, for our border check en route to Yakutsk.

Landing in Yakutsk

A Sakha couple held up a sign with our names and a bouquet. Kolya and Lenora were a gracious couple who welcomed us to their apartment. Kolya is the endearing form of the name Nikolai. They had a translator who helped bridge the language gap for both of us. We knew only a few words of Russian, and our hosts knew only a few words of English. Kolya was an agricultural researcher, and Lenora trained school teachers. Our hosts followed the Sakha religion. We had prayed for the Lord to place us where He wanted us.

We were within a block of the prestigious Yakutsk State University. Their apartment was on the fourth floor of a typical five-story complex with no elevator. Yet, we were awakened by roosters crowing by the log homes outside our window.

The first night, our hosts took us for a walk around several city blocks. At one point, Kolya informed us, "Watch out for the manhole in our sidewalk; it's missing a cover." Someone had stolen the cast iron covers to sell for scrap iron.

We felt like we were taking a giant step back in time to the 1930s! All the large government-building projects were stopped due to the fall of socialism. They had run out of their own and other people's money.

Sunday Worship Service: David

We found the evangelical church building. I recognized the entrance from a picture our Fairbanks friend Roger Wheelock had shown me. A telephone pole was leaning against the nondescript building with no church sign. The next Sunday, they met at a *dacha* (summer home) 10 miles out of town. We enjoyed the singing and preaching, and we had the opportunity to introduce ourselves and share briefly with the group using their translator.

How the System Works

Our hosts took us to wonderful ethnographic museums, the geology museum, the Permafrost Institute displaying a complete baby woolly mammoth, the park, the post office, and rides on the public bus system, which were free at that time.

They introduced us to the Russian way of shopping where one had to look at the price of an item, then stand in the cashier line to buy a coupon for that price, and wait in another line to present the coupon for the purchase. Stores were limited in the quantities available. There were many stand-in-line routines. A milk trailer was parked in the street. People brought their own containers for fresh, whole milk.

Souvenirs and Shamans: David

Numerous souvenir shops were available. I noticed a shaman's mask in one area and deliberately stayed away from there. To our surprise, after we arrived back at our host's apartment, we were presented with that mask as a gift. I didn't know what to say, with our extremely limited Russian, except to thank them and take it back to Alaska.

We decided to burn the shaman's mask soon after returning home. We did not live in fear, but we did not want to have anything promoting Satan in our home or have a mask available to influence any of our family or visitors. The Christians did this in the book of Acts and so did we. "Also, many of those who had practiced magic brought their books together and burned them in the sight of all. And they counted up the value of them, and it totaled fifty thousand pieces of silver. So the word of the Lord grew mightily and prevailed" (Acts 19:19-20).

Our hosts invited a scientist and a shaman to the apartment. One was the leader of the International Conference of Shamanism. With a limited understanding of English, we shared our faith, used their translator, and gave them Christian literature. We trust that the Lord's timing for our meeting demonstrated His power to these people.

To the Lena Pillars: David

Our sister city connection included a three-day cruise on a luxury liner *Demyan Bedny*. The food was great! Many cultural dance groups performed. We hiked to the top of the Lena Rock Pillars Nature Park and enjoyed the view.

On Sunday, I asked if we could hold a short service and give out free literature. My message was short, using one of their translators, to 20 Sakha and Russian travelers. They gave us time and space in the "musical bar" with the disco lights turned off. We set out literature which was immediately taken.

Our northernmost stop was at the mouth of the Vilyuy River, which was polluted from nuclear blasting. At a stop along the way, several of us walked to the Sakha village of Rassoloda. It was another opportunity to distribute Christian literature. I gave a tract to a Sakha man on his horse as he asked in astonishment, "Where are you from?"

On the Edge of Nowhere: David

A couple of Sakha women came to our suite on the boat and asked for a souvenir piece of clothing from Alaska. I took off my T-shirt and gave it to one of them. If you ever see a young lady in Siberia wearing a T-shirt with the words "On the Edge of Nowhere—Galena," know that it came from me. She gave me a Sakha fly chaser made from a horse's tail. We were both happy.

Guarding the Money: David

Back in Yakutsk, we needed to exchange dollars for rubles. A sister city assistant told us to go to a specific bank. We found the bank, looked through an open side door, and saw the tellers at their windows. We started to walk in. Instantly a Russian guard holding an AK-47 rifle stopped us and sternly said, "No." I pointed to the tellers, but again he said, "No!" We backed up into the corridor. Someone pointed upstairs, so we went to a large, bare room with only a small table and a bank agent sitting by it. I laid down 100 dollars, and the agent gave me rubles and a slip of paper. I said, "Спасибо" (spa-SEE-bah) "thank you" and left without counting the rubles.

Village Life

Our hosts took us by car to the city of Pokrovsk, where we went to one of their relative's dachas on the bank of the Lena River. Then we stayed overnight with the head Sakha gynecologist, whose department delivers an average of five to eight babies a day. No wonder the Sakha population was growing!

We drove farther to the Sakha village of Ulakhan-An, with 1,000 people, 15 miles beyond the end of the pavement. The dirt road section was terrible, but the people, houses, and gardens were great. Our translator was the local English teacher, who had never spoken with American speakers.

Our daily diet included fresh cucumbers and tomatoes. Their log house was 30 feet square and crafted out of larch, almost a foot in diameter. These timbers are harder than the white spruce used in Alaska. A few of their homes were more than 100 years old.

Everywhere was "open range," so one had to dodge the fresh cow pies on the roads or sidewalk areas. Both cows and horses are part of their main diet. They gave us *kymys* to drink, which was fermented mare's milk from their horses. We had trouble swallowing that but developed a taste for it after moving to Yakutsk. Raw, fresh fish, six inches long, was the main dish. They showed us how to pop off the heads and eat the fish. We politely took one, leaving the rest for them.

Is There a God?
Our village hostess told us that her mother, a former dedicated communist, believed Jesus had lived and died on the cross for her. Our school teacher translator said, "I don't believe in God, since that was what we were taught in school." Their Russian Orthodox Church building had been destroyed during the Communist Revolution of 1917, and only the bell was left.

"Come Over and Help Us"
We met with Sergei, the leader of Light of the Gospel Mission. Their desire was for American missionaries to come immediately. God had thrown wide open the window of opportunity to spread the gospel among so many hungry souls after 70 years of communism. Many cities of 10,000 or more had no church whatsoever. City leaders who were not believers requested a pastor or missionaries to help fill the moral and spiritual vacuum. The Lord impressed the words of Jesus to His disciples on our hearts: "Do you not say, 'There are still four months and then comes the harvest'? Behold, I say to you, lift up your eyes and look at the fields, for they are already white for harvest!" (John 4:35).

Return to Alaska
Our charter plane stopped in Magadan so we could clear Russian customs. We do not recommend the agents in Magadan. They took our

samovar, telling us, "We will send it to your host family." We knew that was a lie, but what can one do? They let us keep our smaller souvenirs.

That was a busy two weeks in Siberia! We had much new and valuable information to process. We decided to prepare to go but were willing to stay in Alaska and continue with the Koyukon people.

Russian for Missionaries

Six weeks later on September 12th, we flew to Albuquerque, New Mexico, to enroll in a special three-month Russian language course for missionaries. The teacher was Marc Canner, who was writing his workbook during the program, so the later lessons were written the previous night. It was intensive, with five hours of classroom work five days a week, plus homework. In the end, we weren't fluent speakers, but we learned more and memorized 10 Bible verses.

Calvary Church, Albuquerque

We attended the second of three Sunday services at the Calvary Church in Albuquerque, New Mexico. Skip Heitzig continues to be the senior pastor. He taught through the Bible book-by-book. The building was full, seating 2,500 people. During every service, people would come forward, making their decision to follow Jesus Christ. A young couple who attended generously offered us a room and fellowship while studying Russian.

A Lot of Hot Air

The Albuquerque International Balloon Fiesta was an event like no other. We walked among 650 hot-air balloons while they were being inflated. The sky over the city displayed decorative balloons of every shape, color, and unique description, such as a cow, a shoe, a truck, a stork, a parrot, or a polar bear.

The Flat Christmas Tree: David

Back in Fairbanks, spruce tree branches from our yard were tacked in the shape of a Christmas tree on our log wall. When the family came over to celebrate, grandson Jon Henry said, "Grandpa, you have a flat Christmas tree!" He was right, but it didn't take up much space.

1993
Time with Grandchildren

We watched Steve and Cheryl's three young boys while they went to a retreat. We kept busy, but it sure was fun! Granddaughter Melissa came over and spent one night with us. Then Dan and Donna hosted separate young adult slumber parties, so we were delighted to have their two girls overnight. Kara wanted to sleep on the floor in front of the TV. We missed our children and grandchildren when we moved to Siberia but we thanked the Lord for all the fun times we had in Alaska.

The Life of Jesus Christ: **David**

I flew to Yakutsk on a two-week trip in February with Bill Chesley. We needed to jumpstart the printing of 50,000 copies of *The Life of Jesus Christ* in the Sakha language. We added the way of salvation, using Jesus as the bridge at the back of the book. The book's original title was *Jesus and the Early Church* by David C. Cook. The publications had realistic drawings in full color.

Sergei, the director of our Russian partner mission, introduced us to a professional Sakha translator, Aita, a new convert to Jesus Christ. Bill drew up a simple agreement. Now the translator had work that would pay. Others in that era, shortly after the fall of communism, had work but with long-deferred payments.

Aita, the Sakha translator, and Bill are looking at a print master sheet.
Bill is holding a copy of the same book in Russian.

We hired two Sakha believers, Aleksei and Kidona, to handprint all the translated texts in the speech balloons or designated areas. They did this on large masters, approximately two feet by three feet, containing 16 pages of the future book. They also needed to keep the print masters clean. This was an intense, tedious, and time-consuming job that took several months.

Fast-forward: In the end, Aleksei and Kidona got to know each other much better and proceeded to get married. Working closely together has benefits!

More Major Decisions in Life: Kay
While David was gone to Siberia for two weeks, I had the honor of leading a young Indian lady to trust in the Lord Jesus Christ. This was wonderful, but it gave me mixed emotions about leaving these close friends and going to Siberia.

More Language Learning: David
We enrolled in the second year of the Russian language course at the University of Alaska Fairbanks. The intriguing Professor Serge Lecomte was well-qualified, with classes five days a week. It was challenging and helpful as we studied in a non-Christian atmosphere with 25 younger students.

The professor asked the class, "Would you memorize the Lord's Prayer in Russian if I gave it as an assignment?" The overwhelming majority of the students responded, "No."

Another time he asked, "Is it грех if we run out of beer? What does грех mean?" No one answered, so I raised my hand and replied, "Sin." Then he shot back, "How did you know that word?" I responded, "I took a class called Russian for Missionaries."

Feeling like Another Old Couple
We identified with Abraham and Sarah in the Bible when they left the security and safety of their home country and followed God's calling. "By faith Abraham obeyed when he was called to go out to the place which he would receive as an inheritance. And he went out, not knowing where he was going" (Hebrews 11:8). Only God knew what would happen next.

It was challenging to learn Russian in our late 50s. The language was our most significant barrier, but we kept learning a little here and a little there.

We decided to strive to be witnesses for the Lord right from the start, not waiting until we learned Russian well enough to speak clearly about Him. People read our hearts and saw our lives in spite of our limited Russian. With all the uncertainties in the country, we did not know how long we could stay, and we wanted to share the Good News of Jesus Christ as much as possible.

Sent to Siberia
~ The Sakha People

Khabarovsk (1993)

Sendoff to Siberia

A prayer of dedication for us was given at Bethel Baptist Church in Fairbanks. Our luggage was loaded, and we headed to the airport. We said our goodbyes to all our children, their spouses, and our grandchildren at the airport on Sunday, April 18th, before our 45-minute hop to Anchorage.

We went directly to the Aeroflot Airlines counter, the national Russian airline, concerning our six-hour flight to Khabarovsk (Kha-BAR-rovsk), Siberia. It was on time, so we checked our baggage. Bill Chesley, our Russian field director, and Doug Prins, a Bible teacher, flew on the same flight. We would serve as a liaison for more InterAct staff and teaching teams coming over. It appeared that all of the Alaskan staff was at the Anchorage airport to see us off. We were the first resident InterAct missionaries to move to Siberia.

Expensive Underwear: David

Kenny Hughes Jr. privately asked me a critical question for anyone going to Siberia, "Where do you hide your money?" He was safe to tell, so I replied, "Kenny, I'm wearing the most expensive pair of underwear I've ever worn!" Kay had sewn a pouch out of thin material the size of a dollar bill. I attached it to the inside of my underwear band, so it was also under my belt. There are other places to hide money, but that was what I did on this trip.

Maxed Out!

This was the only time in our lives we maxed out our credit card! Normally, we charge things and pay them off by the end of the month since we do not like paying interest. We were convinced it was the Lord's will for us to go to Siberia—and we took a leap of faith. The card was paid off before the end of the year.

Khabarovsk, Siberia

Our Aeroflot plane touched down in Khabarovsk, a bustling city of 600,000 people, equal to the total population of the state of Alaska. It was an important city in the Russian Far East, serving as a major transportation hub for the Trans-Siberian Railroad, river traffic on the busy Amur River, and air travel. We could see China, 21 miles away.

Soap Operas: David

Pastor Fyodor and members of his congregation met and cared for us. Our first three weeks were spent with a widow and her two daughters. They became Christians three years ago.

Part of their daily routine included viewing the soap opera, "Santa Barbara." The family and any guests present were all glued to the tube when it was on. They asked for our opinion, so I responded, "They show us how God does not want us to live." We had never heard of it, but neither did we watch the soaps.

Russians would ask, "Where do you want to go when you get to America?" The obvious answer was, "Santa Barbara because I already know everybody there!"

Name Change: Kay

In Siberia, I quickly realized I needed to change my name. When they heard my American name "Kay," they would typically say "Kate," which did not appeal to me. The easy choice was *Katya* (KAT-ya), a popular name in Russia.

"David" was a biblical name, but they stressed the second syllable instead of the first as in English. So over there, he became "Da-VID."

Our Choice of Words

To be consistent, we will use the following terms for locations and people:

Russia—The Russian Federation includes the old Union of Soviet Socialist Republics (USSR), extending from Europe through Asia, almost touching Alaska, and spanning 11 time zones. It existed from 1922 to 1991, until communism fell.

Siberia—The word makes us think of long, harsh winters, frozen wastelands, Soviet prisons, and labor camps. It is a vast geographical region, including central Russia and the Republic of Sakha (Yakutia). Prison camps were terrible in Yakutia, with isolation and extreme cold. We will use "Siberia" when referring to this region, including the Sakha Republic, but "Russia" for the entire country.

We usually use the name "Russian" when referring to the people. There were other nationalities, but we lumped them together as Russian for readability.

Sakha—The Russian government and many people use the word "Yakut," but we decided to use their preferred name, "Sakha," (Sa-KHA) out of respect for the national people, who are outgoing and outstanding in every type of activity. They comprise 55 percent of the population of their republic and are increasing.

The Republic of Sakha/Yakutia—This northern republic is the homeland of the Sakha people. Yakutsk is the capital city of the Republic of Sakha.

Measurements

All temperatures in Celsius have been converted to Fahrenheit. To make reading easier, all distances in kilometers have been converted to miles.

Phoning Home

To place a call to the USA, we went to the International Telephone Office, which had 20 non-soundproof booths. We placed a call through one of their operators, paid in advance, and waited 45 minutes for the connection to go through to our son Dan in Alaska. This was our only option for international calls.

As runaway inflation kicked in, we were shocked! When we moved here in April, the phone rate was 4 dollars for 10 minutes, but by July, the same call was up to 18 dollars. They also eliminated the reduced rate for off-hour calls.

Russian Evangelical Church Services: David

We attended all the church services to listen, learn more, and improve our Russian. A typical service lasted two hours. It included two or three sermons and several women who recited Christian poems. I preached several times using a translator. They liked to sing, and we followed along in our Russian hymnal. Six church members taught Sunday school in their respective neighborhoods during the week.

An Apartment for Us

Pastor Fyodor drove us to a member's apartment on the fourth floor overlooking a green zone in Khabarovsk. It was a nice one-bedroom older place belonging to a Russian widow named Alexandria who was 76 years old but still very active. She asked, "Would you like an apartment like this?" We said, "Yes," as she gave us a note to deliver to her pastor, who was waiting for us in his vehicle outside.

Then it dawned on us—she wanted us to rent her place. She was on a small state pension, with barely enough income to buy food and pay her bills. She needed the extra money to buy a winter coat and boots. So we agreed. We slept on the living room couch, which easily opened into a bed. We left all her belongings in place and did not use her bedroom while she stayed with her daughter nearby.

The Price of Being a Christian

Our landlady, Alexandria, trusted Jesus Christ in 1941 after being invited to a Christian meeting. Then at her factory, she stood up during a company meeting, stating, "There is a God!" This was during the dark days of communism. She was soon relieved of her job and income. Later, she boarded the Trans-Siberian Railroad and headed to Moscow on a six-day trip. Life was controlled and stringent, so she took one dried fish and four loaves of bread to eat on the way. Her goal was to meet with government officials about getting a job.

Her daughter, Alla, was friendly and appreciated us renting her mother's apartment for income. She said, "I wanted to go to church when I was young, but my mother would not let me go!" If the godless communist system had caught the little girl going to church or in any religious meeting, they would have taken her away to reeducate her. Now the daughter and her children are no longer interested in knowing about God. We asked the Lord to open her eyes.

Phoning Home from Our Apartment: Kay

The telephone system was updating and allowed phoning from a private dwelling, so I decided to call my parents by dialing 07. The local operator answered, "*Allo!*" (Hello!) I responded, "*Allo!* I want to call America." Click! She hung up on me! So I tried again, "*Allo! Allo!* I want to call Michigan, America." A different lady responded, "*Da!* (Yes!) What town?" I replied, "Vassar!" The operator said, "Vassar!" "*Da!*" Then I gave her the long string of numbers for my parents and hung up. I waited for two hours before the operator called back to our apartment. She had made the connection, and I had a short visit with my mother.

Inflation: David

We realized inflation was growing every time I exchanged dollars for rubles. People openly complained about it. A pair of socks cost one ruble three years ago, but today they cost 167 rubles. No sugar was available in our city for a week, and Russians put a lot of sugar in their cups of tea. Shortages of essential goods and food caused people to sell or barter their material possessions to meet the basic needs of life.

Alexandria was on a pension of 8,000 rubles a month equal to 10 dollars, but it was not enough to keep up with inflation. Prices for goods and services were artificially low since they were heavily subsidized under communism.

Early Ministry—Mokhsogollokh: David

Flying north, we arrived in Yakutsk. Then we bounced along with Russian Christians in a chartered bus for two hours to Mokhsogollokh (Mokh-so-gol-LOKH). This was the beginning of close friendships with our new brothers and sisters in Christ. The Evangelical Church, which started in 1991, was the only Christian witness available in their town of 6,000 people.

Pastor Valeri and his wife, Valentina, were actively praying for and doing the Lord's work. Written in big, bold letters across the front of the church was: "Christ Jesus came into the world to save sinners" (1 Timothy 1:15). Seventy-five people met in a converted bar. Several members said, "We used to drink in this same building; now we are praising the Lord." This reminded me of my childhood church in Illinois which met in a converted bar.

Our ministry team included Anya, our translator, Bill and Robin Harris, and Alan Jordan. We prayed for the surrounding villages and then traveled there to hold meetings and distribute Bibles and Christian literature.

One site was at an alcohol rehab camp where several hundred men were confined for a year or more. I told Anya this was a wonderful opportunity to share the gospel with all these men. She responded, "Great for you, but you don't understand the raunchy, sexual comments from the men!" I felt bad that our translator was put in that awkward position.

We ministered with the Russian believers in eight surrounding villages. One of those villages had a Russian Orthodox Church building converted into a museum. Most had never heard about Jesus Christ except in swearing.

Dancing *Babushkas*: Kay

We rode on a fast commercial passenger boat with 60 people for four hours up the Lena River to Sinsk. Ten singing *babushkas* (grandmothers) dressed alike, and two Sakha people welcomed us. One *babushka* gave each of us a piece of bread with a pinch of salt, representing the two staples of life. A Sakha lady gave us a sip of fermented mare's milk from a traditional wooden goblet. Then Anya helped me say a few words. One Sakha lady sang a song accompanied by another person playing a Sakha mouth harp. What a fantastic way to welcome missionaries to your village!

The *babushkas* sang more songs and invited the ladies to join them, so I (on the far right) danced with them.

A feast was served as we mingled with the people. After that, we led a service in their House of Culture. The locals went for free Christian literature. Under communism, every town had a House of Culture to provide a place to proclaim that God did not exist. We used the same building to tell them that God does exist and what the Bible had to say to them. Sinsk was a town of 1,000 Sakha and Russian people, with no believers. The only way here was by boat in the summer or on the winter "road" along the shoreline. There was no airport.

Praying Ahead: David

The spread of the gospel was preceded by prayer. Pastor Valeri took me to the bank of the Lena River, where he frequently prayed for those across the river. We saw the distant, flickering lights of six Sakha villages at night. We prayed for them. All the believers lived on the river's north side, including Mokhsogollokh. The Christians continued to pray for and reach out to those across the Lena River.

While driving to Mokhsogollokh, we noticed a rundown prison on the edge of town. Under communism, several had been held there for living out their Christian faith. The benefit to believers in prison was that they and their families prayed for the gospel to spread freely in prison and this region. This was part of the reason for the present spiritual growth in this area.

Teaching Trips in Yakutia
Missionaries with InterAct and American pastors taught Christian seminars and led concentrated training sessions in Yakutsk and outlying towns. The door of opportunity was wide open. We were glad to assist with what God was doing. The subjects included: Christian education, running Sunday schools, developing teaching materials, pastoral help, and general biblical topics. This brought practical Bible school education to the local missionaries while bringing them together for the sessions. The time together was as important as these special seminars since they discussed and applied the learning to their situations.

Trips to the Airport
Back in Khabarovsk, we made trips to and from the airport via the electric trolley on Karl Marx Street. It was a one-hour ride each way to welcome or say goodbye to somebody and receive personal mail. A few times, we went to the airport to look for a likely American stranger who could carry a couple of letters with stamps on them to drop off in America.

Street Sellers
Many individuals, especially *babushkas,* sold small quantities of vegetables from their gardens, fresh-cut flowers, clothing, greeting cards, or whatever to make a few rubles. A few also sold a bottle or two of vodka. Times were tough!

***Babushkas* in Shorts!**
American tourists started visiting Russia, especially the larger cities. Modest Russian *babushkas* wore head coverings and long dresses that extended below their knees. When Russians saw Americans, they laughed because they saw chubby *babushkas* in shorts. It was outside of their cultural norm!

Working on Documents: David

There were no courses for teaching Russian to missionaries. A lady who spoke English and ran a small store offered to teach us. She had never taught before but was helpful. When our documents needed to be extended and stamped, she connected us with the right individuals whom she knew. The man stamping them was happy to see Americans and practice his limited English while giving us another month on our visas.

As we neared the deadline on our visas, we were informed such changes could only be made during the last 15 days. We were thankful to extend our visas until November. It took more trips to our local *militsiya* (police) and paying a small fee.

InterAct needed to be officially registered with the government in the Khabarovsk region. Lena, our Russian translator, was knowledgeable and helped me with the documents. We went to the regional White House to connect with the head lady. We went back and forth for several trips. Lena told me, "Sit here and don't say anything. Let me do all the talking." The lady said to Lena, "Your client is very boring; he doesn't say anything!" I was praying, with my eyes open, for patience and wisdom. Finally, the mission was registered.

The Wrong Word: David

One ministry trip included traveling on the Amur River by hydrofoil with 120 other passengers. Our team of Russian and American Christians stopped at eight villages. People were open and sought spiritual answers. We continued to a town on the Pacific coast. In one city, we used a local translator who wanted to work with us. When I spoke about Noah and the ark, she didn't know the word "ark," so I simply said, "Big boat." Unfortunately, a person in our group shouted *teplokhod*, which meant a diesel-powered motorized ship. The Russians in our group all burst out laughing, but the translator was embarrassed!

Cross-Country Bus

The only way to access certain villages from towns on the Amur River was to take a Russian "bus" cross-country. These buses were military 6x6 trucks with three axles and a canvas cover over the passengers. We piled into the back with others and sat on some rough boards.

It was not made for tourists, but the oversized tires gave us a smooth ride over the rugged terrain.

Cruising along, we made good time until we had a flat tire. The spare had not been fixed since their last flat. He raised the axle with a chain, took off the tire, threw it in with us passengers, and drove off a little more slowly on five wheels.

On the Pacific Ocean

Our bus ride ended at De-Kastri on the Pacific Ocean. Five locals had repented of their sins during a previous gospel trip, prompting us to come here. Unfortunately, they were not meeting together. We met with these believers and encouraged them to meet in somebody's home for fellowship and to read Scripture.

Two 14-year-old girls, Anya and Oksana, wanted to practice English with us. They had a Russian grandmother in Khabarovsk, so we invited them to visit us when they came to the big city. Everywhere we went, people fed us and wanted to talk. One even brought freshly smoked shark meat.

In a nearby town, there were no believers. A local person, fluent in English, wanted to translate for us. She was very interested in the gospel and asked us many questions in English until midnight. Then she continued talking with our pastor in Russian until after 1:00 a.m. The Russians wanted to know more about Jesus Christ.

Our Money Stash

We encouraged our landlady to stay in her apartment while we were gone. Returning from our river trip, she confronted us. She had been looking at our books where we hid 700 dollars within the pages of a theology book. We thought nobody would ever look in that book, but we were wrong. She exclaimed, "I almost had a heart attack when I saw so much money!" We planned to use it for our living and travel expenses for several more months.

A Major Problem: David

On another river trip, we were all enjoying our 12-hour ride on our hydrofoil, heading back to Khabarovsk on Saturday, July 24th, 1993. Suddenly, an announcement was made over the loudspeaker, and the people all gasped in disbelief. Our American translator didn't fully understand, so we had to wait until we returned to our apartment to know what had happened.

Alexandria, who knew no English, told us the sad news and helped us understand what was happening. We often conversed with the assistance of a Russian dictionary. The Russian Central Bank had abruptly announced that all ruble bills issued before 1993 would not be honored the following Monday. Foreigners were told that each person could exchange 15,000 old rubles (about 15 dollars) for new ones. The bank decree was to curb galloping inflation, stabilize the ruble's value, and eliminate counterfeit bills.

Monday morning we headed out to find the right bank to exchange our rubles. Several banks had no activity, but one had a long line extending along the sidewalk for 100 feet. That was the right bank. Inside, we were directed past 20 tellers to a room where they exchanged our rubles for new ones after checking our passports and visas.

The Russians had two weeks to exchange up to 35,000 old rubles (about 35 dollars) for the new 1993 bills. Anything over that amount was deposited in a state savings bank, but the funds could not be withdrawn for six months. The new rubles slashed three zeros off their currency. I wondered if this could happen in the USA.

No Small Change: David

I tried to buy two Russian ice cream cones on the street. The lady gave me three because she had no change. The government had not made coins available. When I went to buy bread, the clerk gave me money and two small boxes of matches. I didn't want them, but that was a common way of making small change. Even the big stores did not have change. This helped us realize how people lived and why they were so frustrated with their government.

Repenting: David

We followed the Russian way of talking about new Christians by saying, "He repented!" In America, we would talk more about the believing aspect of salvation and say, "He believed!" Russians emphasize recognizing the sin in our lives and thus the need to repent of our sins. As Jesus said, ". . . unless you repent you will all likewise perish" (Luke 13:3). Russians liked it when I raised my hands in a surrender form and said in Russian, "I'm an American sinner."

Repentance means we change our mind about the direction we're going—away from God. Then we stop, turn around, and start going toward Him. Everybody needs to repent.

A stronger emphasis needs to be placed on discussing sin as the Bible does. In America, we have "easy believism." Just say you believe, or pray the "sinner's prayer," and everything will be fine. This can be false security! The believing part must involve both the head and the heart, resulting in a life-changing difference in our lives.

The Gospel via Train: Kay

The night train from Khabarovsk headed northeast to Komsomolsk-on-Amur. We shared a small sleeping cabin with two others for a good night of sleep, leaving at 11:00 p.m. and arriving at 7:00 a.m. Fourteen Americans, Russians, a Japanese, and a Russian translator traveled on our third major gospel trip. We sang "This Is the Day" and "How Great Thou Art" in Russian and English along with preaching.

Then we traveled farther on a day train to four more Russian towns for meetings. The Good News of Jesus Christ was new in three of these towns. People repented of their sins and accepted Jesus Christ into their lives in each location.

On one train ride, I checked on part of our group in another area of the car. On my way back to my seat, I said "Hello" in Russian to a lady. She asked me to sit down by her, so I did. Soon we were talking about spiritual things, so we went to where David was sitting to get some literature. Our group sang the Russian hymn "Russia, Russia," which was the theme song for our trips. The key lyric line was: "Russia, Russia, you have forgotten God, but He has not forgotten you."

The lady I "happened" to meet, or, rather, the Lord planned for me to sit by, was a longtime believer. Her father and grandfather were believers, and a book was written about her father. She asked, "Why are you passing by my village? Come, have a service." She belonged to a house church with 12 others. We prayed, talked, and decided to stop there on our return trip.

Christian Hospitality: Kay

We returned to her village, and the Christians were delighted to have our group. They fixed a snack of tea and honeycomb cut into small squares to eat with a spoon. They drove us to the country to see their honey operation with 100 hives and a guard to keep the bears away.

They generously gave us a three-liter glass container of honey. Three liters equals a little less than a gallon.

The Music Man
The man who played music at church and on this trip was almost blind, but the Lord gave him talent. He had a bag for his heavy electronic keyboard plus some accessories for the rest of us to carry. A few in our group complained, but the music sounded great.

Paying the Bills: David
I held the money bag and paid all our bills for the InterAct-sponsored trip. Before our travels, I went to a store with our musician. He told the clerk what we wanted for our trip since all the food was behind a counter. I paid the bill, and we carried several boxes to a waiting car.

At one 30-minute stop along the river, Pastor Fyodor said that we needed bread, so Kay and I went to the people selling food on the street, but there was no bread. I decided to buy rolls instead. Upon returning, we were scolded for not getting real bread. Bread is the "staff" of Russian life.

A Drunk Driver
We attended a Sunday school rally in a village 40 minutes from Khabarovsk. We rode there in a private car but returned by bus while our landlady rode in the car. A drunk driver hit the car head-on! Alexandria ended up in intensive care. Her heart problems added more concern, but she recovered.

A New Missionary
On another trip to Vladivostok, we met Vika, who planned to go to Plastun, a seaport with a population of 5,000, as a missionary. She took us to visit patients and their families at Vladivostok's Children's Cancer Hospital. Lots of sad cases, but a great opportunity with the freedom to witness and encourage them.

Returning on the Train
We took a night train to return to Khabarovsk and were in our compartment with four beds. You never know who will be sleeping in the other two beds. Our roommates were two ladies, and one spoke English well. Then they checked their tickets and discovered they were in the

wrong room. They invited us to their neighboring berth, and we talked about spiritual things for several hours.

Seeing Grandma in the Big City

The two young girls from De-Kastri, Anna and Oksana, came to Khabarovsk to see their *babushka*. They regularly attended the Russian Evangelical Church and participated in the activities. Once school started, one decided to return home, but the other stayed in Khabarovsk.

We visited their grandmother, who asked us, "Do you believe in one or two gods?" meaning God the Father and God the Son. We had a hard time explaining that with our limited Russian. But she started attending church with her granddaughter and received answers to her questions.

Another Blessing from De-Kastri

Sveta, from De-Kastri, had a severe liver problem and came to Khabarovsk for further treatment. She had started teaching Sunday School in her hometown and needed help. Both her mother and her daughter were new believers. Her husband now believed God existed, so that was one giant step for him. We gave her a flannelgraph set and other Christian materials to help her teaching ministry. We remembered Jesus' words, "I will build My church" (Matthew 16:18). We saw progress.

Planning Ahead

At this time, one could book flights only 15 days in advance on any airline within Russia. Insecurity made planning difficult. Our Russian friends often said, "You Americans say you are going to do such and such in the future, but we don't know the future." As Americans, we don't know the future either, but we act and plan as if we do. Only God knows the future of America or Russia.

Our Sixth Grandson: Kay

Joshua Mark Rathbun, our sixth grandson, was born in October in Fairbanks to Debbie and John. He was named after Joshua in the Bible who led the Israelites into the Promised Land.

Now we were a balanced family with four children, two boys and two girls, and 12 grandchildren, six boys, and six girls.

I'm reading to Danielle and Joshua Rathbun,
who were wearing the sweaters I knitted for them.

Back to English-Speaking Fairbanks

We returned to Fairbanks at the end of November. Our Russian visas had expired and could not be extended. It was great to be welcomed by family and friends and understand every word. We had many stories to share.

My Brother's Funeral: Kay

I felt so sad about losing my brother Thomas Robert Temple. I attended his funeral in Greensboro, Georgia, with my parents. He died at 50, leaving a wife and four beautiful girls. This ended his successful coaching career at Green County High School. Everybody knew him, and he was well respected.

I grieved with my parents and family. I still have many "whys," but God doesn't always explain reasons when something like this happens.

Christmas Activities

We joined our family for all the Christmas activities. Elizabeth's and Steve's children were in the program at Bethel Baptist Church. Kara had a piano recital at the Pioneer Home. She was also in the program during Sunday School.

First Sakha Christian Book (1994)

1994
Learning Russian for Free: Kay

There are benefits to being a grandmother. I enrolled in the Russian class at the University of Alaska Fairbanks to audit the course. Since I was over 60, the university waived the class fees.

To the "Bush" of Alaska

We flew to the "bush" of Alaska to visit three villages in January. In Huslia, we were encouraged by the new log church they had built. We saw the maturity and growth of believers in Galena and the loving care of those in Tanana.

Native believers encouraged us with gifts of fresh and dried meat and money to help reach the people in Siberia with the gospel. Several people were interested in going to share their testimonies with the Sakha people. It would be a giant step of faith for Alaskan Natives to go to Siberia, but we were praying to that end.

Making Progress toward Printing: David

On January 25th, Bill Chesley and I returned to Yakutsk to pick up the print masters for *The Life of Jesus Christ* in the Sakha language. Over the past 11 months, Aita had finished the translation while Aleksei and Kidona hand-printed 252 pages of text. The book had been checked for accuracy and clarity. The next step was to get the masters safely to the printer.

Where Are We?

In mid-March, we moved to Yakutsk, Siberia. The city is north of North Korea close to the Arctic Circle in Siberia. If you played the classic board game Risk, you would know where Yakutia was located; otherwise, you may need a map. The Lena River flows north by Yakutsk and empties into the Arctic Ocean. The river is 2,668 miles long, which is 688 miles longer than the Yukon River. Yakutsk is six time zones west of Fairbanks and six time zones east of Moscow.

Built on Ice

Yakutsk is the capital of the vast Republic of Sakha (Yakutia), which is twice the size of Alaska or four times the size of Texas. Yakutsk is the largest city in the world built on ice (permafrost) and in the coldest inhabited region in the northern hemisphere. It was colder in the winter than in Fairbanks, with an average January temperature of around minus 35 degrees. The population was 200,000 and growing.

Looking for a Home

We left Fairbanks on March 13th spending nine hours in the air plus clearing customs. All of the flights were uneventful as we had prayed for.

The church in Yakutsk placed us with a Christian family while we started looking for a permanent home. We wanted to find a widow needing help with her expenses or a family with an extra room so we could hear and use more of the Russian language.

After three weeks, Misha, the director of our Russian partner mission, invited us to move into a bedroom on the second floor of their Yakutsk office. The Evangelical Church started here five years earlier but soon outgrew the building. It was like Grand Central Station, with a constant flow of mission guests and others, so we met a lot of people. It was near the Yakutsk International Airport. The housekeeper, Anna Pavlovna, prepared the food and ran the household affairs. She was a blessing to be around.

The Evangelical Church

The packed Evangelical Church met in a converted warehouse near the center of town. Almost 200 people attended, and the number kept growing.

Filling the Vacuum

Seventy years of godless communism and socialism created a colossal vacuum—people were longing to learn more about God. The system hardened many hearts. For example, very few veterans were interested in spiritual things. Others could not let go of their past indoctrination and still firmly believed in communism, even though their government had debunked it. A faithful contingent of Communists always marched in the annual parades, displaying a photo of Lenin.

What a difference Jesus Christ made in the lives and hearts of those who turned to Him! It was inspiring to observe the love and joy these people had in their changed Christian lives. It was refreshing because most were not humdrum about their faith but excited to know God.

One Sunday, 10 people prayed aloud, one right after another. There wasn't even a pause between prayers because everyone was anxious to pray. Six people repented in that service, including a girl staying at our "boarding" house. She was like a sponge, listening to people talk about God and singing hymns. She borrowed our tape recorder to listen to more sermons.

Lenin Pointing the Way: David

Every town in Siberia had a Lenin Square. The bigger cities had full statues or busts of Lenin. We were more conscious of his cold, silent presence as foreigners. I told people in Yakutsk, "Look, his outstretched hand is pointing the way to the church!" It was only two blocks away.

Gospel to the East (GTE)

Light of the Gospel was formed in 1987 with its home office in Rovno, Ukraine. Many limitations, including the distance of six time zones, existed between Ukraine and the Republic of Sakha. Relations between Ukraine and Russia had been rocky since World War II and continued to sour. In 1993, the parent mission encouraged the formation of Gospel to the East (GTE) within Siberia with Misha as the director. The name fits their location in the Russian Far East. Their main work was in the Republic of Sakha with 20 missionaries. Several more ministered in neighboring Kamchatka.

A Ministry Van: David

Less than a week after moving to Yakutsk, we took a commercial bus from Yakutsk to Mokhsogollokh to do more outreach with Pastor Valeri and members of their church. They prayed for a vehicle to use for spreading the gospel in their surrounding area. A friend answered that prayer by donating a four-wheel-drive military van. We would have recommended it for retirement. None of the doors shut properly, and it burned oil. The church's skillful driver, Kolya, was great, but he drove like a race car driver.

We rode in their van to the Sakha town of Ulakhan-An. The ruts in the mud near the village were a foot deep, which didn't bother the free-range cows and horses. I was designated to sit in the front seat. A six-year-old Sakha girl sat in my lap. She soon fell asleep in spite of the rough road. We were deep in thought and prayer for these people.

This van was a bit like us—still going after many miles with lots of jolts, ups, and downs in life. We wished we were 30 years younger, but the Lord gave us strength. It was satisfying to know we were serving where the Lord wanted us.

The Service in Ulakhan-An

Stas, a promising teenager studying under Valeri, traveled with his mother, Shura, and younger sister, Marianna, on this trip to Ulakhan-An. We visited here two years ago with the sister city program, so we found the house where we stayed overnight and invited them to the meeting. The wife and son came. We were delighted to continue our friendship with them.

Our team shared the gospel in their House of Culture with 80 people. Stas led the Sakha service; the literature was quickly taken, and one lady repented.

Two people from Ulakhan-An went to Mokhsogollokh earlier that winter for medical help. They also received spiritual aid when they heard the gospel in the hospital, repented of their sins, and believed in the Lord Jesus.

Roof Repairs: David

While in Mokhsogollokh, I worked with the young men to dismantle part of a building. Valeri and his family were living in the end apartment of the single-story six-plex. The opposite end of the building had burned beyond repair. The city gave Valeri permission to tear down a third of the building and use the materials to repair his apartment. The building was slated for demolition, but the city had no money to do anything. I spent several hours on the roof in the sun. My red face looked like I had sunburn, but it was from all the heat. The Siberian summers had hot spells!

Teaching English: Kay

I accepted the position of teaching English as a second language for three weeks. Two classes were with students nine to 11 years old, and the other included teenagers. All the students loved looking at our family picture album. One day, I shared the puppet story about the big round cookie and thanked God for our food. The repetition in the story was like many of their stories.

What Stops Your Church Service?

What does it take to bring your church service to a screeching halt? In Siberia, we've seen services in Yakutsk and Mokhsogollokh stop when someone publicly repented for their sins and turned to Jesus Christ. Usually, the individual would come to the front of the church, so the pastor and the one repenting kneeled. First, the repentant sinner would cry out to the Lord, seeking forgiveness. Then the pastor would pray for that person.

One time, a pastor stopped his message twice as people came forward. Others repented during the announcements or prayer time. Once, during the dedication of a church building, a teenage boy stepped forward, and the service paused. A few people were repenting for the second or third time, but that was also necessary. Only God knows if they spoke empty words or genuinely from the heart. A changed life will reveal who has repented. The church is full of forgiven sinners.

Russian Cuisine: Kay

Anna, the cook at the mission house, had heart problems, so she asked me to take her place while she got medical help. I enjoyed helping her prepare meals, but doing it on my own was difficult. One Russian entree I made was *piroshki*. I made the shell with raised bread dough and filled it with whatever was available, such as cooked cabbage, mashed potatoes, hamburger, fish, or rice. Think of it as a crescent-shaped hand pie since you pick it up in your hand to eat. I could not make borscht, a Ukrainian beet soup, but they all liked my American-style pizzas.

At the Doctor's Invite

Maya is a city of 7,000 Sakha people with no believers across the Lena River. We drove there with Dr. Leonard Rascher from Moody

Bible Institute, Jack Vogt from Canada, Lena, a Russian translator, and Elena, a Sakha believer.

Final Choices in Life: David

Elena's friend was the head doctor at a large regional hospital. We met with a group of elderly Sakha patients. We spoke in English, Lena translated into Russian, and, finally, Elena translated into Sakha. It took longer to go through three languages for better comprehension.

The doctor asked us to visit two cancer patients who were nearing the end of their lives. Elena accompanied us as I used my limited Russian. A few times, she gave a longer explanation and filled in vital information. The first critically ill patient was Goya. He asked, "If I become a Christian, will I be healed?" I answered, "That is up to God. But I can promise this—Jesus Christ will be with you to the end of your life, whether He heals you or not." "For He (Jesus Christ) has said, 'I will never leave you nor forsake you'" (Hebrews 13: 5b). Goya repented and accepted Christ.

Then we visited the other patient who was near death, but he was not interested at all in talking about God. We left some literature and prayed for him after leaving his bedside.

The Sakha Cultural and Religious Center

The following day, we rode on a rough, winding "shortcut" through the woods to visit Soto's Cultural and Religious Center. It was a site from which they promoted the Sakha religion throughout the Republic of Sakha.

Sakha Pastor Ordained

Aleksei was young, but he had been preaching for more than a year. I participated in his ordination at the Evangelical Church in Yakutsk. Several leaders and I laid hands on him as we prayed for God to bless him in the ministry.

Conference in Palmer

We flew back to Alaska in July to attend the annual InterAct Conference in Palmer, Alaska. This included a special meeting of the growing Siberian team. More Alaskan missionaries were interested in serving in Siberia.

Ordained: David

Bethel Baptist Church in Fairbanks officially ordained me. Pastor Paul Holmes, the senior pastor, wrote about the church's decision:

"This letter is to inform you that on September 29th, 1994, by the unanimous vote of the Elders and Deacons of Bethel Baptist Church, David Henry was set apart by ordination for the work of the ministry. This action was based upon a faithful track record of many years of Christian service. We appreciate him greatly in the Lord and commend him to you in God's name."

AWANA at Bethel

We spoke at the AWANA program at church. We were thrilled that several of our grandchildren were involved. The previous week, the children brought many small McDonald-type toys for us to take back to Siberia.

First Birthday Cake

Debbie fixed a train cake five feet long for Joshua's first birthday. All the grandchildren were amazed!

The three taller ones: Kara Henry, Cheryl Cogan, and Melissa Cogan holding Joshua Rathbun; the shorter ones: Danielle Rathbun, Jon Henry, Curtis Henry, Milo Cogan, Wes Henry, Suzanna Henry (toward the front), and Andrea Cogan

Debbie had a piñata. Each child tried to break it open using a large metal spoon, starting with the youngest up to the oldest. For round two, they used a heavier piece of wood. Tim Henry broke it on his turn, making everybody happy.

Back to Yakutsk

We returned to Yakutsk on October 21st. The temperature moved down to minus 40 degrees before the end of the month. Our temporary housing situation was still at the GTE mission house. Several people were helping us look for an apartment, but nothing was available.

Registering with OVIR

As foreigners, we had to register with the local OVIR office within three days of arriving in the city where we intended to stay. This is an acronym left over from Soviet times for the Passport and Visa Agency. One does not make mistakes when dealing with government agencies. Misha kindly registered us. We visited their office every three months and reregistered ourselves.

Nicodemus: David

Our Sakha neighbor was Nikodim—Russian for Nicodemus, a good biblical name. He was a dedicated Communist. I reminded him that the Communists had eliminated 1,500 Sakha shamans throughout the Republic to stamp out their religion. He still backed them for doing what they had to do. He asked if we had any videos about God, so I loaned him our set of Moody science films in Russian.

The Four Gospels in Sakha

The four gospels of the New Testament were first printed in the Sakha language in 1898. The Sakha Outdoor Museum in Sotinski displayed an original copy under glass. We were there with a Sakha Christian lady and asked her to read it to us, but she was unable to do so in the old-style alphabet. The current Sakha alphabet was established in 1939 based on the Cyrillic Russian script, which added five letters that are distinctly theirs.

The Welcome Dinner: Kay

Bill and Greta flew from Khabarovsk to participate in the dedication of *The Life of Jesus Christ*. Before the celebration, we gathered at the mission office. I assisted Anna in preparing cabbage salads, fried fish, borscht, rolls, cabbage leaves filled with hamburger, thin waffles rolled like crepes, and pizza.

After feasting, they sang several songs accompanied by a guitar. Then we Americans sang songs for them and shared how we came to know the Lord.

The Book Is Here: David

The Life of Jesus Christ in the Sakha language was printed in Japan. The shipment of 50,000 copies filled a large metal container which kept them safe and dry. I signed official documents for InterAct so they could get to Yakutsk. I also signed a waiver giving a pastor in Neryungri the right to accept and transfer the container to a trucking company. All the paperwork had to be completed with the official stamps before distributing the books.

Foreign and local believers were praying for these books. The colors were vivid, and the handwritten text reflected excellent penmanship. God provided 70,000 dollars for the printing and shipping through numerous generous Christian donors.

Dedication of *The Life of Jesus Christ:* David

Excitement hung in the air for the first-ever Scripture portions in the modern Sakha alphabet. On Saturday, November 19th, the public dedication of *The Life of Jesus Christ* was held near the center of town in the huge drama auditorium of the Children's Palace, which was filled with people.

A hush fell over the crowd as Sergei stepped from behind the curtain and warmly welcomed everyone to this historical event. There were speeches and singing in Sakha and Russia.

The administrator in charge of the facility had asked that no sermons be preached. Instead, several speakers showed strong emotions as they spoke about this first piece of Christian literature in their language. Aita, who translated the book, was deeply moved as she shared the history of the project. Prayers for God's blessing were spoken in Sakha, Russian, and English.

As the bright stage lights blinded my eyes, I spoke in Sakha, "Greetings. My name is David. God loves you." Then I switched to English with Vera as my translator, telling about our first trip to Yakutsk two years ago when our Sakha hosts took our pictures in front of the monument of Manchaari. Every Sakha person knows about their famous "Robin Hood." We want everybody to know about Jesus Christ, who is much greater.

First Sakha Christian Book (1994)

Distribution Followed Dedication

The next step was distribution. The goal of InterAct was to make one copy available for every Sakha household. There were 380,000 Sakha people at that time. Most of them lived within the Republic of Sakha, with a significant number living in Moscow and Saint Petersburg. Ninety percent of the Sakha people spoke their language.

Sakha belongs to the Turkic language family. The Sakha people fled north to this region when Genghis Khan went through Mongolia and northern China 800 years ago.

They were well-educated, outgoing, and took much pride in speaking Sakha. Their language was taught in more than half of the public schools throughout the republic. Many secular books were written in their language along with other media such as newspapers, radio, TV programs, songs, poetry, and dramas. It was the "in" language at the Yakutsk State University.

Everyone who attended the dedication received a book. The missionaries with GTE distributed them through their churches. Copies were given to Sakha officials, libraries, hospitals, and prisons. The National Library in Yakutsk planned to distribute *The Life of Jesus Christ* throughout their 500 libraries within this republic.

Galina and her three children displayed their copies.

Fast-forward: Galina's youngest daughter, Vika, became a missionary in a Sakha town. In 2015, we visited her while she served in Nizhny Bestyakh.

The Day After: David
On Sunday, the day after the dedication, Misha drove to Mokhsogollokh and Pokrovsk with 300 copies. We shared in both Sunday services. Misha had room for only Bill and Greta, our translator, Vera, and myself. The trunk was full of books, and we had more on our laps.

The following day, Misha drove back with more books and Lena, a Sakha singer, to visit the new fellowship in Oktyomtsi, population 2,000. This was also the hometown of Mikhail Nikolayev, the president of the Sakha Republic.

The Administrative Center: David
Misha, Vera, and both of us went on another trip. After stopping at the church's warehouse in Yakutsk and loading several hundred copies, we headed to Mokhsogollokh to get Pastor Valeri and his guitar. The task is easier when you work with believers who want to serve the Lord.

Then we went on to Pokrovsk, the administrative center, with a population of 15,000. Pastor Valeri started the Evangelical Church here shortly after beginning the work in Mokhsogollokh. Twenty-five believers had been baptized, and 40 attended their regular services.

The main administrative building was similar to the county seat in the US. We checked our fur hats and coats in the cloakroom as we entered. Then we met Olya, the official secretary, who was an active Sakha believer. It was good to have Christians serving in high places.

Olya led us to Fyodor, the head administrator for this region. I commented to him about the World War II memorial out front and how the US had sent warplanes through our town of Fairbanks, then to Yakutsk, and ultimately to the war front in western Russia. Now we were helping people in Siberia learn about God.

At 2:00 p.m., we entered the general hall, where 50 government employees were seated around the four sides of the large meeting room. We were overwhelmed! They were all Sakha. A man from the local TV station recorded the event. Pastor Valeri, Misha, and I spoke briefly. The others in our group sang several Christian songs. Then we handed *The Life of Jesus Christ* and other literature to everyone.

Can You Imagine This in America?

Can you imagine such a meeting at your city hall in America? The head administrator, who is not a believer, sets up a formal meeting with 50 employees. Our meeting lasted an hour during the busy time of their day. Immediately after our presentation, one employee repented. If we were bolder for the Lord in America, more Christians would pray for our government workers, and more people would repent.

Then we were hustled into a small side room for an interview that aired on TV. Next, we returned to Mokhsogollokh for their daily two-hour evening church service. How appropriate to close the service by singing all eight verses of "How Great Thou Art!" Consequently, we didn't have time to watch ourselves on TV. The station offered to air Christian programming for free, so I sent videos to Pastor Valeri for them. This was a long, productive day.

Bible Correspondence Courses

GTE had four new missionaries who started a Bible correspondence course, especially for seekers in the Republic of Sakha. When they answered their request, they included this new book.

New-To-Us Apartment

Shortly after the dedication, we moved into a new-to-us, typical concrete apartment on the fourth floor without an elevator near the regional bus depot. We lived there with two new single missionaries in three rooms with adequate space.

Outside our window, across October Street, was the old Nikolskaya Russian Orthodox Church built in 1852. The Communists turned it into a book depository, and the building deteriorated under 70 years of their rule. Now it was being restored to its former glory. We watched Workers apply gold leaf to the roof.

The Mundane Things of Life

A large shipment of toilet paper finally arrived in Yakutsk. Practically every store had it. We were glad because we had brought a roll with us and were rationing it. When we first came to Siberia, the tissue, if available, was rough.

Total Commitment (1994~1995)

The Lada That Kept on Going

We rode with Misha and Vera for a 250-mile trip to Khandyga. The Russian Lada was small but dependable. It depended on prayer and constant mechanical assistance. It took us 14 hours of travel over rough roads on which we would not have wanted to take our car. We had only one flat tire and several exercise breaks to push the car out of the snow. Parts of the roads were one lane, so when we met a vehicle, one of us would usually get stuck as we tried to let the other pass. Misha drove faster than we would on these roads, hitting one big bump that surprised everyone. Later, Misha discovered that the car frame had bent. After every trip, he would spend a day or more repairing the vehicle.

We distributed books everywhere, praying for open hearts all along the way. Roads were poorly marked, and we were traveling in dense ice fog since it was negative 45 degrees. We missed one turn and continued straight. The usual thing to do was to stop and ask somebody for directions. God led us to share the Sakha books in a village off the regular route, where Christians would not normally go. We thanked God, backtracked, and continued on our way.

Poster People

We noticed that the posters advertising our upcoming meetings included mention that Americans would be there. We both felt uneasy about this. However, our Russian and Sakha partners assured us it was a drawing card because people wanted to see real Americans during our early years in Siberia.

An Active Church

We rode with Misha on another 10-day trip to Khandyga with a population of 9,000 people. The city was predominantly Russian, but the surrounding area was Sakha. The town was heated with coal. The locals stated, "The snow is black, but our homes are warm." Unfortunately, all the coal dust presented serious health problems.

Khandyga had a strong evangelical church started by Light of the Gospel missionaries in 1990 with 80 members. Their permanent meeting place was in a wing of the public library. A Christian meeting was held nightly, and they were actively reaching out to every settlement in their region.

A Sakha Church Planted: David

God mightily used one middle-aged Sakha couple, Valeri and Zoya, to plant a church in their hometown of Krest-Khaljai, population 2,000. They both repented the previous year through the witness of believers from Khandyga who continued to disciple them. Fifty people attended this growing church which met twice a week in the auditorium of the music department.

Our team met at their public school, where 150 attentive students packed into the largest classroom. Others were in the hall just outside the door. We shared for more than an hour and handed out literature.

Valeri and Zoya's total commitment to the Lord was evident in their lives, their family, and the Lord's work going on in their hometown. On December 14th, we participated in a public meeting in their House of Culture with 100 adults listening in rapt attention. The heat was set on "super economical," so you could even see your breath. Everybody wore their heavy fur coats and boots for the whole two and a half hours. Colder than minus 40 degrees outside did not stop anything.

The Christians wanted Valeri to be ordained as their pastor. So we met with 40 Christians, as well as those very interested in spiritual things, in a separate, warmer room for another hour and a half to sing, pray, and explain the ordination process. Valeri questioned if he should become the pastor since not all of his six children were believers. We replied that he was responsible only for those living at home since the oldest lived elsewhere.

Both Valeri and Zoya kneeled and prayed. Then Misha, Nikolai, Dimitri, and I laid hands on him as we prayed and set him apart to become their pastor. Valeri had been the only man in this church, but things changed as they reached out and added more believers. Late that night, we returned to their apartment for more tea and lots of food.

Sakha Hymns

Several Sakha Christians translated hymns into their language and gave us copies. They knew we had worked closely with Native people in Alaska to produce literature. God used these early songs to prompt us to install five special Sakha letter fonts on our computer. Soon we printed their songs for initial testing and sharing with more people. We repeated this process many times. These were the first steps toward producing a standard hymnal in their language, a huge project that continued until we returned to Alaska.

E-mail Improving

In May, we set up an e-mail account through a local system. Communication continued to be frustrating but was improving. Our initial contact was with Dave and Mary Ellen Wurm in Fairbanks, who passed our information on to our children. Our e-mails took two hours to be sent through the system to Fairbanks!

Phone Problems: Kay

We dialed directly from our apartment to my parents in Michigan, paying five dollars and 52 cents per minute. The cost was two dollars and 72 cents per minute to call us from Alaska.

It was difficult to hear. On one occasion, I kept saying *Allo, Allo* (Hello, Hello). I thought I heard a familiar word, so I asked, "Do you speak English?" My dad responded, "Kay, it's your father. I speak English."

There continued to be problems with the telephone in our apartment. Other areas of town were without any phone service for a month or two. Sometimes our phone would ring in another person's apartment.

American Christmas Celebration

A Christmas celebration at our apartment with 28 Sakha, Russian, and American friends on December 25th was dubbed "American Christmas." It was a great time of fellowshipping, singing, and sharing. If more came, we would have made room—like they did on our city buses!

1995
Teaching English: Kay

At the School of International Languages, I taught English to 50 Sakha high school students from outlying villages during a special one-week training session. I had seven classes with different age groups. Since many American holidays on our calendar are based on our Judeo-Christian heritage, it was a good opportunity to share our culture and the influence of Christianity on America.

Private Russian Lessons: David

A church member coached me and included Kay in the sessions. Galina was a good Russian teacher. Her goal was for me to preach without a translator. Kay helped her granddaughter with English. Galina encouraged Kay to write a letter to our grandchildren, as shown in Appendix B.

We were also thankful for Vera, who met with us three days a week for Russian language sessions. We frequently needed to discuss cultural issues and local news.

Ice in Our Entryway: Kay

Cracked water lines were common in Yakutsk, especially under buildings. Leaks were stopped when they reached a critical point. An iron pipe in the only entrance to our concrete apartment building had sprung a leak. The cold floor turned the water into solid ice. People kept adding boards for a walkway on top of the ice and water to keep our winter boots dry. I would walk on the narrow board behind David holding onto his coat. With a foot of ice underneath, we had to duck our heads as we walked in or out! Fortunately, the two inner wooden doors in the entryway were frozen open. The exterior steel door was still usable, and our apartment was warm.

Russian Scrabble

Our children gave us a Russian Scrabble game with Cyrillic letters. It was very challenging. We even bent the rules for ourselves by using the dictionary if we had a word in mind.

Eight Days Up North

We flew with Vera on an eight-day trip to Kular, close to the Arctic Ocean. Our direct flight from Yakutsk was on a twin-prop plane that could hold 40 people and took three hours.

After landing, we discovered why we needed our visas officially stamped to buy our tickets and board the airplane. Everybody exited the plane while we waited for a special agent to check our visas and passports. She told us, "If your documents were not in order, you would have had to wait on the plane and go directly back to Yakutsk on the return flight."

We assisted Sergei and Sveta, who served with GTE. Sergei was the first believer in this region, and he became their missionary pastor. Their family had been broken and lived elsewhere, while Sergei worked in Kular for five years. In April 1993, Sergei accepted the Lord Jesus into his life, and then both his wife, Sveta, and their daughter did. The Lord brought them back together as a young Christian family in the fall of 1993.

The Kular Airport was 12 miles from the town and farther from the bigger city of Severni. Thirteen people had been baptized in these towns, and they planned to baptize six more. Fifteen attended the regular Saturday meetings in Severni in their House of Culture. Since Vera was with us, we met in the "disco" hall, where she played the piano. Each week, they had meetings in two homes and a Sunday Shool class with 15 children.

Witnessing in the Area: David

We spoke with the Severni Hospital staff and toured their facility. They showed us two birthing rooms, one for live deliveries and the other for abortions. We shared freely with a high school class and the staff. One girl muttered that it didn't matter whether her soul went to heaven or hell and left the room. Then the principal said, "There's a shy girl here who has a question." Her question: "How can I receive Jesus Christ?" Later, Dima, a new believer in Severni, asked me to pray and dedicate their 15-day-old baby, Katya, to the Lord as he took a video.

Where's the Gold?

The government was demolishing the whole town of Kular since they discovered a vein of yellow underneath. The gold in the area meant tighter restrictions on flights to their region. Consequently, the population of 7,000 dropped to 100! The number of people in Severni had declined to 2,000. Russians and Ukrainians were moving out of the area, but more Sakha people moved in from outlying areas.

Helicopter Ride to Khayyr: David

The Kular Airport served eight surrounding towns. Sergei asked if we would like a helicopter ride to the reindeer-herding village of Khayyr.

Sveta, Vera, Dasha, and Kay before boarding the large Aeroflot helicopter which could hold 20 people.

We attempted to put on our seatbelts, but they had not been used for years! Mine was jammed under the grime and reindeer hair. We made it a tourist flight as we moved about, peering out different windows, and even spotting a herd of 500 reindeer.

Upon landing, we distributed copies of *The Life of Jesus Christ* as we were greeted by 50 children and many adults. They loaded several sick people and threw in a whole frozen reindeer carcass. Then we realized this was a medical emergency trip. They had not flown for over a month due to serious financial constraints in the country. The reindeer was for feeding relatives. I gave our pilots American postcards and Christian literature as I thanked them.

A Unique Practice: David

Sergei and Sveta started a unique practice. Every night at 10:00 p.m. they and others in the church would stop what they were doing and pray for people in their churches and villages. Many people in Ukraine followed this practice. We prayed with them every night in Kular, except for two when we had unbelieving guests. Praying is so essential.

A Sakha City of 10,000: David

The churches in Krest-Khaljai and Khandyga actively continued their outreach to the needy people around them. Misha coordinated the outreach with them. He took us to the Sakha city of Ytyk-Kyuyol, which is five hours from Yakutsk. Fifteen members of the Khandyga church choir sang in Russian. Pastor Valeri, Zoya, and a lady from Krest-Khaljai were also there.

There were no believers in Ytyk-Kyuyol, a town of 10,000 Sakha people. We usually bought gas here on our way to Krest-Khaljai, but we always prayed for this town and many others we passed while traveling. We met in their House of Culture on Sunday at 11:00 a.m. Two hundred people came to the two-and-a-half-hour service. Pastor Valeri preached in Sakha, and then the three from Krest-Khaljai sang hymns they had translated into Sakha and shared their testimonies.

Pastor Valeri informed us that more people had repented in his hometown, and now 60 were attending their meetings. They continued reaching out to the surrounding villages. Ministry in Yakutia was like living in a modern-day book of Acts. Great interest in the gospel continued with people asking many questions.

We arrived home at 4:00 a.m. Misha liked to drive at night as he put the pedal to the metal.

Getting Dollars: David

We had to be creative to get dollars, which I exchanged at a local bank for rubles. When we traveled from Alaska, we carried enough money for our expenses to live on for several months. Later, I would order five pictures of Ben to be sent in an envelope with our hand-carried mail. Ben referred to Benjamin Franklin's face on a 100-dollar bill. They needed to be crisp, with no tears and dated 1990 or newer. A warm iron smoothed wrinkled bills. Russian banks were fussy.

Surviving Siberia (1996)

Some Background

Last year, we flew from Yakutsk to Magadan, where we cleared Russian customs for our return to Alaska. A missionary friend traveling with us planned to catch a connecting Alaska Airlines flight to Khabarovsk.

The flight was two hours late, so our friend asked us to pray that she would make her connection. As we landed, the Alaska Airlines plane was on the tarmac with its loading door closed and its engines revved up for takeoff. She ran out and kneeled in a praying position on the tarmac facing the plane.

Outside the terminal, we stood watching. Finally, the entry door dropped down under the tail section, and our friend ran up the steps into the plane. The plane retracted the ramp and took off. We breathed a prayer of thanks to God as we were ushered inside by security.

At the Alaska Airlines counter, we thanked the agent for letting her board. He snapped, "She did not get on the plane! The man in charge said they would not take her." At that moment, the pilot called in, "I just picked up a passenger on the tarmac. I thought she was an American."

The Rest of the Story

Now to connect the story pieces. Last November, Bill and Greta were flying from Khabarovsk to Yakutsk. They met a pilot who spoke to them in broken English on their flight. He gave them his address in Yakutsk. Bill and Greta didn't have time to visit the pilot at home, so we did.

We looked for the pilot's place, but the weather was frigid with heavy ice fog. We tried another day and were warmly welcomed into their apartment. He wanted to talk with Americans and improve his English. They were leaving soon for a long vacation in Ukraine. The conversation turned to spiritual things, and we gave him a small booklet about God.

Because of his interest in spiritual things, we wanted to give him a New Testament, but we didn't have one with us. That evening when the doorbell rang, he was at our door. He asked questions about the booklet we had given him. We handed him a New Testament in English and Russian.

With His Parents in Ukraine

Several weeks later, we received a letter from Ukraine with the sad news that his father had passed away three days after their arrival. He wrote that he had been reading the New Testament every day with his family. It was a comfort to them during their time of sorrow. He wanted to know what happened to his father's soul and where was he now. What an opportunity and challenge to write briefly, clearly, and in simplified English to answer hard questions.

The Revealing

We visited each other several times after he returned from Ukraine. During one visit, he related an experience in Magadan when a young lady wanted to catch another plane. The head pilot didn't want to help foreigners, but he was the co-pilot and wanted to assist her. He radioed Alaska Airlines and asked them to help her. The Russian head pilot disgustingly asked, "What are you doing?" He responded, "I was practicing my English." You can imagine our excitement as we connected the dots and said, "That was our friend!" God had used him to answer her prayer. This was another step in his journey to Jesus Christ.

Fast-forward: Both the pilot and his wife accepted Jesus Christ as their Lord and Savior and are now living in Israel.

From My Diary: David

Occasionally, I kept a diary; it included these entries for a five-day ministry trip:

Friday: Misha T. and Victor picked us up at 8:00 a.m., left Yakutsk, and crossed the Lena River on the ice. Stopped at a GAI checkpoint (State Automobile Inspection site) near Amga, where all vehicles are checked for proof of ownership and vehicle registration. Misha gave them *The Life of Jesus Christ*.

We stopped in many Sakha villages to post signs about the Bible correspondence courses and distribute books to everyone in sight. A Sakha man was riding his horse near the road. Misha stopped, and the man came over. Misha handed him *The Life of Jesus Christ,* which he studied in amazement.

We arrived in Onnyos, a sleepy Sakha village of 700 people. Seven believers emerged out of nowhere and immediately gathered for tea and fellowship. Their leader, Vladimir, accepted the Lord while in prison, where he heard the gospel. Their house church met three times a week.

We left more books with them. Yuri, who was partially paralyzed, came with us. Misha's little Lada was crowded, but he was glad to help a brother in Christ. Gassed up in Amga. Their roads had many bumps, and we felt every one. Today we passed more villages and towns, but none have a church except little Onnyos. We Met another Misha and Lyuda with GTE in Ust-Maya. More tea, then to bed at 1:00 a.m.

Saturday: Traveled to Kyuptsi with both Mishas and Yuri on the winter road through thick woods and the frozen Aldan River for two hours. Three believers lived in this town of 5,500, but they were not meeting together. We need more Christian workers!

We gave books to people on the street. Posted signs about the Bible Correspondence Course on the outside bulletin board of the library and entered the building with a case of books. The librarian was pleased. People kept coming for a free book. Two Sakha men had a serious discussion with Misha about black and white magic powers. One man who received a book tried for another, but the librarian physically retrieved it. I gave the librarian a thumbs-up sign as he smiled back.

We dropped Yuri off with his sister and declined tea to save time. Crossed the Aldan River to Eldikan, population 6,000, with three or four believers but no missionaries. More help is needed!

Returned to Ust-Maya, and went to a one-room Sakha home measuring 12 feet by 16 feet. Great time for sharing and eating. Twenty of us crowded together. Met Marina, a Sakha believer and English teacher fluent in Sakha, Russian, and English. They discussed a passage of Scripture and asked for my comments. Marina translated for me and was delighted to do so. We prayed about getting into their public school on Monday, but the principal did not want us. Marina was the first local believer. Her 16-year-old son was a believer, but her parents were Communists. Marina searched for real peace and found it in Jesus Christ.

Other believers in Ust-Maya had repented in Khandyga through the ministry of GTE 215 miles away. Misha, originally from Khandyga, married a missionary, joined GTE, and moved to Ust-Maya in 1991.

Banya (steam bath) time. The wife of the man in charge of the road system was a believer. We enjoyed their banya plus another big dinner. We heard that the public bus to Amga had been discontinued due to bad road conditions—we understood why, but it also hindered gospel trips.

Sunday: Walked across town for the church service. The only church in the city of 4,500 people. Thirty-five of us met in a large room in Nikolai's house. Six of them were Sakha. We both spoke during the service. Sunday school met at 1:00 p.m. with 15 children. We spoke without a translator. One child commented, "We had real Americans in Sunday school today!"

We returned home to rest a little. More than 30 people came to ask spiritual questions, sing, and hear Misha T. preach again. Ate at 11:00 p.m. and prayed again about going to the public school.

Monday: Day started slowly for us as both Mishas went to Kyuptsi to pick up Yuri.

We went to the public school at 2:00 p.m. and met in a large classroom with two classes of enthusiastic students eager to listen. After five minutes, one unhappy teacher charged in and marched out with her students. We continued to share freely as we had done in many other schools. The bell rang, and our translator had to leave for an appointment, but we continued without her. All the students wanted to stay after school to hear more—so we stayed. One Sakha girl asked, "Can you help me? I'm bothered by evil spirits in my bedroom." Jesus Christ is stronger than all evil spirits, and we need to turn to Him for protection.

That evening, 35 people came to Misha and Lyuda's house. A newspaper reporter from Kyuptsi interviewed us and asked questions. Marina translated for us in Sakha.

Misha started preaching. In the middle of his message, Lyuda entered the living room and quietly informed Misha, "The police want to talk to you in the kitchen." Misha went out and talked with the two uniformed officers. Then they called for us Americans. They wanted to see our passports and visas.

I returned to the living room to get our documents. Everybody in the living room was either kneeling or standing as they quietly and fervently prayed. They knew exactly what to do and Who to turn to for help in times of need! They understood the situation better than we did.

Our documents were all in order. The officers were satisfied when Misha informed them we would leave their region tomorrow at 5:00 a.m.

Who told the police? I thought the school principal did. Kay thought the reporter did. Who do you think reported us? This was another vivid reminder that our days in Siberia were numbered, and the door of opportunity may swing shut again. Our goal was to do all we could to spread the gospel of Jesus Christ while the door was still open.

Tuesday: We left at 5:00 a.m. for Onnyos to take Yuri home. We arrived in time for an early supper. Planned for our evening meeting in their House of Culture. First-ever public Christian meeting in this village with 40 people. First time I shared publicly without a translator. Misha filled in the gaps or added necessary information. We distributed more of *The Life of Jesus Christ*. One unbelieving man came with us to the house and asked more questions. Ate another big meal followed by a time of singing hymns.

Prayed at 10:00 p.m. and started back to Yakutsk. An hour out of Onnyos, we landed in a snowbank for an unwanted "exercise break." After an hour, we were on the road again. We prayed, "Lord, give Misha physical and spiritual strength." Arrived home dead tired at 5:00 a.m.

Wake UP!: David

At one home prayer meeting, we all kneeled to pray. Pretty soon, I jerked to attention as I heard myself starting to snore. A Sakha believer looked at me as we both flashed an understanding smile and bowed our heads again. Since then, I have asked the Lord to help me stay focused while praying. Jesus also had to remind His disciples to stay awake and pray.

Easter Celebration: David

We spent Easter in Mokhsogollokh. The church was packed with 175 people. Their regular Sundays had 75. Many sat up front, and a number had to stand in the back of the church. People came from Pokrovsk and Oktyomtsi to celebrate together. We noticed a significant increase in the number of Sakha people attending.

A Sakha lady, Rhaya, started a Sunday school class this winter for junior high students and brought 21 of them on a chartered bus to church. Rhaya accepted the Lord two years ago and was growing spiritually.

She and another lady had stayed with us in Yakutsk last month. She showed us her notebook with lists of members and those baptized. The book included her long prayer list. Starting at 5:00 a.m. each day, she prayed for several hours. That would be a good list to be on! Her church planned to baptize 13 people in Pokrovsk as soon as the river opened up.

The newspaper *The Lena River Lighthouse* highlighted an Easter article on the front page about Pastor Valeri with GTE and their work which started in 1990.

On Easter, Pastor Valeri greeted the congregation in Russian with "Christ is risen!" The audience responded, "He is risen, indeed!" This was based on "He is not here, but is risen!" (Luke 24:6) and "The Lord is risen indeed!" (Luke 24:34).

Russian Calendars

It was hard to get used to the Russian calendars. Monday was the first day of the week, honoring work. Their word воскресение (Sunday) means "Resurrection" (of Jesus). We liked the meaning of that special day, but most Russians lived as though it was just another day.

Not Telling Too Much

E-mails were like giant postcards that could be read. Our view was that encrypting messages would raise a red flag, challenging a person to look at them. Instead, we limited our news flow by not telling everything we did or knew, especially concerning the number of baptisms or what certain people did.

Where Was Our Church Meeting?

The meeting place for the Evangelical Church in Yakutsk kept changing. One Sunday, while riding the city bus to the service, we spotted our group in front of the State Theater of Opera and Ballet. We jumped off the bus at the next stop and joined them. Our part was to encourage Sakha and Russian believers to follow the Lord, with or without buildings. We recommended home churches scattered around the city, which would increase the number of preachers and leaders. But the Russian believers saw things differently. They wanted everybody to meet together.

Cooperating with the University: David

Yakutsk State University cooperated with the University of Alaska Fairbanks (UAF). They offered an intensive eight-week course, "Summer in Siberia," which highlighted the Russian and Sakha cultures. Dr. Stephen MacLean, from UAF, wondered if we would complete the entire course at our age. We told him, "Yes, we are dedicated to continuing to live and serve here." We participated as exchange students from Alaska.

Nine of us met five hours a day, five days a week. It was a unique experience taking a course with American college students. Several girls wore shorts in class. One did not sit in a lady-like position, which did not fit the Yakutsk college culture and greatly upset the teacher.

Summer Solstice—Sakha Style

The annual Sakha celebration of the summer solstice was very elaborate and a national holiday in the Republic of Sakha. Most of the formal activities took place in a large area with a horse racetrack.

The event began with a shaman dedicating the event to the sun and nature, which they considered to be alive or animated with a spirit. Unfortunately, they honored and worshipped creation rather than their Creator. As the Bible says: "who exchanged the truth of God for the lie, and worshiped and served the creature rather than the Creator, who is blessed forever. Amen" (Romans 1:25). Several small smudge fires burned where people could waft the "cleansing" smoke onto themselves and drive evil spirits away.

Everybody was dressed in their finest national costumes for this joyous time. Many dance ensembles were dressed alike and wore silver (white gold) jewelry. Other events included various sporting competitions, horse racing, national clothing displays, drinking fermented mare's milk from wooden goblets, playing mouth harps, and remembering

World War II homecomings when many people did not return. Near the end, several large circles were formed with invitational dances of more than 100 rotating clockwise, singing their traditional chants.

More Workers

Our new Russian field director, Mike Matthews, and his wife, Kathleen with their boys moved here to direct the ministry. Rhaylene Abbey, who had served in Kodiak, Alaska, came for three months. Yasue Ibaraki, a single Japanese missionary, also worked in Yakutsk. Bill and Robin Harris and their children moved to Yakutsk and temporarily stayed with us.

Amga English Camp: Kay

We rode the commercial bus to Amga to teach at a special English camp for a week. They were excited to have real American teachers speaking English. Many schools and higher levels of education taught the "proper" British English, but most people wanted the freer American style.

The 44 Sakha students were in junior and senior high. We taught four classes each day. The students were divided according to their comprehension of English. I had the two highest groups in the morning, while David had them in the afternoon. We played games like Uno, Memory, and a variation of Scrabble during our free time. We also had good opportunities to share our faith.

Rear-Ended: David

I drove Misha's little Lada on a busy four-lane main road. While stopped at a red light, I saw a medium-sized truck coming up behind me in the rearview mirror. He didn't even try to stop! Fortunately, a Russian friend riding with me jumped out and got the offending driver's passport. He held on to his passport until the police arrived. Misha's trunk wouldn't open, but they worked out a deal between themselves. I had a sore neck from the whiplash but recovered after several days.

Wet Cement

The lives of young children are shaped early, affecting their entire lifetime. They copy others, including us, which creates lasting impressions. This is similar to molding wet cement which hardens as shaped.

While we were in Siberia, our children had a family picnic on July 8th, 1995, at our house. Dan poured cement for our garage floor and saved enough for the sidewalk. Elizabeth coordinated the project. Each grandchild pressed one of their small hands into the wet cement and signed their name. It was unique and a "forever" memory for them and us.

Everyone notices the 12 handprints by our door. Younger children put their hands into the molds and compare sizes. We want to see all of our grandchildren grow and be shaped to do what is right in God's eyes. Parents have the most influence on their lives, but grandparents do, too. "Train up a child in the way he should go, And when he is old he will not depart from it" (Proverbs 22:6).

Continuing to Grow

The church continued to grow. We rode the bus to Mokhsogollokh for their fifth anniversary. Inside the building, there was a tree with paper fruit listing the names of people who had come to know the Lord. Seventy pieces of fruit were hung on the tree!

Some fruit was still immature, others had moved away or even fallen off, but most were still on the tree and maturing. No other church was within 60 miles of this populous area when they began. The active Mokhsogollokh congregation started daughter churches in Pokrovsk and Oktyomtsi, plus they reached out to Bestyakh and Chupayeva in their region.

To Sinsk and Beyond

After traveling with Pastor Valeri and several others for five hours, we arrived at the village of Sinsk. When the road ended, we followed the shoreline along the Lena River, which was smooth most of the way. We bounced along in a Russian UAZ, a four-door military vehicle manufactured in Russia during World War II.

Two ladies from Sinsk were baptized in the frigid river. They did not waste any time in the water. These were the first baptisms in this village. They were wrapped in blankets after emerging from the water as we sang two hymns and prayed over each of them. Later that afternoon, their first snowfall of the season began. An unfinished 20-by-24-foot cabin was donated for a church. The obvious need was for a spiritual leader or missionary pastor.

The following day, we drove another hour along the shoreline to the Sakha village of Yedey. Students in the public school heard the gospel for the first time.

Official Stamp Needed: David

We looked forward to Christmas in Alaska. Our baggage was loaded onto our Aeroflot flight in Khabarovsk, but we could not board. Kay's exit visa was not stamped; only the exit date was written down! The officials told us, "David can go, but Kay has to stay here." We were shocked! My immediate response was, "We'll both stay here." Then we desperately took off to a customs office in town to get the visa stamped, but no official would help us.

We stayed with our former landlady, praying and wondering what to do. She insisted on taking us around town over the next three days to see certain officials, such as the military, police, and government agencies. Every official knew about our situation in advance, but no one would help. Our landlady told us, "You need Russian experiences so you will understand. This is normal here."

At one point, her heart bothered her, so she popped a nitro pill and cured that problem on the spot as we prayed. We passed a building where 700 Russians had been gunned down for refusing communism right after the 1917 revolution.

I flew back to Yakutsk and visited the visa office as soon as it opened. The lady was apologetic, realized the mistake, and immediately stamped Kay's exit visa. I sincerely thanked her but did not relay any of our frustrations, extra plane tickets, expenses, or the "lost" 10 days. We learned the importance of having one's documents correct in Russia. I took the next flight back to Khabarovsk.

Be Still, and Know: Kay

While David went back to Yakutsk for the exit stamp, I went on a gospel trip with our landlady's church to another village. She explained how to eat bread and sausage during hard times. She put a round of sausage on a piece of bread but kept moving the meat away from her mouth for the next couple of bites. That was how she stretched her prosperity and ate the sausage I bought for her.

I noticed the Russian poster we had left hanging on her wall: "Be still, and know that I am God" (Psalm 46:10a). The Lord used this time to teach us this verse in a more meaningful way, like when we say, "Lord willing, I will do so and so."

Visiting My Parents: Kay

Shortly after returning to America, I visited my parents in Michigan. My father was recovering from back surgery and was thus unable to help my partially paralyzed mother. Uncle Ed and Aunt Leola lived in the same town and spent a lot of time helping them. Elizabeth flew out to assist, and we celebrated Thanksgiving dinner together.

Fun in Alaska

Kara was in a choir concert and looked nice dressed in a black blouse and floral skirt. There was a picture of Suzanna in the newspaper with a school dignitary fixing her scarf before she went out to recess. We took in the school programs for Elizabeth's and Steve's children.

Slides of Siberia were shown at the Fairbanks Native Bible Church. The people were interested and asked questions. They gave us gifts to pass out in Siberia.

Twelve under Twelve

Twelve grandchildren under 12 years old. Back row: Kay Henry, Melissa Cogan, Andrea Cogan, Cheryl Cogan, Tim Henry, David Henry; middle row: Jon Henry. Kara Henry, Danielle Rathbun, Curtis Henry; front row: Joshua Rathbun, Milo Cogan, Wes Henry, and Suzanna Henry

Christmas with Our Family

We thoroughly enjoyed celebrating at our house with all of our children. The grandchildren were growing and seemed more active than usual. This made us wish we could come home often to be with them.

Africa Will Be Warm

Our younger son, Steve, and his family were preparing to go to Ethiopia in June. Wes, their middle son, said, "We're five missionaries going to Africa." Children can help advance the cause of Christ.

They served with Mission: Moving Mountains in community development and discipleship. Steve was a civil engineer with the State of Alaska and would use those skills and others to meet the spiritual needs in Ethiopia.

1996
Back to Siberia

On January 16th, we left Fairbanks. Siberia would be our home for a year or more, with our visas needing to be renewed every three months. This time, Alaska Airlines flew eight hours from Anchorage to Khabarovsk via Magadan. Arriving in Yakutsk, with thick ice fog and negative 40 degrees, we were welcomed back to a "three layers of clothing" day.

Staying Warm in Our Apartment

The central steam heating system was cheaper to operate in the big city, but all the apartment buildings with thick concrete walls and loose-fitting windows like ours were hard to heat. They controlled our temperature, so we had no thermostats. We wore our sweaters indoors. Our living room registered only 57 degrees, but the bedroom and kitchen were warmer. It was minus 55 outside!

Dealing with Glasses: Kay

Out on the street, people put their mittens over their noses to prevent frostbite. Many people preferred this method rather than wrapping a scarf around one's face as little children did. If I took my glasses off, I couldn't read the numbers on the buses. David often removed his glasses while out in the cold so he didn't have to defrost them before coming indoors.

Our *Khrushchyovka*

All the buildings around ours were the same unpainted, concrete-grey color. *Khrushchyovka* was the unofficial name for our type of low-cost five-story apartment building. They were developed during the early 1960s when Nikita Khrushchev was head of the Soviet Union. There were no elevators since the building was under six stories. Every day, we got our exercise by walking up and down five flights of uneven stairs.

The floor plan was configured the same way, with four dwellings on each level. Each section had 20 apartments and included a series of three entry doors to keep out the cold. Our building complex housed 80 living units which butted up to the next building complex.

Close Neighbors: David

Our apartment and our neighbors on the fifth floor faced the front of the building, while the other two apartments on our floor faced the opposite direction. One time, our neighbors forgot their key. The husband asked if he could go out on our balcony and crawl over to theirs to get in. I sometimes took chances but did not plan to take that one. I told him to go ahead but to be very careful. Five stories down would be a long way to fall!

A thin concrete wall separated our apartments, so we could hear which channel our neighbors were tuned into on their TV. We knew when things got a little rowdy but kept such information to ourselves. This cautioned us that they could also hear us. Fortunately, they were good neighbors.

"Now You Know How We Live!": David

The lights in our hallway were very dim or completely out. On the fourth floor, I bumped into a neighbor in the dark. She spouted, "Now you know how we live!" We only knew her by face, not by name. But she and everybody in our apartment knew we were the Americans on the fifth floor. She was merely expressing her frustration at their living conditions. The electricity was not on continuously due to aging equipment and lack of money for repairs.

When our water stopped running, I followed others with buckets or containers to one small spigot out by the street. I waited for my turn to draw water. We rode the same overcrowded buses. We stood in the same food lines. We experienced everyday life in Siberia.

People worked at their jobs but with diminished interest. It was essential to pay into their "Social Security" or pension fund. Hopefully, the government will have money to pay out in the future.

Mystery of the Missing Meter: David
Our electricity was off one morning in January. The apartments across the courtyard had lights. We opened our entry door and the hall lights were on. Looking at our main electric breakers in the hallway, I realized our meter was missing! What was going on? My first guess was that the electric company took it for not paying our bill. I didn't have the official electric payment book, so I couldn't make a payment and get it credited to our apartment.

Vera was coming over soon, so we waited in the dark until she arrived. She called the electric company, but they did not take it. Somebody stole our meter during the night. The company sent an employee to hotwire it. The following day, they sent another person with a new meter.

Changes and Uncertainties
Changes and uncertainties are the norm in Russia. School teachers, doctors, store clerks, and other employees were months behind in receiving their pay. The Communists made more gains in the last election. There was much uneasiness about the upcoming presidential elections in June, the first in post-Soviet times. We prayed that the government would continue to be open to Christians.

One woman expressed genuine concern to us about the Communists winning. She was a former party member and knew the system. Several people told her, "Remember what we did to Christians like you in the past!" But we advised her, "Continue living for the Lord, expect more restrictions, trust God, and leave the consequences up to Him."

Buying Food: David
The *Krestyanskiy Rynok* (Peasant's Market) had indoor and outdoor vendors, no matter what the temperature was. It was only two bus stops away. Our market had almost everything inside. A few people were

always anxious to sell you frozen fish or meat outside, even at minus 40 degrees!

Sakha ladies dressed in heavy clothes also kept moving to stay warm.

They wore gloves when they gave me the change. I could buy fish or vegetables inside, but I wanted to reward these frozen workers for their labor.

Meat Eaters "Heaven"

All kinds of meat were available at the *rynok*, especially horse and beef. They preferred the tender meat of a foal. Other meats included reindeer, pork, chicken, and whole rabbits. Occasionally, we found geese, ducks, and ptarmigans.

"Bush Legs"

In 1990, US President George H. W. Bush signed a formal trade agreement with USSR President Mikhail Gorbachev for frozen chicken hindquarters to be shipped to Russia. Thousands of "Bush legs" were sold all over Russia. They became trendy and enjoyed by all. They thought Americans wanted to save the white meat for themselves. We bought Tyson chicken hindquarters from the USA by the case and stored them on our "outdoor freezer" (our balcony).

Grandchildren in AWANA

It was encouraging to receive a photo of our grandchildren at an award program with AWANA.

Joshua wasn't quite old enough to get a Cubbie's vest like Curtis and Danielle wore.

Curtis Henry, Joshua Rathbun, Jon Henry; Milo Cogan, Wes Henry, Cheryl Cogan, Danielle Rathbun, and Andrea Cogan received awards.

Baptisms at the Second Church

On Sunday, we witnessed 13 people at the Second Evangelical Church being baptized by Misha. Eight of them were Sakha. An indoor pool was used, so the water was pleasant. More than 50 people attended.

1040 Tax Forms

We gathered additional information to fill out our 1040 tax forms, which weren't available until after leaving Alaska. This was awkward, but we told our daughter where to find the information. Our income was low when you factored in our extensive ministry and travel expenses, but submitting the forms kept Uncle Sam happy.

K-X for a Month: Kay

We traveled to Krest-Khaljai (K-X) for a month. Valeri and Zoya insisted that we stay in their bedroom while they and their four younger children slept in the living room, which had a hide-a-bed and two chairs that folded out into beds. We all got along well. They had a 16-year-old daughter and three sons, ages 13, 8, and 3. They asked many spiritual questions, which I did my best to answer.

Olga, Zoya, Ivan, Valeri, Anton, and Andrei are in their living room.

Their older son, Andrei, had a birthday. Zoya asked me to make a cake using sweetened condensed milk and to just add a few things. There was no recipe.

Zoya taught English in their public school, so I helped prepare the meals. We also assisted in Zoya's English class. The students asked many questions like, "Can you prove there is a God?" We answered that their ancestors' word for "nature" showed that they believed in a Creator God. People also saw and lived with Jesus Christ, who claimed to be God, 2,000 years ago. We believe what they wrote about Him in the Bible.

We spoke six times about Indians, Alaska, and the Lord in their public school. We shared about Jesus on their local TV station, in their hospital, and with the director of their collective farm.

We enjoyed participating in their church services. They rented a spacious room in the Music School plus two smaller rooms for Sunday School. The adult attendance was 20.

Eating Well: Kay

The Sakha diet was primarily meat-based, like the Native diet in Alaska. Along with beef and horsemeat, we sampled moose, but there was little wild game. A Sakha breakfast was extra special with a bowl of fresh whipped cream mixed with wild berries. We broke the bread into chunks, put it on a fork, and dipped the piece into the whipped cream. We didn't count calories! I cranked their cream separator to separate the raw milk from the rich cream. They also made cottage cheese, so we ate many fresh dairy products. Three cows were kept in a small Sakha-style barn with slanted walls overlaid with cow manure for extra insulation.

The Fear Factor

Fear was a major problem. Afraid to turn to the Lord. Afraid to follow the Lord. Last year two women repented and then became mentally "disturbed." They turned back to shamans who "healed" them and warned them to stay away from Christians. The world of evil spirits is alive and active. It was similar to the contest between the Egyptian magicians and God's prophets, Moses and Aaron.

These people had been held in darkness for many years by their religion of animism. This system included the belief that spirits were animated or alive in everything in their world (trees, mountains, storms, animals), and only a shaman could deal with them. The whole system was built on fear. People have also been indoctrinated into communism, another belief system based on fear.

Many Russians were involved in extrasensory perception (ESP). One prominent ESP worker came to K-X while we were there. She told the people that the Christians were causing them to get sick and to stay away from them. We were thankful to have Christians praying for us and His work in Siberia.

We identified with God's purpose in life for the Apostle Paul: "to open their eyes, in order to turn them from darkness to light, and from the power of Satan to God, that they may receive forgiveness of sins and an inheritance among those who are sanctified by faith in Me" (Acts 26:18).

Needing a Refrigerator

Back in our apartment, our old refrigerator motor ran constantly but didn't cool anything, so we unplugged it. It was not easy looking for a fridge, since we also had to find somebody with a truck to haul it and able bodies to carry it up five flights of stairs. It took several weeks to replace it.

Looking for a Church Building

The Sunday after Easter was the last time our church could meet in the auditorium of the Navigational School. Can you imagine an established church of 200 with no place to gather and no Sunday school rooms?

Churches needed to be near popular bus stops since the majority of the people, including us, rode public transportation to and from the services. The Sunday buses were not packed like sardines! Our congregation had only two or three private vehicles parked outside.

Outdoor Service in Chapayevo: David

I drove Misha's van to Oktyomtsi and then to Chapayevo to participate in their weekly meeting at 2:00 p.m. They sang hymns in Sakha and fellowshipped around their picnic tables during the outdoor service. Vera translated while I spoke.

Men's Missionary Conference: David

GTE sponsored a men's missionary conference. Forty missionaries and pastors from throughout the Republic filled their building. The theme was "Building the Foundation of the Church." How to reach people for Jesus Christ in other cultures was my topic with examples from Alaska and the Republic of Sakha. The teaching and interaction benefited everyone.

Forty Years and Counting

We wanted to do something memorable for our 40th wedding anniversary. The Lord gave us the desires of our hearts when a leader of the church in Neryungri, south of Yakutsk, invited us to their daughter Tanya's wedding to Vadim. Misha was our driver, and Vera translated for this seven-day trip. We stopped to see missionaries in five cities and towns on the way to the wedding.

Making Do with What You Have: David

Our trip to Neryungri took 15 hours over rough gravel roads. We had four flat tires on the way. Other vehicles were experiencing tire problems. Misha dug out his jack, but it did not work. So he used his God-given ingenuity. He placed a large stone as a fulcrum, and with a long pole we jacked up his van. A new method of carjacking! It worked.

Misha tied the pole on top of the van as we had to use it three more times before we arrived in Neryungri—where I purchased a jack.

Vera and Misha held down the pole, while I changed the tire. This worked faster than a jack but required several people.

Their Big Day

In Neryungri, the bride and groom had a civil wedding at 2:00 p.m., which was lovely and meaningful as they signed legal documents. Then they conducted the Christian ceremony in their church at 6:00 p.m. The service included two preachers, choir numbers, poems shared, skits, and personal greetings. The ceremony presented an excellent opportunity to tell people what the Bible said about love and marriage. They appreciated hearing from this American Christian couple who had been married for 40 years.

Elections Have Consequences

A Sakha family took us with them to cast their votes. Just outside the polling room was a large poster describing how to vote. Another poster showed the 11 candidates' pictures and brief information about each one. Inside, it looked like an Alaskan polling station, but each booth was surrounded by a red cloth. After filling out the one-page ballots, voters dropped them in a red box. In Russia, you must be 18 and show your passport for positive voter identification.

The countrywide election took place on Sunday, June 16th. Boris Yeltsin received 36 percent of the votes while Gennady Zyuganov, the Communist candidate, was a close second with 33 percent. By law, this forced a runoff election on July 7th. Christians continued to pray and encourage others to vote for Yeltsin.

Dr. David Jeremiah said, "By staying silent, believers are allowing non-believers to decide who will run our government and create our public policies." We all need to pray, think, and vote.

Never Give Up: David

Our official residency permits for foreigners gave us many of the benefits citizens received. We could fly on Aeroflot Airlines for the same cheaper price as Russians, travel to many places freely within the country without getting prior permission, and stay in their hotels at their prices.

To qualify, we had to have a permanent physical dwelling and temporary permission documents to live in Russia. To get those documents, the applications, qualifying pictures, receipts from a bank showing the fees were paid, and an official letter from our partner mission, GTE, was turned in. The final process to obtain the permits was spread over several months and took numerous trips to different offices.

One lady glued in our photos and said, "Now, all you need is the official stamp." When I returned to pick up the permits, another individual informed me, "Don't call us, we'll call you," so we waited for 10 more days. On my final trip to the office, another person brought the permits to the head lady to be signed, which she then presented to me. I thanked her and gave her *The Life of Jesus Christ* in Sakha. We jumped through all the hoops to function more freely in Russia. The Lord rewarded our persistence with official Russian residency permits effective until March 23rd, 2010.

Young Again (1996~1997)

Summer of Service (SOS)

Four students from Moody Bible Institute, Peter and Rhonda Dahlin, Jason Rogers, Melissa Coult, plus Jenna Miller from Cornerstone Bible School, arrived for Moody's Summer of Service (SOS). Zhenya, a great translator, came to Yakutsk for these ministry trips. It was a full-time job taking care of the team and keeping things moving in the right direction. Our first challenge was getting government permission for the students to travel to specific villages.

Everybody had to pass an AIDS test given within Russia before travel permission could be granted. This new law went into effect the day the team arrived! The American AIDS tests were not acceptable. We took them to the clinic and waited for the certificates. Then we took them on an overnight trip to a Sakha church and a Russian one to start orienting them to the local way of life.

Riding on the Rockets: David

At 4:00 a.m. on July 18th, Misha picked up the 10 of us plus our baggage and took us to the Lena Riverport. We boarded two different hydrofoils called Rockets, holding 75 passengers each. Yasue led one group on the Aldan River with Peter, Rhonda, Jenna, and Zhenya.

Traveling in the opposite direction, we led the team with Melissa, Jason, and Vera up the Lena River. We passed five towns where GTE had started churches as well as more towns without any knowledge of the gospel. We traveled 400 miles on the Lena River for 12 hours.

In Olyokminsk, we helped Gennady and his wife Vera with GTE. Various projects needed to be completed for their main church building, such as chinking logs and painting the interior of rooms. One of their classrooms was available for a small church gathering. I was asked to share what the Bible teaches about the church. Every night we gathered for two-hour church services.

Young Again (1996~1997)

The Paddle Wheeler: David

Vera was a gifted singer and a delightful individual, so the students dubbed her Mary Poppins. Vera tried to buy first- or second-class tickets for our team's trip to Lensk, but no space was available. So she had to purchase third-class tickets under the main deck of the diesel-powered paddle wheeler. It was an open living area with eight bunks, but we could walk around and visit up on deck.

Jason climbed halfway out of a small porthole to experience the fresh air and the churning water. The large paddle wheel was 30 feet away, turning and propelling our boat. Vera was shocked! If he had fallen out, the wheel could have hit him. Oh, to be young and fearless again!

On the deck, I offered one woman *The Life of Jesus Christ*. She laughed and said, "I already have that book. Two ladies gave it to me when I was in the hospital in Yakutsk, but I don't know their names." Then Vera dug out a photo of them. She was delighted and glad to hear more about Jesus Christ.

After 36 hours, we arrived in Lensk at 1:00 a.m. The believers met and took us to a home where we were served a hearty Russian dinner, sang songs, and went off to bed at 3:00 a.m. It would have saved time to have fallen asleep in our clothes since we were up at 7:00 a.m. for a two-hour trip to Sovhoz Novy.

Lensk was without a pastor, but Zhenya, a young leader, drove us in his boss's UAZ, which rode like a horse. We stopped near our destination for a big lunch with a Sakha Christian family. Then we headed to the baptism site.

Vera, David, Pastor Sergei, and nine people dressed in white robes for the baptism.

Mirny (Peace): Kay

Next, we drove to Mirny for a communion service followed by a feast with Pastor Sergei, his wife Anna, and 15 believers. On Sunday, we participated in a regular service with 70 people. I spent special time with new believers and seekers, encouraging them to follow the Lord.

Mirny was founded in 1955, with a population of 40,000, following the discovery of diamonds. We peered over the edge of Mirny's quarry, the second-largest mining hole in the world. A steady stream of trucks hauled plain-looking clay spiked with diamonds.

On Our Way Home: David

Returning to Lensk, we hoped to go on the Rocket to the Sakha village of Peleduy with four believers. But there was dense smoke from regional forest fires. After three weeks on the river and road, we ran out of time.

It was impossible to purchase five airline tickets, so Jason and I flew one day while the three ladies came the following day. We were all tired from our adventures, but we wanted to reconnect with the other team members. This was part of a busy summer. If God wasn't in it, we were wasting our time.

Sharing the Good News: David

Back in Yakutsk, I preached on Jesus' parable about the prodigal son's father and how he showed his love and patience in waiting for his son to repent. The following Sunday, a lady in the church relayed that her husband went to their "prodigal" son and told him how much he loved him. We were delighted to see people applying the message.

Mushrooms Plus: Kay

The youth group invited us to go mushroom-picking with them. At 8:00 a.m., we caught the bus heading out of town. Then we walked for six hours along a dirt road, through the woods, and over rough terrain while trying to avoid water. We stopped for breaks to eat and lighten our backpacks, but we finally made it.

Searching for mushrooms, talking, laughing, and singing together was fun. The young people were very good at identifying safe ones to pick. We showed them a mushroom book from Alaska with pictures, and they commented, "We have the same ones here." It was a long way back to catch the bus. We enjoyed mushrooms and the youth, but that was a lot of work.

The next day, one of the girls celebrated her 17th birthday and invited us to her party. Most of the nine mushroom pickers were there. We ate, sang, and played games. The game Gossip required one person to go out of the room while everybody else whispered a word in the ear of the person on their left. We had to remember that word. When the person returned to the room, he asked each of us a question, and we answered using the word our neighbor had whispered. I mispronounced a word that came out very silly! Oh, the joys of learning a foreign language!

Remember Your First Car?: David

We remember our first car as newlyweds. It was a 1949 Ford sedan. Owning a vehicle was a learning experience.

Riding on the hydrofoil from Yakutsk to Krest-Khaljai, we enjoyed all the autumn colors. Upon arrival, we asked some bystanders where Valeri lived in their summer home. Before I could say his last name, the stranger filled it in for us and stated, "They are Christians." People recognized them as believers. At the dock, we waited for an hour until word got to them. This gave us time to share with a Sakha lady and give her Christian literature. Delays aren't time wasters.

A large dump truck arrived and drove us to Valeri's place. A few hours later, Valeri came home with much excitement as he drove into their yard. Valeri was beaming! None of his seven brothers owned a car. Valeri's children went and sat in the car every day. Zoya, his wife, commented, "Our unbelieving relatives can see that God is blessing us." Valeri planned to travel to surrounding villages with the gospel in the winter. Their roads were impassable during the summers, especially when it rained.

The little Moskvitch (meaning a native of Moscow) was a brilliant green. "If you want your husband to stay home, buy him a Moskvitch!" Then he would need to keep working on his car. This was a common line with a lot of truth. They bought the used car when Zoya received five months of back wages.

Time with Women: Kay

After returning home, I went to Mokhsogollokh. The church had a big heart for continued growth and outreach. They hosted a women's conference in September for missionaries, pastors, and singles serving the Lord in the Evangelical churches throughout Yakutia and the Kamchatka Peninsula. Sixty women participated.

"Women—Serving the Lord" was the theme since they serve in their homes and through witnessing, teaching, praying, encouraging, and outreach. I taught several sessions on serving the Lord in our family and the church.

The key verse was: "that you may walk worthy of the Lord, fully pleasing Him, being fruitful in every good work and increasing in the knowledge of God" (Colossians 1:10). Everybody went home with a lot of information to put into practice.

A Bad Nose Job: David

Mike Matthews and his family moved into a house near the GTE mission office. A dog that came with the house jumped up and bit off the end of Mike's nose! They retrieved the piece of his nose and had it sewn back on. Then he flew to Canada to see if the doctors could aid in the healing. Mike appointed me to be in charge of the work in Siberia while he was gone.

A Long-Term Student: Kay

We welcomed Marianna, a 16-year-old Sakha girl from Mokhsogollokh. She had repented at 13 and was growing spiritually. We had traveled together with her brother and mother on our first gospel trip here.

Marianna lived with us while studying in a four-year program at the Music Department of the Yakut Pedagogical College. She was a lovely girl and asked our advice about various things, including boy-girl relationships. She was very diligent in her studies and wanted to have all 5s, the equivalent of all As. Since a piano came with our apartment, she was learning to play, and I enjoyed helping her.

Bible Study Time: Kay

On Friday nights, I taught a ladies' Bible study group about serving God in our marriages. I enjoyed it but felt inadequate to help the ladies with the common problem of alcoholic husbands since I had never experienced that.

Young Again (1996~1997)

Teaching Kindergarten: Kay

Galina asked me to teach English two days a week for several months at a special kindergarten where she taught Russian. She used flannelgraph and told Bible stories in Russian. We worked together on the English songs. One student carved a bone which we used in class. The children liked playing "Doggie, doggie, where's your bone?" Another favorite was "Button, button, where's the button?" It was amazing how fast they learned English. We sang "Jesus Loves Me" and several other songs. I also taught parenting classes to help their parents. Family problems seemed to be overwhelming. God gave us many opportunities to speak for Him.

A Call from Ethiopia: Kay

We were delighted to get a phone call from Steve and Cheryl in Ethiopia. We knew it was expensive, but it thrilled us to hear their voices and learn more about their work.

Celebrating Thanksgiving in Russia: Kay

We celebrated Thanksgiving at Mike and Kathleen's place. Mike had just returned from the home office in Oregon. His nose was healed and looked normal. He brought a smoked turkey for our team to sample. I made a "faux" pumpkin pie out of carrots. One has to improvise.

A Delivery on Christmas Day: David

A postal worker knocked on our door and delivered a small package. I profusely thanked her and gave her a Christmas ornament as a present. She was pleased. After I closed our door, it dawned on me that the 25th was my traditional Christmas Day, but it was just another workday in Russia.

The "New Year's" Tree

Lenin Square had a huge, beautiful "New Year's" tree with a big red star on top. It was not the star of Bethlehem but their traditional communist one. Standing by the tree was Grandfather Frost, a bearded figure similar to Santa Claus. He delivered gifts on New Year's Eve with the assistance of his granddaughter, the Snow Maiden.

We could see more than 100 apartments from our windows, but only one had Christmas lights. The manger scene in our apartment reminded us of the birth of Jesus. The Grinch and the Communists tried to steal Christmas!

Welcoming the New Year with a Bang

New Year's was big in Russia—with a tree, drinking, food, and celebrating. Throughout the day and past midnight, fireworks were going off everywhere in the city. As a friend told me, "The Chinese are happy. Look at all the money they are making." This year we did not stay out until midnight. It was interesting to lie in bed on the fifth floor and even have fireworks hit our bedroom window.

1997
Be Prepared: David

The church was packed! We were enjoying a special service with many others remembering the birth of Jesus on January 7th, the day Russians celebrate. Misha made his way to where we were sitting and whispered, "Pastor Valeri just arrived and wants to take you to Oktyomtsi. Can you go right now?"

As quietly as possible, we made our way to the back of the church and the overflowing pile of heavy winter coats. We dug in and found ours. We had no idea how long we would be gone. I asked to go back to our apartment and grab my Russian sermon notes since no translator would be available. There was no time to pack a change of clothes.

We rolled down the road in the church's UAZ. The hour-long trip gave us time to catch up on the news. Would we return home after the service, go to Mokhsogollokh, or what? Those are logical questions, but we knew the Christians would take good care of us; the Lord was with us, and there was no need to worry.

Warmly Welcomed: David

A horse was tied up with a sled outside the Oktyomtsi Church. We still think of dog sleds and snow machines. Inside, 35 Sakha people warmly greeted us, all excited and feasting on Christmas dinner. We also saw several new faces.

They quickly washed a few dishes and silverware while preparing a place for us. Their program included Sakha hymns and gift distribution. Children were called forward one by one to sing a song or say a verse and receive a present. Near the end, we had an opportunity to give a personal Christmas greeting to them. We had not expected to be guests for such a joyful occasion.

The dinner and service took longer than expected, so it was too late to go to the church in Pokrovsk. Instead, we piled into the UAZ and brought a Sakha couple home. We asked them if they were Christians. They responded, "We are the only Christians in our village, and the church meets in our home." One Sakha believer told me he wanted to evangelize in that village two years ago, but a strong shaman, who opposed the gospel, lived there.

Our Five-Day Outing

The Christmas Day trip turned into a five-day ministry time with Pastor Valeri. Our last service was on Friday night. We expected Misha and Mike Matthews to show up for the meeting, but they arrived during the closing prayer. So the two-hour service with six speakers turned into a three-hour service with eight speakers! Then we ate at Pastor Valeri's apartment. Valeri and Misha discussed plans for reaching out to more Sakha people. We left at 2:00 a.m., arriving home shortly after 4:00 a.m.

Designated Driver: David

I became the designated driver for the GTE mission van. InterAct had given the van to them, and I was listed on their official document for the vehicle title. The most important thing was that I had my passport in my possession. Misha's passport had been stolen, and it took time and jumping through hoops to replace it.

I was not excited about driving anybody else's vehicle, especially in Siberia. Their "rules of the road" were different. So when driving, I usually gave the other guy the right of way. My Alaska driver's license had a three-dimensional hologram, which I showed when stopping at police checkpoints. The police and others liked Alaska. One policeman asked if he could show my license to his buddy in their office, so I waited.

You're Just a Bug

The Russians have a saying, "Without a piece of paper (document), you're nothing. You're just a bug!" And that was very true. Without documents, you are not a person, just an insignificant bug! A passport must be presented for everything. It is needed to drive a vehicle, buy a plane or train ticket, leave or enter the country, exchange money at the bank, get married, vote, get a library card, or pick up your package at the post office. We carried them everywhere because they were our official identification papers.

Frostbitten Toes: David

I drove the mission van to minister in three Sakha villages. In the first one, a man had just repented, which was a highlight of this trip. All three towns were doing well, although one had only women believers. I have nothing against the ladies, but they needed local men to lead their church rather than depending on Christian leaders from a neighboring town.

The van heater barely gave out heat, and it was cold, but we stopped and warmed up in each village. Our return trip was colder. The Toyota van from Japan had the steering wheel on the right side. This meant the gas pedal was by the door, farther away from the heater. My toes were cold! Thankfully, the frostbite was minor and only left me with some red, tender toes. The steering mechanism was cold and stiff. I used the back of my comb to scrape the frost on the inside of the windshield. It took me a while to warm up at home.

Brother's Conference: David

Fifty-five Sakha and Russian pastors and men from various churches in our republic met in Mokhsogollokh for the Brother's Conference. They shared what the Lord was doing in their area. We prayed together and discussed their problems. Certain issues people faced greatly hindered their outreach, but it was encouraging to see the Spirit of God working.

The conference included a "Walk Thru the Bible" with the same hand motions as the English version. The average speed to review the whole Bible was two and a half minutes. Hermeneutics, how to interpret the Bible and other helpful subjects were presented.

The Eighth of March: David

Several people asked us if we have March 8th in America. I said, "Yes, and we also have the ninth and 10th." Seriously, the eighth was celebrated as a major Russian holiday for International Women's Day. Kay received a big box of chocolates from the kindergarten where she taught English. We both appreciated their kindness.

All Work and No Pay

We left Yakutsk on March 14th, a day before the airline and airport workers called a strike since the employees were way behind in receiving their pay. Other sectors of Russia's economy were a year or more behind! The government had no money! The delayed paydays lasted for more than a year, which impacted air travel in other areas of the Sakha Republic.

Better than Christmas: David

Upon arriving in Fairbanks, Danielle told me at the airport, "This is better than Christmas!" That made my day. We appreciated being in our house where everything was so convenient. We enjoyed the ice sculptures with our children and grandchildren. The park had several ice blocks hollowed out to sit in and one that could spin on an ice base. One child climbed inside, and the others gave it a good spin.

Helping My Parents: Kay

For three weeks in April, we were in Michigan helping my parents. My mother required much care after her stroke, and my father needed a well-deserved rest and a change of pace. On Sundays, we visited several supporting churches.

What Is That?: David

After finishing our stay in Fairbanks, we flew to Khabarovsk, Russia. Going through customs, all our baggage was x-rayed. The official blurted out with a puzzled expression, "What is that?" I quickly answered, "Coffee mugs." He accepted my response, and I did not have to open up any boxes. We had prayed there would be no problem with customs. Seven dozen coffee cups were packed in our luggage. That was a lot for one American couple!

A company in Fairbanks made 144 mugs for us with a picture of the famous Yakutian Rock Pillars. We included Psalm 121:2 written in both Sakha and Russian. "My help comes from the Lord, Who made heaven and earth." None of the believers in Siberia had coffee cups with Bible verses on them.

Once we arrived in Yakutsk, we gave a mug to Marianna. She exclaimed, "Every time I drink tea, I will think about this verse and the Lord." That was exactly what we wanted her and other Sakha people to do.

No Problem!
We returned to Yakutsk on June 4th—the day our Russian visas expired. We strongly recommend not doing this, but we couldn't do it differently due to unforeseen circumstances. There were no problems in renewing our visas, as the official signed and stamped them, giving us freedom of movement while living here. We planned to live and work here as long as the Lord gave us good health, an open door, and we were being productive for our Lord.

Drink Your Milk!
The best milk is country fresh from a cow. The church's summer camp on the outskirts of Pokrovsk needed fresh milk for their campers. Would you walk your milking cow 12 miles to camp? A Christian loaned her cow to the camp for July. She also provided fresh beef from another one. What a great way to give to the Lord and His work. People had little cash! Those with jobs were still months and months behind in receiving their pay.

We drove to the camp and delivered urgently needed supplies from Yakutsk. We transported the camp director and head cook as they plunged into the work. Many helpers, handymen, cooks, dishwashers, and counselors came. The last group of 47 campers kept everyone engaged. Christians from the five surrounding towns actively participated with prayer, food donations, and hands-on help.

It reminded us of Nehemiah when the Hebrew people rebuilt the broken walls of Jerusalem in record time. "So we built the wall . . . for the people had a mind to work" (Nehemiah 4:6). It was their camp, and it was being used for the Lord's glory.

The setting was rustic but functional and met all their basic needs. They converted two older log houses into dormitories, one for boys and the other for girls, with outdoor plumbing. Their dining hall had a couple of mosquito-net walls, while the roof was covered with plastic. The archway leading to the camp, with John 3:16 and Acts 16:31 written on it, identified the camp. The campers' schedule called for activities from 8:00 a.m. to 11:30 p.m. Back in our apartment, these two former campers remembered to pray for them as we put milk in our coffee and on our cereal.

At an Elder's Home
Lyudmila Maksimovna, a medical doctor, actively served the Lord in our church. She asked us to visit her lonely mother, who had difficulty getting to church because of foot problems.

Lyudmila's grandmother lived in southern Russia and told her mother about God every time she visited. Her mother turned to the Lord in 1941 when communism was strong in Russia. Later, her husband turned to the Lord and grew in his faith. Her grandmother was one of three women—Yevdokia Pavlovna, Vilma Rikhardovna, and Maria Nikolaevna—who were the first known Christians in Yakutsk.

A Touch of Alaska
We enjoyed visiting our friends' dachas during the summers, especially when it was hot. They usually had two greenhouses, one for tomatoes and the other for cucumbers. Many older dachas had raised beds three feet high, measuring three feet by eight feet.

Beautiful flowers grew on their property. We brought delphinium seeds from Alaska to share with friends. One family called their plants "David and Katya" because they were entwined around each other.

More Baptisms: David
Sunday afternoon, August 3rd, was an exciting day in Yakutsk with more answered prayers. Nineteen young people and adults were baptized near the GTE office by White Lake. Six were men, including an older man who used his cane to walk into the lake. Two of the men were Sakha. Two Sakha college students were baptized. One of them volunteered to help us enter more Sakha hymns on the computer. She was the only Christian in her village.

Where's My Oscar?: David
Two men from Holland came to Yakutsk to produce a movie in Dutch featuring the Sakha Bible translation project. I was the van driver in one scene when two men flagged me down in Yakutsk to get a ride. We haggled about the price but finally agreed.

I drove for four hours to an airport near Berdigestyakh along with the Sakha translator, Sargylana, and her daughter, Kaya, to catch a biplane to their village farther north. Several times, we stopped along the gravel road and let out the cameraman. I backed up and passed him at normal speed while he filmed us. At one point, he hung out the rear van door while filming the translator sitting in the van's middle seat.

Sasha rode with us in case we had any mechanical problems. On our return trip, we talked for several hours with two Sakha young men who were interested in hearing the gospel for the first time. No believers lived in this region, but we distributed literature. I arrived home at 3:00 a.m., exhausted. No Oscar nomination for me!

On the Hydrofoil Again

In September, we rode on the hydrofoil to Krest-Khaljai. Their Sakha church has continued to make remarkable growth since our last visit. One of our top projects with them was to have more Sakha hymns entered on the computer so we could print a trial edition of the hymnal. But our main job was on-the-spot discipling.

Joyful Sakha Singers

The list of hymns and choruses had risen to 40. Larry DeVilbiss came to Yakutsk to record these new songs and make them readily available throughout the Republic of Sakha. He recorded individuals and groups of singers.

Sakha Home Bible Study: David

On Tuesdays, I met with Valeri and Volodya as we prayed and studied a portion of the Gospel of Mark together. We answered two basic questions: "What does this passage mean?" (Interpretation) and "What does it mean for us today?" (Application). This was a time to train Bible teachers.

The home Bible study in Yakutsk was a new and exciting development. They met every Friday night, and the two men did all the teaching. I was there for moral support and to answer any questions, but I had to do it in Russian. After completing Mark, we went through Acts. The weekly study grew to include 12 people. Five Sakha people repented during these studies. It met a basic need for Sakha believers from the two evangelical churches.

First Modern Missionaries in Yakutsk: David

A Baptist church in Odesa, Ukraine, presented the spiritual needs at the ends of the earth—in Yakutsk. Anton and Lyudmila accepted the challenge. In 1980, they moved to Yakutsk to start a church but found only two believers. People in Ukraine faithfully prayed for Yakutsk and gave them names and addresses of relatives exiled to the many prisons in the Sakha Republic. The current evangelical movement in this remote part of the world started with these missionaries.

This is a condensed version of an article I wrote for *World Pulse* by the Evangelism and Missions Information Service of the Billy Graham Center. It appeared in the February 5th, 1999 issue. I wrote two other articles for *World Pulse* about the Sakha people.

1998
The Top Five Questions: David

We reflected on the top five questions we've been asked over the years. Everywhere, people were asking questions and seeking answers. I wore a wristwatch and was often asked, "What time is it?" My answer, given with an accent, confirmed their guess, "Yes, he's an American!"

Top Question Number One

What do you think of our country? Do you like living in the Sakha Republic?

We like the people; they are friendly and kind. Life is easier in Alaska, but God wants us to live here.

Top Question Number Two

What are you doing here? Where do you work?

We work with the Sakha and Russian people in the Evangelical Christian Church. We travel around the Republic to visit other churches because we want people to know that Jesus Christ loved and died for them. We also teach English.

Top Question Number Three

What is the difference between our Russian Orthodox Church and your church?

We use the same Bible, and some things are similar. People can go straight to God through Jesus Christ, and He will forgive their sins. The

Orthodox Church takes people to God through rituals, icons, candles, and crossing themselves.

Top Question Number Four

Is President Bill Clinton a Christian? We hear he is a Baptist.

He says he is a Baptist and a Christian—but it seems in name only. God knows his heart. He is married, but he likes other women. (This latter aspect didn't bother most.)

Top Question Number Five

Where did my relative go when he died? He was a very good person, but he did not know about Jesus Christ. (This was the HARDEST question.)

Almighty God is perfect, and He will judge each of us after we die. He is fair and he knows our hearts and everything about us. The real problem is that we are all sinners. There is no sin in heaven with a perfect and holy God. That's why we have to repent and go to God through Jesus Christ in this life. It all depends on what we do with Jesus Christ.

To Strike or Not?: David

We spent nine days in Ust-Maya with their church. We couldn't fly due to the air traffic controllers' strike, so we drove for 10 hours. The local people were concerned about problems closer to home, such as not receiving their wages.

They asked me to speak on serving the Lord through giving. I gave them a biblical perspective, but there were practical considerations: "We want to give, but how can we when we haven't been paid for a year?" I suggested giving produce from their gardens and spending time helping at church. I spoke with the men each night, using character examples from Scripture. Kay spoke with the women each afternoon about serving the Lord.

On Sunday, the first preacher was Sasha, who had not been paid for more than a year at his full-time job. His boss told him to strike. He said, "I pray, and I know the Lord is with me whether I'm paid or not." What a testimony to God's grace!

I was the second preacher, so I spoke about praying for and living a godly life before our unbelieving wife or husband. We were concerned about the low number of men in many churches.

How do people live who are not paid for a year or more? The church family and others generously assist as they are able. People received bread or a limited number of monetary coupons, which were good at specific stores for certain food supplies. They take gardening, picking berries, gathering mushrooms, and fishing very seriously. Bills go unpaid, or they don't buy things that require cash.

The town's heating system was at the absolute bare minimum to keep the pipes from freezing, but not enough to heat their apartments. They could use electric heaters, but those bills had to be paid, or else their electricity would be cut off. Schools had been closed for lack of heat. Life was slipping into survival mode for many people. But what a difference it makes when Jesus Christ is the center of one's life! The lights of their lives were powered by our Lord. Growing Christians kept working hard, having good attitudes, and sharing with needy people.

Pastor Misha and Lyudmila moved here from Khandyga in 1991. Their four children reminded us of our own grandchildren. They called us *Dedushka* (Grandpa) and *Babushka* (Grandma), making us homesick for our children and grandchildren.

Bending and Flexing: David

Being flexible should be a normal part of the Christian life, yet we need to be inflexible when it is not the Lord's will. Bible doctrine was not open to change, but our lives and ministry were.

We planned for an eight-hour trip with Misha and Vera to Khampa and Lyokyochyon. Newly converted elderly believers Trofim and Pasha lived in Khampa. Their daughters were new believers and traveled with us. The gospel needed to be established here.

Part of the main highway was a regular two-lane gravel road. Other parts were a glorified one-lane road with two main ruts. The wavy sections of the permafrost with no gravel foundation made us feel like we were surfing. I told Misha, "In America, highways can be boring!" Several vehicles had gone into the ditch. We finally took our turn. The Lord sent a kind driver in a 6x6 military truck with the power to pull us back onto the road.

Khampa Home Meeting

Pasha's daughter was also named Pasha. The daughter asked the local officials about using the House of Culture for our meetings. They were skeptical. "First, have a home meeting; if there is a lot of interest, then you can have a public one."

Forty people came for the Khampa meeting. They listened intently as we shared the gospel and sang hymns in Sakha. After two hours, we distributed literature.

Two More Cities

Misha asked us, "Would you be interested in going to Nyurba and Mirny?" We gave an instant "Yes" since we were already halfway there. Nyurba was four hours farther down the road, and Mirny was another seven hours away.

Nyurba was a growing Sakha city of 15,000. We found one church member at 10:00 p.m. and arranged to meet with them on our return trip.

We arrived in Mirny bright and early Sunday at 5:00 a.m.! Fortunately, Pastor Sergei and Anna were forewarned and happy to see us. We ate a hearty meal and rested for a few hours. It was exciting to be in a healthy church with 70 members reaching out to surrounding towns. This was a diamond mining town, but the believers were the real gems in God's eyes. They gave Misha gas money for our return trip.

In Nyurba, we attended a special Monday night service with 30 people. Pastor Yemelyan led the two-and-a-half-hour service. Many Sakha people learned and sang their Sakha hymns for the first time.

More Sakha Opportunities

The next stop was Khampa at 4:00 p.m. for a home meeting. They listened and learned more about following Jesus Christ. We slept there for the whole night and left the following morning.

The family of the main Sakha Bible translator, Sargylana, lived in Lyokyochyon. Her parents were open to the gospel, so they planned a village meeting in their House of Culture. Forty people attended; they heard the gospel for the first time, asked questions, and received literature.

We enjoyed working with the Sakha and Russian believers and advancing God's work in their lives. Flexibility required us to heed these Bible verses:

"For though I am free from all men, I have made myself a servant to all, that I might win the more; and to the Jews I became as a Jew, that I might win Jews; . . . to the weak I became as weak, that I might win the weak. I have become all things to all men, that I might by all means save some. Now this I do for the gospel's sake, that I may be partaker of it with you" (I Corinthians 9: 19-23). And to the Sakha, I became as a Sakha so that I might win Sakha people.

Russian Grandparents—Sakha Granddaughter

We rode the hydrofoil to Krest-Khaljai to arrive before the annual Sakha celebration on June 21st, the longest day of the year. A few people got up at 4:00 a.m. "to meet the sun" as they prayed to the spirit of the sun. We did not observe this opening event.

Later in the day, we watched sporting events such as wrestling, running, various jumping events, and playing mouth harps. Exciting horse races stirred up clouds of dust! Lena, the pastor's oldest daughter, accompanied us while we pushed their youngest, Ruth, age one, in a stroller. Lena told us, "I overheard people asking, 'How do those Russian grandparents have a Sakha granddaughter?'"

Wild Lilies: David

Sardaana (wild lilies) are the beautiful national flower of the Sakha people. Valeri's daughter showed us fields of flowers. I photographed them to make a poster. The mosquitoes were also numerous—I had more than 50 on one pant leg!—so we hurried back to the house.

That evening at a Bible study, Kay shared about the *Sardaana* flowers. Jesus reminded us: "Consider the lilies, how they grow: they neither toil nor spin; and yet I say to you, even Solomon in all his glory was not arrayed like one of these" (Luke 12:27). All the Sakha people enjoy the creation God made for them.

Staying in the Background: David

We were delighted to witness seven Sakha adults being baptized in the Aldan River. Twenty believers were on the riverbank with us, while several bystanders peered from farther away.

Pastor Valeri asked if I wanted to baptize them. We knew they would be glad to let me have the honor, but I declined. We wanted them to recognize Valeri as the spiritual leader. Americans needed to stay in the background and not hinder the growth of the Sakha church.

Four Bottles of Vodka: David

Four bottles of vodka stood in the middle of our table overflowing with excellent Sakha food. We were special guests of honor, along with Pastor Valeri and a dozen Sakha people. I thought about Jesus' eating with publicans and sinners!

Titus moved to his new home in heaven. His family and special friends, all unbelievers, remembered his life and departure according to their custom one year after he died. His aged mother sat directly across the table from me.

The host opened a bottle of vodka and started filling the shot glasses for everyone. Pastor Valeri spoke in Sakha and turned his glass upside down. We followed his example. Then he stood up and spoke about the Lord in Sakha.

At my turn, I spoke in Russian about how a hunter could fool ducks with artificial decoys and a duck call, thereby deceiving and shooting them. Satan had deceived Titus until he repented four years ago. Our host had shown me his beautiful hand-carved duck decoys while others were preparing the dinner.

Udarnik was small, with only 100 people. It was a collective farm with falling production and only token help from the government. Our host had not been paid for more than a year. Their gardens, cows, hunting, fishing, and gathering berries were critical to their survival.

It had taken us two hours to travel 18 miles to this little village. The "road" was terrible, with huge dips and water up to the floorboards of Pastor Valeri's little green Moskvitch.

Little Udarnik was forgotten by the once-powerful Soviet system. However, their Creator had not forgotten them. God used Pastor Valeri where his father had once lived—to share the gospel with Titus' family and friends.

My Children Need Jesus

Varya, a Sakha believer in Krest-Khaljai, invited us to her home to talk to her children about the Lord. After eating, she gathered her four children, from 11 to 20 years old, and had them sit on a bench facing us. They listened intently for an hour while we explained the gospel and answered their questions. The number one decision you can make is choosing to follow Jesus Christ. The one you marry is the second most important decision in life. We continually prayed that God, who is all-wise, would use our lives and our limited use of Russian for His glory.

How to Get to Heaven: David
A little Russian gospel tract titled "How to Get to Heaven" was at church. No copyright, author, or organization's name was written on it. After getting the Bible verses and key phrases translated into Sakha, I made trial copies. Several were given to Galina from Pokrovsk to review and get her feedback.

Galina attended a seminar about Christian witnessing taught by a gray-haired missionary in his 70s. Gus Matero had worked with Child Evangelism Fellowship in Norway. During the class, he asked, "Are there any gospel tracts written in Sakha?" Galina said, "Yes," and held up a copy. Gus laughed and responded, "I wrote that tract. Thousands have been printed in Russian."

As soon as I found out, I met with Gus. He had designed and printed the tract in Russian during the communist era. The absence of an address prevented anyone from finding the source. He just wanted to get the Good News out and knew a printer who shared the same goal.

Gus contacted his friend, who printed 10,000 copies in Sakha for us. We thanked God for Christian friends, writers, and printers who made the Good News available in the Sakha Republic. Later he sent us another 15,000 copies.

English Camp
Yvonne Betters brought seven high school students from Far North Christian School in Fairbanks to Yakutsk. They taught English in a state-run camp. Her husband, John Betters, had made several trips to Yakutsk to assist in the public educational system. They came to our apartment for dinner. We were excited to see Christian youth from Alaska traveling to the Sakha Republic and sharing their faith.

Alaska Again
Flying to Alaska at the end of July, we saw Steve, Cheryl, and their boys, Jon, Wes, and Curtis, before they returned to Ethiopia. We worked on our technical needs and made the usual trips to the dentist and for eye care. We audited the Russian language course at UAF. The excellent teacher wanted her students to do well.

Doctor's Advice: David

I had minor problems with my knee. The doctor in Fairbanks did arthroscopic surgery to check things out. Even though I had no other physical problems, their equipment showed that I had skipped several heartbeats in a row during the procedure. I was not sure I needed a pacemaker but decided to take the doctor's advice. Consequently, they installed one that kicks in when my heartbeat drops below 50 beats per minute.

Telling True Stories: Kay

Suzanna and a friend stayed overnight, so we played a game and then started telling stories. At first, they told scary stories, but then I told a true story. They wondered why they had bad dreams! Soon they were telling true stories, too.

Sakha Scripture Calendars

Making calendars with Bible verses became our new trend. The first one was completed in 1999 with a picture of the Lena River Pillars, a natural rock wonder in the Sakha Republic. Genesis 1:1, "In the beginning God created the heavens and the earth," was printed underneath in Sakha. We made extras for Christians to distribute. During this era of financial problems, people were delighted to get a free calendar in their language.

Stopover in Khabarovsk

We flew on Aeroflot, which was the only airline flying from Anchorage to Siberia at that time. Our flight to Yakutsk was delayed for three days until the smoke cleared enough from the forest fires.

Since we had lived in Khabarovsk, we stayed with a wonderful young Russian couple, Kostya and Natasha who were expecting their first child. Kostya had been ministering each week in another town and also holding Bible studies in their home.

Seeing "Old" Friends

On Sunday, we attended the "Green Church," the only registered Baptist church that existed under communism and government surveillance in this city of 600,000. The small log building painted green was easy to find. When the church outgrew the building, they traded their building and valuable land for another area approved by the city and built a larger concrete church.

We were delighted to meet Sasha and Galina again. They had planned to be missionaries but had drifted away from the Lord. Three years ago, they returned to fellowship with God. It was encouraging to see those who came back to the Lord and were moving forward in their Christian lives.

Russian tradition is to have a big dinner on the ninth and 40th days after a person dies. Our former landlady's daughter prepared a meal for us in honor of her mother. This was another opportunity to continue being a witness to her daughter.

Out of Gas at 30 Below Zero! (1998)

Church Dedication in the Diamond Region: David

The western portion of the Sakha Republic is the diamond-producing region. Members of the Mirny Evangelical Church had renovated a former grocery store. The fresh smell of lead-based paint still lingered in the air. Pastor Sergei hung a new sign outside the building that read: "*Dom Molitvy*" (House of Prayer). Many Evangelicals in Russia use this title for their churches based on what Jesus taught. "Is it not written, 'My house shall be called a house of prayer for all nations'?" (Mark 11:17). Bible verses were stenciled in front of the sanctuary.

The church dedication was packed with 200 Russians, Ukrainians, Sakhas, and Americans, including their 80 members. New Testaments were given to 26 first-time visitors for answering a brief questionnaire. The service was extended to three hours. My message compared two preachers, one whose pants had worn-out knees while the other man's seat was worn very thin, with a challenge to "pray without ceasing" (1 Thessalonians 5:17). Russians usually stand or kneel to pray. The service included singing special songs, relating testimonies, reciting Christian poems, and hearing three other preachers.

Two eternal diamonds glistened during the service when a man and a woman came forward to receive Jesus Christ. This brought tears of joy to a number of us. One church member had been praying for this man for six years!

In closing, a dozen pastors, leaders, and I stood in a row facing the audience while stretching out our arms over the congregation, praying for their future success in upholding Jesus Christ and His Word in this church and region.

A Brother's Conference: David

The church dedication was followed by a two-day brother's conference with 40 men. The first day's theme concerned revival, as they wanted to reach out to more unbelievers. Prayer is key to revival, so we all got down on our knees and prayed for God to revive us. God's generous grace versus man's legalism was the next day's theme. Our world system teaches us to be legalistic and work for our salvation by following specific rules, but God's grace is free for everyone who believes.

Lights Everywhere

A brilliant display of northern lights filling the entire sky welcomed us to the city of Svetly (Light). The Fire Department declared the church building "unsafe" since their small structure had only one entrance, so we gathered in an apartment. The pastor lost his job at the Fire Department because of his faith and was forced to move since the Fire Department had provided his housing. Two other church members had also lost their jobs for being Christians. They paid a price for following Jesus Christ.

Traveling in the Region: David

As we traveled farther to Almazniy for an evening meeting, we met Maria who shared that she had a strong burden to reach her people. Vera and Maria sang several Sakha hymns. She was overjoyed to discover that 48 hymns were already written in her language!

At midnight, another car picked us up for the return trip to Svetly, but the heater had problems at minus 30 degrees. We were packed in the car, which kept us warm. We continued to Chernyshevsky, 57 miles from Mirny. A massive dam and hydroelectric plant were built on the Vilyuy River to supply electricity to the region for diamond production. The cheap electricity benefited everyone.

We stayed with a young man who had lost his wife after giving birth. His mother-in-law took the baby in place of her daughter who had died, and she moved elsewhere. Now he lived alone. He had been saved shortly before his wife died, and we witnessed his baptism two years ago. He sang, played the guitar, and wrote songs.

That night, we spoke at his church service, which started at 7:00 p.m. and ended at 11:30. The members asked many questions about God, our lives, family, and Alaska. We went to bed while Vera and our host walked to the pastor's house with others. They sang and fellowshipped together until 4:00 a.m. We were thankful for this three-week opportunity to serve the Lord and see what He was doing in Siberia.

No Gas for Sale

Misha borrowed the UAZ from the Mirny Evangelical Church for another gospel trip to a new area. Tas-Yuryakh is an isolated Sakha town of 500 people with a group of believers. We passed two 24-four-hour filling stations along the way with signs reading: "No gas." The discovery of oil put this region on the map, along with a new gravel road and a small refinery.

Pavel rode with us back home to his wife and their three children. His whole family followed the Lord! Pavel was Sakha and the leader of their house church. While in Mirny for a month of job training, he benefited from the fellowship and discipleship at the church.

Misha left the UAZ running during the service since it would have been hard to start after sitting in the extreme cold for several hours with a defective starter. It was a practical vehicle and it even came with a hand crank.

Face-to-Face Fellowship: David

We were excited to have a face-to-face Christian fellowship in Tas-Yuryakh with six adults and their children. The members had repented three years ago but needed help and more contact with other believers. We sang hymns in both Sakha and Russian. Misha and Vyacheslav preached. I compared sleeping bears in their dens with the rich fool in Luke 12:16-21. The bears think they are safe in their underground home, taking a long, uninterrupted winter's sleep. Indians in Alaska hunted bears in their dens. The rich fool depended on himself and thought he was prepared for many years. Both deceived themselves!

After the service, we had tea, which included a rich soup, meat, salted cabbage, salads, fish, *piroshki* filled with hamburger and mashed potatoes, various pastries, candy, and cured pork fat.

The Trip Home

The waiting UAZ was warm and toasty. We began our four-hour trip back to Mirny. First, we headed to Almazniy to drop off Vyacheslav and his daughter who came to teach the children. Now we had more wiggle room.

Out of Gas

We came to an abrupt stop along a deserted section of the road. Misha pointed to the gas gauge, exclaiming, "We're out of gas!" Vera was not feeling well, as the three of us got out of the vehicle. It was minus 30 degrees! We did what we had to do. We started walking and praying as we headed toward Mirny. The night was crystal clear and beautiful as we trudged along the side of the road. There was no traffic.

A brilliant display of meteorites showered all around us. Then a bright light burst directly over us, causing us to stop and look around in amazement. It was a bit too nippy, though, to enjoy God's beautiful creation. Besides, it was 3:30 a.m.!

The light-colored sky over Mirny in the distance was a welcomed sight. We were thankful for our heavy coats and *untees* (local horsehide or reindeer boots with one-inch-thick felt soles). But it was cold, so we kept moving.

We walked several miles before Misha pulled up, all smiles. We gladly climbed into the warm UAZ. He had purchased enough fuel from a trucker to get us to Mirny.

Why?

Why would anyone be thankful to serve the Lord under these circumstances? Why did we have to run out of gas in the middle of nowhere on a frigid night? We felt at peace because we knew we were doing the will of God. We depended on God to take care of us as we did our part.

1999
Two Thousand Years Ago

Two heavy-duty Russian Ural Army trucks arrived in the Sakha Republic. On their sides was emblazoned "2,000 Years since the Birth of Christ" for everyone to read and consider, with a route map extending from Moscow to Pevek farther east. They left Moscow on December 20th, 1998, and arrived in Yakutsk in the middle of January. It was an obvious way to present Christ to the masses across Russia and encourage believers.

The Light of the Gospel mission from Ukraine sponsored the trip. Their expedition leader, Leonid, traveled with six full-time participants. They spent two weeks with GTE churches in the Sakha Republic. Much of their route was accessible only during the middle of winter when numerous rivers, lakes, and marshy lands froze and became "roadways." Four new missionary recruits joined them, making a two-year commitment to this part of the Lord's harvest field.

Goodbye, Mother Dear: Kay

My mother was nearing the end of her life. To see her one last time, I flew by myself to Michigan. First, I flew into Domodedovo Airport south of Moscow, and then I took a taxi for 53 miles through Moscow to Sheremetyevo Airport on the north side of the capital for my connecting flight.

Three days before my mother breathed her last on this earth, I joined my family. We gathered around her bed, sang hymns, and read Scripture. It was a special time of comfort for me.

My father told her, "See you in the morning, Vera," as she fell asleep. Ten years later, he reunited with her in heaven. My mother, Vera Mae Temple, died on April 14th, 1999, in Vassar, Michigan. She was 89 years old.

I remembered how she had read the Easter story from the Bible to my sisters and me when I was eight years old. Then I accepted the Lord. Before she died, I thanked my mother for leading me to Jesus, exemplifying the Christian life, and helping me grow spiritually.

Sakha Church Growth (1999~2001)

A Bigger Trial Edition

A major trial edition of the Sakha Hymnal with 59 songs was published. Once Sakha believers saw the book, they were inspired to write more songs.

Christian Lyrics for the National Anthem: David

Pastor Valeri asked, "Do you think it would be good to write lyrics to our national round dance to praise our Creator rather than creation?" Immediately, I responded, "That's a wonderful idea!" His newly composed words for the Sakha national anthem, honoring our Creator, were an instant hit among believers. Several others followed his example and wrote Christian lyrics to go with the round dance melodies.

First Sakha Christian Conference: David

The first-ever Sakha Christian Conference met in Oktyomtsi. The meetings united the believers and helped them grow spiritually. The new hymnal was enthusiastically received. Maria gave me 40 Sakha poems to turn into songs. Members of their little church were kept extremely busy—joyfully serving the Lord.

While in Oktyomtsi, I went to the local hospital to visit someone. My helper and translator thought speaking in Russian would be sufficient. I spoke to some patients but got a strong rebuke, "Speak in our language, not Russian," so the translator switched to Sakha.

Put to Music: David

I encouraged Valentina to use Maria's best poems as lyrics for new songs. She worked with her friend Raisa, who played the guitar. They sang one of the new songs together in church. Soon, Valentina gave me a recording of two more poems with original music.

Anti-American Tensions: David

Yugoslavia was continually in the news. One common question: "Why is the US, a big, strong country, picking on such a little country?" Under President Bill Clinton, the US bombed the Chinese embassy and a busload of civilians.

While enjoying dinner in a Sakha village, a man burst into the room and angrily shouted, "Why is America bombing Yugoslavia?" I replied as calmly as I could, "We're against that war." Then I read from the Bible: "Where do wars and fights come from among you? . . . You lust and do not have. You murder and covet and cannot obtain. You fight and war . . ." (James 4:1-2). I silently prayed he would not return with a gun, since he was so angry. Decisions made in the White House affect all of us.

Sakha Church Dedication

We rode the hydrofoil to Krest-Khaljai for the dedication of the new church building. It provided space for the services and Sunday school classes.

Nikolai, the local administrator, attended the service and spoke in Sakha: "I was a Communist, and I still am a Communist. But I believe there is a God. I also support your church. I know you will not cause problems in our town by drinking, stealing, or fighting."

Pastor Valeri called for a lunch break. It was a celebration feast with various Sakha specialties: fish cooked, smoked, and raw; horse meat cooked and raw; beef, salads, eggs, pastries, and a few local dishes we couldn't name or explain, but they were tasty.

When the meeting continued, two older women in their 80s repented. One told how her 90-year-old husband, a lifelong Communist, had repented a year ago. She had seen the change in his life and wanted God's help.

To close the three-hour service, everyone formed a circle and prayed, committing the building to the glory of God. The church continued to reach out to the surrounding villages. Pastor Valeri baptized five local people, then six more in a neighboring town the following Sunday. God was growing His church.

A Stone Hit My Window: David

About midnight that evening, we were drinking tea with Pastor Valeri. A young Russian lady knocked on the door and wanted to repent.

She and her husband were visiting Krest-Khaljai. Her husband had repented in another town along the river.

What an exciting day! Four people repented. Then the sharp ping of a stone on the windowpane by my head made me jump! One couple was angry, falsely accusing the pastor of something. Several people hated him for following Jesus Christ and abandoning the Sakha religion. The committed Christian life includes spiritual battles.

Alaska Again

Before our visas expired, we flew back to Alaska. Debbie, Dan, and Elizabeth continued to live active lives with their families in Fairbanks. Steve had moved his family from Ethiopia to Kenya due to the dangerous border war with Eritrea.

Visa Delays

After a four-month wait, we finally received our one-year visas. Now we could plan our trip back to Yakutsk, leaving Fairbanks in September. We flew to Anchorage and then on Aeroflot Airlines' once-weekly flight to Khabarovsk. On the long flight, we visited with Steve Stevenson and his son Greg. Steve's mission, SEND International, was working in southern Siberia, whereas InterAct worked in the northern part. Finally, we got on Sakha Avia, flying once a week, for another two hours north to Yakutsk.

We thought about our delayed return. When the time isn't right, God says, "Wait." How strange it would be if God answered every prayer at the snap of a finger. If that happened, God would be our servant, not our Master. God would be working for us instead of us working for Him.

In Our Baggage

Our baggage included 4,000 copies of the new Sakha Scripture calendars for the year 2000. "Jesus said to him, 'I am the way, the truth, and the life. No one comes to the Father except through Me'" (John 14:6) was highlighted among a colorful field of yellow crocuses. We had videos of the *Jesus* Film, and ballpoint pens with "God loves you" written in Sakha. We packed a minimal amount of clothes and food items not available in Yakutsk: cornstarch, banana chips, dried celery, chocolate chips, and concentrated lemon juice.

Marianna's Birthday: Kay

Marianna welcomed us back to our apartment in Yakutsk for her final year in music school. We celebrated her birthday, but we also thought of Dan's birthday on the same day in October. I made her pizza, a fruit salad, and a cake. She was delighted with the bouquet. Giving flowers is important in Russia.

Frozen Cabbage Salad

This year, we bought several cabbage heads at a time. A huge wooden frame two feet long with a single blade, finely sliced the heads. For flavor, we added a few shredded carrots and salt before mixing them.

A board on top of the large container held everything at the water level. We poked and turned it daily for four days, then bagged the fermented cabbage in small quantities and froze them on our balcony. We would thaw a bag and mix it with olive oil, finely diced onions, and dill for a quick, tart, easy salad.

Obstacles in Life: Kay

Our lives are full of obstacles and opportunities. One issue affecting our daily lives was how to get to the closest bus stop. We depended on public transportation like the majority of people. Workers had removed our footbridge spanning several large steam pipes, two feet in diameter, and a deep ditch. After that, everyone had to climb over them.

We need these pipes to provide centralized heating in our concrete jungle. They crisscrossed the city of Yakutsk, carrying steam from the central heating plants to apartments and businesses.

From our fifth-floor apartment, we could see grandmas and grandpas working their way over the pipes. Younger people carrying babies climbed over cautiously. Young ladies in their finest, dressiest coats and boots gracefully navigated the obstacles while young boys simply jumped over. We often took the long way to a less convenient bus stop.

After a month, they replaced our bridge before it got too cold. I did well on the obstacle race at Bryan College, but not so well at this stage of life.

Missing Potatoes

Most people have a potato bin just outside their door in the hall where it is cooler. These wooden boxes were two feet high and 10 inches wide. The sideboards had space to allow air to flow freely. A friend gave us a big bag of potatoes, so we stored them in our bin with a padlock. An unknown person pried off a couple of sideboards and stole all our potatoes! Thankfully, it was only potatoes.

A Steel Door

The potato thief caused us to think seriously about the security of our apartment. We ordered a metal security door with a peephole to be added outside our wooden exterior door. It came with a lock system in which the key turned 360 degrees twice, going twice as far into the wall. Now we were like our neighbors.

Who's There?: David

About midnight, we heard a knock on our door, so I called out, "Who's there?" The stranger wanted somebody we didn't know, so I did not open the door.

Then he violently and repeatedly kicked our steel door. Our old wooden door would not have held him back. It took half an hour until the police came and escorted him away. I heard him give his friend's address at apartment number 40, the same as ours, but in a neighboring complex.

We thanked the Lord for nudging us to buy that steel door and for His protection. "Some trust in chariots, and some in horses (*and even in steel doors*); But we will remember the name of the LORD our God" (Psalm 20:7).

Where's My Hamster?

We enjoyed showing hospitality to many guests. Paul and Lori Lokke and their daughter, Amanda, age six, were visitors. They found a hamster for Amanda but couldn't find a cage. She put her new pet in a plastic bucket, and we did our business in town.

While we were gone, the hamster escaped. We looked for her pet under the couch and a bookcase. Finally, we found it under the piano as he ran out into Amanda's waiting hands.

Sakha Singing Group

The Sakha hymn singing group met weekly. They invited us to sing with them, even though we were poor singers. One of our main projects was promoting the Sakha hymns. The group practiced until they were satisfied with each song.

The Teen from Amga

Anatoli, a Sakha teenager from Amga, heard about us through a lady selling clothing at the outdoor market. We had given her Christian literature. He traveled four hours to Yakutsk to find us. He had been an exchange student in California, and his hosts introduced him to Jesus. He was fluent in Sakha, Russian, and English. Anatoli was excited because he wanted us to tell his people about the Lord.

New Year's Eve

Fireworks lit up the sky as we looked forward to the new century. Listening to President Boris Yeltsin's New Year's speech, we were shocked when he announced his resignation. He promoted Vladimir Putin to be the acting president of Russia for a fresh start in the new century.

Y2K

The Millennium Bug, Y2K, was a worldwide scare where computer users feared their technology could be adversely affected. We had interruptions before Y2K with our heat, water, electricity, and phone, but our older systems were not computerized. The abacus would still work in the stores. Life goes on!

2000
McDonald's in Russia

How far away is your nearest McDonald's? In Siberia, our closest one was six time zones or six hours away by jet. McDonald's in Moscow was jam-packed with numerous short lines. They sold good burgers, were clean, friendly, and had reasonable prices. A Big Mac costs 39 and one-half rubles (one dollar and 38 cents). They kept prices low by buying most of their food in Russia. Students were employed starting at 20 rubles (70 cents) per hour.

The grand opening in 1990 of the first McDonald's in Russia happened a year before the fall of communism. A record-breaking 38,000

Soviet people lined up for hours on that first day for a taste of America. They had a 900-person seating capacity, inside and outside—the biggest one in the world. Smashing the previous record of 9,100 burgers, the Moscow McDonald's sold 34,000 burgers in one day!

Christmas Concert in Pokrovsk

The 30 members of the Pokrovsk church organized a unique Christmas program on December 25th, the day the rest of the world celebrates. Only four unbelievers were present in the city's clubhouse.

Then, on January 7th (Russian Christmas), we rode the bus to Pokrovsk for their Christmas concert in the same building. We were overwhelmed to see 300 eager people listening to the two-hour program in both Sakha and Russian! The Christians distributed literature to everyone. It pays to pick the correct date in Russia.

Going the Wrong Way: David

The boxes and bags in the mission van kept shifting on the rough roads while on the way to Nyurba. Three of us sat in the back on a bench seat facing the rear.

Russians like to drive all hours of the day or night. No maps were available, and the roads were not well-marked. We were looking for a village off of the main road in the early morning hours. Watching a beautiful sunrise, I realized that if we were going west, the sun should be coming up on the other side of our vehicle. I commented to our driver, "I think we're going in the wrong direction." He didn't agree, but the driver traveling in the other vehicle had similar thoughts and soon stopped.

The typical response would be to ask someone along the road. Two Russians knocked on doors with lights on, but there was no response. Then Kay and Ksenia got out and found a Sakha person to ask. We were thankful to have a stranger redirect us.

In Nyurba, their church held two evangelistic meetings in the city auditorium with a large stage. More than 300 people attended as we remembered 2,000 years since the birth of Jesus.

The Sunday after the meetings, 15 new people attended their church. The congregation of 40 had grown and now had a pastor. They were an outstanding church in praying for and reaching out to others.

On our way back to Yakutsk, we gave a hitchhiker a ride. Ksenia witnessed to him and gave him literature. We picked up another hitchhiker a little later, and Ksenia shared her faith again. After they got out at their villages, she reported, "Both men said that was the first time they ever heard about Jesus!"

Election Day Winner

Vladimir Vladimirovich Putin, age 47, won the presidential election on Sunday, March 26th. The lackluster presidential campaign was continually on the radio, TV, news media, posters around town, and in our mailboxes.

The following showed Russian humor, with an account of a simple Chukchi person unwittingly providing a salient social tradition. (The Chukchi are an ethnic group in northern Siberia near Alaska.)

A Chukchi was stranded at a remote location in the middle of winter. A small plane managed to deliver an aid package to him containing a large box of food along with newspapers, a radio receiver, and a TV set. One week later, the Chukchi sent a distress message: "I'm hungry, please help me!"

An aid organization radios back to him, "But we just sent you a large shipment of food!"

The Chukchi replied, "I looked at your newspapers, and it was all about Putin. I turned on the radio: Putin again. The TV was the same—Putin, Putin everywhere. I was afraid of those cans." (A joke from RFE/RL Russian Election Report, No. 3 (11), March 2000)

Second Sakha Christian Conference

The second Sakha Christian Conference met in Oktyomtsi in April. Over 100 guests were present, representing 400 Sakha believers scattered across their republic. All were recent believers. Stas led the singing while playing his guitar. The theme was the Christian home and family with practical ways to improve them to glorify the Lord. Stas mentioned our newest project of collecting and printing a booklet of testimonies. That evening, three people gave us new ones.

Sakha Language Course

We took a unique Sakha language course. It was not intense but challenging enough to make us study. The course gave us a boost in assisting Sakha Christians to produce more Christian literature. Several projects were at different stages of readiness.

Sakha Words Matter: Kay

The words we speak, how we express them, and which language we use are essential. On Friday before Easter, I asked a Sakha lady how to say the traditional Russian Easter greeting in Sakha, "Christ is risen!" with the response, "He is risen, indeed!" Fortunately, it was not a complex phrase to memorize like a few others.

On Easter Sunday, Anya phoned and asked where the church was meeting. She informed us her mother would be coming. At church, I used the new-to-us Sakha Easter greeting with Valentina. She responded immediately. We also shared the same greeting with other Sakha people. We wanted them to internalize spiritual concepts in their language rather than relying on Russian for their spiritual growth.

Anya and her parents joined us for Easter dinner. Valentina, her mother, brought a song she had composed based on "Christ is risen!" We were thrilled to receive another original hymn with a catchy melody.

He Popped the Question: Kay

We were traveling toward Namtsi to evangelize in two villages along the way. Out of the blue, Vera, our English translator, asked me, "What should I do? Victor wants to know by next Sunday if I will marry him." We all had missed any signs of particular interest! Vera was shocked as much as we were. Victor had repented five years ago while serving time in prison. He was a growing Christian who played the guitar, and they harmonized well together. His goal was to do missionary work—with Vera in the Sakha Republic.

We were so surprised; it was challenging to think clearly, but everything seemed to be moving in that direction. Russians liked to pray about it and make quick decisions. Victor and Vera announced their engagement the following Sunday.

Happy Fourth of July: David

Today was a happy day in Yakutsk. A Sakha friend invited me to watch the wrestling matches between the USA and the Sakha Republic. The flags of the USA, Sakha, and Russia were flying high in the stadium. The American team consisted of 20 wrestlers from various states plus seven coaches. The local men were great wrestlers, and they won. My friend cheered loudly for you-know-who!

That evening, we saw our first copy of *Jesus Christ Has the Power to Save Us* in Sakha. We coordinated this project with Sargylana and a man from England more than a year ago. Ten thousand copies of this 37-page booklet were paid for and printed by Scripture Gift Mission International. The stories were compiled from the Gospel of Mark and emphasized the power of Jesus Christ to forgive sins, drive out evil spirits, heal the sick, and control storms.

Baptisms in Borogontsy

We took an overnight trip to Borogontsy, a Sakha city of over 7,000, to witness the baptism of four people. The Lord placed these believers in different locations to spread His Word. More laborers are still needed.

Where to Share Next: David

Gus Matero asked me to travel north on an evangelistic trip along the Yana River, which flowed into the Arctic Ocean. However, air transportation did not work out due to gas shortages.

So we changed plans. With Stas as our translator, we traveled to 11 Sakha villages south of Yakutsk. Most of our meetings were impromptu, held in the open on a pile of house logs or at apartment complexes.

In Pokrovsk, we talked with children sitting on their apartment steps. We had to move when someone in an apartment above threw water out of a window at us. Then we visited the hospitals. This trip was an excellent opportunity to share the gospel and distribute literature.

On the Aldan River: Kay

Since David planned to travel with Gus, I became Ksenia's partner on a two-week mission trip. We flew east to Ust-Maya, where we stayed with Marina, a believer who taught English.

A Difficult Job: Kay

The next day, the three of us caught a ride across the Aldan River in a small boat to Troitsk. After supper, when the ladies finished milking their cows, we had a Bible study at the home of a Christian. We stayed with a veterinarian who was not a believer. On Saturday, we picked wild strawberries with the Christians and swatted swarms of mosquitoes.

Sunday morning, our new hostess announced, "Today is a workday in Russia." She assigned all of us jobs: washing dishes, mopping the floor, digging post holes for a chicken pen, and moving it. Marina protested, but Ksenia told her, "We need to show love." So we got busy. First, I washed the dishes, then helped Ksenia dig post holes. A lady came to talk about spiritual matters. Marina dismantled the old chicken pen. We assisted our hostess in constructing the new one.

Later, Ksenia called us over to pray with a different lady ready to repent. Another woman from the Saturday Bible study came to ask questions. Finally, the chickens were transferred to their new home, and our work was done.

We joined our hostess to drink tea and discuss spiritual things in their culture. She asked my opinion about their traditional practice of "feeding the fire" to contact the spirit world, seeking help and protection. I responded that some Natives in Alaska observed this practice, but they were still experiencing fear in their lives. After answering her questions and explaining the gospel, our hostess repented! Joy filled our hearts and there was rejoicing in heaven!

Another Sakha Town: Kay

Next, we found a man to take us to Ezhantsy, but the driver had to look for gas. After fueling, we had traveled only a short distance when the motor made a strange noise and quit. The driver worked on the engine for several hours before he repaired it. We finally arrived at midnight. The village power plant went off at 11:00 p.m., but we drank tea and ate in the dark. We were glad to crawl into bed.

The following day while walking around town, we shared Christ with several people on the street. One older lady, her grandson, and a blind man with severe hearing problems accepted the Lord that day. Ksenia had to shout in his ear so he could hear, "Do you believe there is a God?" He responded loudly, "Yes!" He had been thinking about repenting, but said he kept hearing his grandfather's and great-grandfather's voices telling him not to listen to God. They told him, "If you follow us, we'll give you your eyesight." Ksenia responded, "They're lying. Don't listen to the spirits, but turn to follow Christ." He chose to pray and ask God to forgive him as he rejected the evil spirits.

This man's wife left him previously because he drank so much. She raised five children by herself but had become a Christian. The next day Ksenia suggested that he make peace with his family. At first, he commented that they had separated too many years ago, but then he understood the need to make peace. He asked his wife and two of his older children to forgive him, and they did.

Back to Ust-Maya: Kay

Our host informed us about a barge to Ust-Maya, but we had to leave immediately. We rode on a three-wheel motorcycle to the beach. To our disappointment, the barge carrying coal had already left. Our host dashed to the nearest house and asked him to give us a ride out to the boat.

As we approached the barge, we threw our rope to the crew and proceeded to get our baggage and ourselves onboard. Marina grabbed my arm and shouted, "Katya (Kay), climb up!" as she pulled me over the top of the railing.

The ride back to Ust-Maya took seven hours. We talked to the crew on the boat, drank tea, ate fish soup, and enjoyed the beautiful scenery. I missed not seeing any moose or ducks, but I saw a mounted moose head and caribou antlers on several homes.

Teaching English at Camp: Kay

As soon as I returned to Yakutsk, we started teaching English at Camp Bingo in the city. We and other Americans had been helping at the camp in exchange for using several of their buildings for a Christian camp later in the season. It was summer, and the kids preferred camping more than learning English. We showed them photos of our family, which piqued their interest. They wanted to meet Americans and learn about how we live.

The Children of Asia

The second annual Children of Asia International Sports Games took place in Yakutsk in August. Tuymaada Stadium was just half a mile from our apartment. Near the stadium, the front sides of wooden buildings and fences were freshly painted in bright colors. The stadium was the largest in the capital city. We got tickets to attend a dress rehearsal for their opening exercises, which was an action-packed hour and a half of pageantry. A massive crowd packed the stadium.

The participants numbered 1,178 and came from all over Russia and Asia. The games included three unique and ethnically Sakha ones: an unusual running and jumping event, the greased stick pull, and *Hapsagay* (agility) wrestling. This wrestling event explained why many Sakha young men had misshapen earlobes from rubbing against their opponents.

Fast-forward: Anatoli, the young Sakha believer from Amga had those kinds of "Sakha" ears from wrestling. We thought they were a Sakha status symbol! He told us, "I want to have plastic surgery on my ears so the girls will like me!"

Buying Tickets: David

We attempted to buy tickets for the hydrofoil to Krest-Khaljai. One could purchase tickets only 24 hours in advance. On our first try, I put our names on the list, but they sold out before getting to us. Someone put us on the list the next day, but I didn't see our names when we arrived. They had written in Sakha about the elderly non-Native couple. A Sakha person understood and told us to hurry to the ticket counter. Now we had tickets!

The boat had seating for 60 passengers. We had company at home, so we packed late into the night. Then we had to be at the riverport at 5:00 a.m. We relaxed while traveling for 11 hours on the hydrofoil. The fall colors were beautiful along the calm Aldan River.

Harvest Sunday

The evangelical churches highlighted the harvest of another growing season, thanking the Lord for His abundant goodness to everyone. Beautiful displays from their gardens and the forests adorned the churches.

The service included songs of thanksgiving and praise by all different age groups. After the service, everybody feasted. The celebration dates were staggered over several weeks so one could thank the Lord in other churches as well. We enjoyed three Harvest Sunday services.

They usually had a good crop of staples such as potatoes, carrots, and cabbage from the fields. Tomatoes and cucumbers grew well in their greenhouses. A variety of berries and mushrooms were picked from the forest.

A Bigger Family

Our son Dan Henry married Linda Sue Isaacson in September in a private ceremony in Fairbanks. Their family was enlarged to eight children. Dan had three children: Aaron and Kara (Henry) Fields, Tim and Bambi Henry, and Andrew and Suzanna (Henry) DeCarlo. Linda had five children: Joseph and Rachelle (Isaacson) Warner, Seth and Erin Isaacson, Luke and Rebecca Isaacson, Jared Isaacson, and Patrick and Lynnea (Isaacson) Schuster. They became a large Christian family working together.

The Big Potato

We caught the flight to the "Big Potato," Moscow, leaving Yakutsk at 4:00 a.m. on the last day of October. Halloween was not celebrated in Russia, so we avoided all the pranks. We thanked the Lord that our new visas came in time to fly to America.

We rested in Moscow for two days as we readjusted our internal clocks to the six-hour time change. We ate on the street at the Russian fast-food restaurant *Kroshka Kartoshka* (Little Potato). They served healthy baked potatoes with a wide variety of toppings. They eat lots of potatoes in Russia! We helped their business grow every time we were in Moscow.

Landing in the USA: David

Our next flight was 11 hours on Aeroflot Airlines over the Atlantic Ocean to Newark, New Jersey. We rented a car and started driving on I-95 with three lanes going in our direction. It was so easy to drive in America. We kept up with the traffic, and several vehicles even passed us. I glanced at the speedometer, which read 85, and backed off a little. There were no police checkpoints or mandatory stops like in Russia either.

In Egg Harbor, New Jersey, we joined friends at Faith Bible Baptist Church and shared about the Lord's work in Siberia. Then we drove to Gaithersburg, Maryland, to see our granddaughter, Suzanna, and her mother, Donna. Suzanna kept improving her cooking skills to everyone's delight.

Warming in the South: David

Our next stop was Bradenton, Florida, at Calvary Baptist Church. It was good to be in a church with many older saints serving the Lord. They were actively praying for us and the work in Siberia.

Thank-You Letter

Pastor Valeri and his wife, Zoya, from Krest-Khaljai, sent an unsolicited letter of thanks. Excerpts follow:

"We want to thank you and all American believers for your love to the Sakha people, for your help and prayers. We want you to know that the Bibles and spiritual books printed and sent by you have helped many Sakha people find their way to God and receive His salvation. Most of all, we thank you for our brothers and sisters with your help who come to our country to talk about God and His Son. David and Kay Henry's lives and love for Jesus are shining examples for us. We have known them for a long time, since 1994, and all these years, they helped us with word and action to grow in faith. That's why we say their life is the best testimony of the faith.

"So we thank God and you for David and Kay. Many Sakha believers know and love them. We thank our Wonderful Savior for you all because your faith is being proclaimed throughout the whole world."

Celebrating 90 Years: Kay

Our family celebrated my father's 90th birthday. Family members rented 10 rooms and a conference room at the Comfort Inn in Birch Run, Michigan, for three days over Thanksgiving weekend. This was the first time the extended family had come together for a reunion in 15 years. The celebration began with dinner at Fritz's Restaurant in Richville.

Fifty family members and spouses were present. Five of his children came from around the world: Shirley Spear from New York; Nancy Horwath from Canada; Jim Temple from Tennessee; Sheryl Cavazza from Bali, Indonesia; and myself from Siberia. His eldest son, Tom Temple, was deceased. The other family members included 10 of the 17 grandchildren and 14 of the 24 great-grandchildren. "It meant everything to me, them all coming," said Bob Temple. "I figure it's better they visit me when I'm alive than at my funeral." We made good use of the hotel's pool and conference room with swimming, family skits, time to visit, and playing games together.

We thanked the Lord for Dad's strong faith and health. He still drove his car and he was active at Vassar's First Baptist Church. He taught the adult Sunday school for over 20 years, served on the mission board, and hosted a Bible study in his home. He served for three years as president of the Michigan Gideons and personally visited all 68 state camps. He enjoyed traveling to visit family members, fishing, playing games, and using his e-mail system.

2001
A Near Half-Century

Staff writer Nancy Tarnai wrote a long article about our lives as missionaries, which appeared in the *Fairbanks Daily News-Miner* on January 15th, 2001. The title: "A Near Half-Century of Mission Work." She gave a complete overview of our ministry which spans two continents. Nancy was impressed that we planned to continue missionary service for as long as possible.

A Chilly Welcome: Kay

We arrived back in Yakutsk at 4:30 a.m. in January. Flying three-quarters of the way around the world, we passed through 18 time zones via New York and Moscow. It would have been closer to fly from Alaska via Korea, but the connections and the number of flights were limited and more expensive.

January was not the ideal time to travel to Yakutsk. It was minus 30 degrees when we arrived. I asked the cab driver how cold it was, and he replied, "It's warm!" It had just been negative 50 degrees, so we were glad it was "warm!" We were happy to be back to continue the work God had given us to do in Siberia.

Father's Stroke: Kay

In February, my father, Bob, suffered a stroke at home. Debbie called a few minutes after we returned from a ministry trip informing us. My sisters, Shirley and Nancy, and their spouses were there to help him.

I phoned the hospital in Saginaw, Michigan, and talked with Dad. The doctor asked Dad, "Who is the president?" He answered,

"Bush." Then the doctor looked at him and said, "You're right; I was thinking Clinton."

Rush Hour: David

I received the game Rush Hour from Debbie for my birthday. We had lots of visitors, and they all enjoyed playing the game. More people in Yakutsk were buying vehicles, the roads were narrow, and everybody was rushing to go somewhere, almost like the game.

Life Goes On at Minus 40 Degrees: David

Does life go on at minus 40 degrees or colder? Yes, but we dress like a cabbage—in more layers. If you're curious and counting, we wear three pairs of long underwear on days like this. Minus 40 degrees separates the men from the boys! If we were bears, we would hibernate. If we were summer birds, we would have flown south. However, we chose to live in the far north with the others who stayed there.

We rode the bus with Rhaylene to Mokhsogollokh for a Russia field team meeting at Frank and Barb Emrick's apartment with their children. I sat in the last seat available in the very back by an emergency door for our two-hour trip. Gaps around the door let in the fresh air keeping the window clear, so I could see outside. The larch, spruce, and willows were heavily laden with snow and frost. I thanked the Lord for all the beauty of the cold. Kay and Rhaylene had seats in the middle of the bus, but it was not warm enough to clear the frost off of their window.

Our return trip was colder at negative 50 degrees. We often remarked, "We don't want to travel when it's colder than minus 40 degrees, but we can't stay home all winter!" The ice fog was so dense that we only saw the fuzzy shapes of vehicles and buildings close by. Traffic slowed down, and everybody drove with their headlights on. Trucks and cars spewed out clouds of exhaust fumes. Our apartment was cool but comfortable.

The Wedding in Nyurba

Stas and Nadia invited us to their wedding in Nyurba. It was a long 24-hour drive from Yakutsk. They both loved the Lord and planned to serve Him together. The church went all out for the big celebration.

Before their big day, the church coordinated an evangelistic outreach concert in the large city auditorium. The Christians wanted everybody to know about Jesus, who loved and died for them.

Flood Time

The waters of the mighty Lena River kept rising from the melting snow and runoff from the rivers. We felt safe living on the fifth floor, but others in low-lying areas did not. One young Sakha family—Slava, Shura, and two energetic boys—moved in with us. They lived on the first floor of a frame apartment complex on low ground. God provided opportunities for us to practice Sakha, but especially to live the Christian life before them for several weeks.

People were advised to stay home. Schools closed. Hospitals moved patients from the first floors. Vodka sales were temporarily banned. Everybody was instructed to fill up extra jars with drinking water. A feeling of apprehension hung in the air. In the end, only 200 homes in low-lying areas of Yakutsk were damaged.

Everything Sakha: David

A Sakha Sunday service was started in Yakutsk. The whole service—praying, singing, testimonies, and preaching—was predominantly in their language. I preached several times on themes of particular interest to the Sakha people. There was always good interaction as they applied God's Word to their lives and culture. The attendance ran between 20 and 25. A weekly Sakha prayer meeting focused on prayer and praise.

Meeting in Moscow

We flew to Moscow to meet a single fellow, Mike Harms, at the airport. He had flown nonstop from Seattle. We took him on a tour of the Kremlin and gave him a quick introduction to Russian culture. Since there was a mix-up on his room reservation, we arranged for him to stay in our room at the Hotel Rossiya (Russia), which overlooked Red Square.

At the airport, we had to fight our way through all the people for his next flight. While standing in one long, slow-moving line, our friend Aleksei, with Athletes in Action, showed up and took Mike to the front of the line. He got through that bottleneck in time to catch his plane to Yakutsk. One of the agents was sure he was in the wrong line since very few people flew to Yakutsk. He was part of the SIS (Summer in Siberia) program.

Next Stop, Chicago: Kay

We flew directly from Moscow to Chicago on Aeroflot Airlines and spent time with my father in Vassar, Michigan. At age 90, he was still recovering from a stroke. My brother and sisters had helped him, but now it was my turn for a month.

The New Log Church

We returned to Fairbanks excited to see the newly built two-story, 4,400-square-foot log building for the Fairbanks Native Bible Church. Many local people and summer work crews labored together to complete the structure at 5360 Fairchild Avenue near the International Airport on two and a half acres. David Joseph was the head pastor, and his wife, Marci, taught Sunday school. George and Judy Richardson had a significant role in the project, and he preached there as needed.

Jubilee Anniversary Conference

Every member from Alaska, Canada, and Siberia celebrated together at the 50th InterAct Ministries jubilee anniversary at Victory Bible Camp. It was a great four days to meet members in person and rejoice in God's faithfulness.

9-11: David

Back in Michigan, we spent a few days with Kay's father in Vassar. Then we drove our rental car to Flint to catch our flight to Chicago on that fateful September 11th. We heard on the car radio that the first plane hit one tower of the World Trade Center. We hurried to the check-in counter for our flight. On TV, we watched a plane hit the 80th floor. Then we were called back to the ticket counter.

A reporter with a Flint TV station interviewed us in the lobby. I told him about my view of terrorists and where we planned to fly. Everybody was in a state of shock!

All flights were canceled! So we rented another car and returned to Vassar. My father-in-law had been watching the events unfold on his TV and stated, "I knew you would be back." Our prayers and sympathies went out to the families who lost loved ones. The attack was a huge wake-up call for America—and us who knew the Lord—to live for Him.

First Flight to Moscow: David

Every day, we checked in and rescheduled our trip from Flint to Chicago. Our Aeroflot connection was delayed for seven days. I listened to many recorded messages until I finally spoke with an agent. Our flight was the first one after the attack. She said, "Your plane is overbooked. Come in at least four hours early." We took her advice seriously, standing in the first line for five hours in Chicago. Then we waited in two more lines, over a block long, to pass through the final security check.

In Russia Again

We were delayed in Moscow for several days until we could buy tickets to Yakutsk. There we connected with Sergei, the mission director in the Sakha Republic. He returned on a flight from New York where he had visited his parents, who had immigrated to America.

After arriving in Yakutsk, we flagged down a taxi. When the driver realized we were Americans, he told us in very animated terms, "These terrorists are wild animals—not people!" He was a Muslim from southern Russia. We appreciated his safe driving, sympathy, and another opportunity to witness to someone who needed the Lord.

Everywhere, the Sakha and Russian people offered us their sympathies and understanding for this unbelievable, savage attack. Many people prayed for our safe return to Siberia. They knew what it was like to have a major war on their home soil during World War II.

Scrubbing Detail

Upon arriving home, we helped scrub our common apartment hallway and steps. Signs were posted on every floor about the event, and we were personally informed in case we didn't understand. The cleaning was long overdue. It was fun to work alongside our neighbors.

A Shower in Our Kitchen: David

Early Sunday morning, we were awakened to the sound of water spraying under our kitchen sink. I tried to stop it with duct tape, but the leak got worse. I phoned the apartment complex technician, who responded quickly and shut off the water.

The next stage was mop-up. We dried our floor and realized the water had leaked into the apartment below us. Kay went down to the fourth floor to help the elderly Sakha widow. Her daughter and granddaughter also came over to help. The moisture ruined her wallpaper.

The final stage was to pay for the damages. Water even dripped down to the apartment on the third floor and damaged the lady's phone, which was sitting on a table. We didn't argue but gave her money to replace her inexpensive phone.

Inspired: Kay

I wrote a devotional, "Lost Fruit," about the fruit of the Spirit and the need for patience in one's life. My example was when a water pipe developed a severe leak. It was a huge mess to clean up, requiring patience and dealing with the neighbors below us. Living in Siberia taught us a lot about patience, especially since this was the third time we had water problems. The devotional is found in *From the Heart of a Lion*, published by Bryan College on page 93.

Sakha Church Service: David

Due to our water leak, we missed the 10:00 a.m. service but made it to the 12:30 p.m. meeting in Sakha. I started the service with a few Sakha words mixed with Russian since the regulars were not there. One of the ladies led the music. A Bible school student shared about his recent trip to several villages. Another woman gave her testimony, and then a college student came forward to repent of her sins. I was delighted to assist. The three of us kneeled as the lady prayed, and then the one who had witnessed to her, and I closed in prayer.

Two Calendars

This year we prepared two calendars. One was full-sized with a picture of a Sakha barn and piles of hay to feed cattle. The accompanying Bible verse was Matthew 4:4. The traditional Sakha barn represented their essential food supply of milk, milk products, and beef.

The pocket-sized calendars had another idyllic Sakha scene of a canoe sitting in the grass along the edge of a calm, serene lake. The accompanying verse was John 4:14. The Sakha people enjoy a quiet area. The lake represents pure, clean water that one can drink.

Living by Faith (2002~2004)

2002
All Out for Our Lord

How radical would you go for our Lord? Pastor Valeri went all out. A year and a half earlier, he chose to sell their family cows. He wanted more freedom to travel to surrounding villages, reaching out to others with the Good News of the gospel, but having cows required daily feeding, caring for, and milking them.

Everybody in their town kept cows. Pastor Valeri still had three young children at home and two in college. They entertained many guests. Zoya wanted to work with her husband, not against him. Yet she had doubts. How would a mother provide for her large family without cows? How would she explain this to their relatives and villagers who already thought Christians were strange?

God Provided

Until two years ago, Pastor Valeri was the only man in their church of 22. Now there were six men. All the wives had been praying for and witnessing to their husbands. Last fall, the members decided to take turns praying around the clock for their husbands. Dariya had prayed for her husband, Budimir, and he repented. God delivered him from a lifetime habit of drinking and smoking. His Christian wife had endured much gossip and pain. He was the town drunk, and God transformed him. Now he wanted to serve the Lord.

Dariya's husband was Zoya's answer to her prayers. Dariya and Budimir had two milking cows. They decided to give one of their cow's milk to Zoya and her family. Everybody in town knew how God provided for Pastor Valeri and their family.

Traveled to K-X: David

We were invited to Krest-Khaljai for Budimir's baptism and birthday celebration. Zoya arranged a ride in the school van, but they had too many passengers, so there was no room for us.

Two days later, we received an urgent phone call, "Be ready in one hour!" I gulped and hesitantly responded, "O.K." We rode in a private van to K-X. We were a little disappointed not to have ridden in the more "comfortable" school van, but this was what the Lord provided for us.

After arriving in K-X, we heard that the school van had had an accident while traveling from Yakutsk. The van had rolled over, and several passengers were injured. We felt sorry for them but were thankful that the Lord had spared us.

Budimir's Celebration: David

The temperature in K-X hovered around minus 40 degrees, but on Sunday, it was minus 50 degrees! I wore my new moose-skin fur boots with one-inch-thick felt soles. My feet were warm as toast! My old pair of horse-skin boots couldn't travel any farther.

The church did not have a baptistry, but it had a bathroom with a bathtub. So they filled the tub, and the men at the church witnessed Budimir's baptism. Then we celebrated his 55th birthday with a big cake. His changed life was a great witness to all in K-X.

Praise to God: David

Seven years of devoted work by 22 musically gifted Sakha people—combined with the Lord's help and direction—produced the first Sakha hymnal, *Praise to God*, with 120 hymns. This grassroots project kept growing! The translated songs served as a bridge to encourage the creation of original Christian songs. Several put Sakha Christian poems to music. A few were inspired to compose original prose and music. Others used their national round dances as a musical style with Christian words. The book cover displayed a bouquet of their national lilies. We received 3,000 copies one day before the Sakha Conference.

God wants the Sakha people to make His name great through music in the north and among their people: "'For from the rising of the sun, even to its going down, My name shall be great among the Gentiles; . . . For My name shall be great among the nations,' Says the LORD of hosts" (Malachi 1:11).

Fourth Sakha Christian Conference

The new Sakha Hymnal, *Praise to God*, was the highlight of the fourth Sakha Christian Conference. More than 100 people attended meetings in Oktyomtsi. People brought their hymnals and Bibles to the services. Hearing so many people enthusiastically sing these songs was sweet music to our ears.

Stas wrote a new gospel tract for unbelievers asking fundamental questions about everyday life to get the readers to think. Now there were two tracts in Sakha.

Ksenya sang a solo as everybody listened with rapt attention. The song had been translated from Russian, but the melody was a beautiful, new Sakha melody she had composed.

Lyuda's Original Song: David

Lyuda excitedly told me, "Now I plan to write my original song. And Ksenya agreed to compose the melody!"

Two weeks later Lyuda had written eight verses for her new song praising our Creator. We drove across the frozen Lena River to Ksenya's hometown. She used her God-given musical talent, and they sang the new song together. Kay recorded this song and another new song that Ksenya combined with a Sakha Christian poem about Christmas.

Audio Cassettes

We produced two 60-minute Sakha audio cassettes with a mix of songs, Scripture readings from Psalms and James, and a sermon. We rushed to get these done and made them available one week before leaving for the US.

Honored to Pray: David

At church, Misha asked me to pray over Pasha, who was going to a Sakha town north of Yakutsk as a missionary. Pasha had accepted the Lord after reading *Jesus and the Early Church* and realized she was a sinner.

Winging Our Way to Alaska: Kay

We bought our plane tickets in faith as we waited for our "must-have" three-month visas that would allow us to both leave Russia and return in August. The visas arrived one day before we flew from Yakutsk to Moscow. Then we flew to Detroit to visit my father. We reported to several supporting churches in the thumb area of Michigan before heading to Fairbanks.

Back Home in Yakutsk

We returned to Yakutsk in the middle of August. Delayed in Moscow, waiting for the next available flight to Yakutsk, the Lord gave us unexpected contacts with two key Russian Christian leaders. Then on our flight to Yakutsk, we were seated next to Sasha, who was also going to the pastors' conference. We were thankful to make the right connections at the right time.

Pastors' and Leaders' Conference: David

The leadership conference included 40 men from throughout the Sakha Republic. The two main themes were preaching and being culturally relevant in the ministry. Using Philippians 2:5-8, I spoke about Christ's giving up many divine rights to reach people effectively. We also had given up some of our "rights" to reach the Sakha people more effectively, such as leaving the comforts of Alaska and speaking English.

Seventieth Jubilee

We went to Oktyomtsi to join in the 70th birthday jubilee for an elder in their church. His wife was Russian but a fluent speaker of Sakha and an outstanding witness for the Lord. The little church went all out to honor him. The best part was witnessing to 20 unbelieving friends who attended the gala event. The Sakha Christians used every opportunity for evangelism and showing Christ's love to others.

They All Knew Us: David

What should we do? We needed tickets for the hydrofoil to Krest-Khaljai. The little window, six inches by eight inches, was closed. We knew the process from past experiences, but that did not excite us. Now we silently prayed again while trying to figure out our next move.

We found a lady carrying around the list and asked her to sign us up for going to K-X. She smiled and instantly responded, "Evangelicals." We were surprised and said, "Yes." She continued, "You're going to see Valeri." Again we responded, "Yes." We had never seen this woman before. She had never attended church but knew Valeri was the pastor in K-X, where she lived. Another Sakha lady standing nearby flashed a big grin, saying, "I saw you on our local TV three years ago."

It was nice to be known as believers, but how do they all know us? We do not live in an isolated monastery or under a rock waiting for the Lord's return. The Apostle Paul wrote about the believers in his day, "You are our epistle (letter) written in our hearts, known and read by all men" (2 Corinthians 3:2). We were being watched 24/7. People know our actions and words and guess our thoughts.

The supply of tickets ended just ahead of us. We were number 24 on the list, but many in front of us were buying multiple tickets. One person told us, "You're Americans, go to the top person and ask for a special privilege." It was a temptation, but we replied, "We are ordinary people; we will wait."

Returning the next day, we waited several hours. This time, we were able to buy our tickets. The agent reminded us, "Be here at 4:00 a.m. for check-in."

We set our alarm for 3:00 a.m., and a taxi picked us up at 3:30 since the city buses do not start running until 6:00 a.m. The taxi driver asked us out of the blue, "Where is your church located?" I responded, "On Pilotov Street," as the driver filled in the address number. He was an airline pilot making ends meet by driving a taxi in his spare time. I gave him a gospel tract and invited him to church. He responded, "I don't have time for God." At least he knew where the believers were in his city of 250,000 people.

We lined up for registration at the riverport and weighed our baggage. It was heavy but just under the limit. An elderly Sakha lady with a wrinkled face spoke to Kay, "I saw you in my church in Khandyga three years ago." We remember faces better than names, but we did not recognize hers. Yet she knew who we were.

The hydrofoil was full, so we scrambled to find two seats together. Every seat was eventually taken for our 11-hour trip to K-X. During one of our stretch breaks, another Sakha lady stood up and caught Kay's attention. Kay recognized her face. They had met in the Khandyga church three years ago!

With the Believers in K-X

It was encouraging to minister and worship with the Sakha church in K-X for 10 days. The church was well-known, even in the surrounding villages. The Christians lived for the Lord with their words and everyday actions.

A motorcycle pulled up to Pastor Valeri's home one morning at 6:00 a.m. The stranger from another village introduced himself and relayed his dilemma. He was seriously considering suicide. His wife and children had left him because of his drinking and inappropriate actions. The young man prayed with Valeri, "If there is a God, please bring my wife and children back to me." The Holy Spirit led him to the pastor for spiritual help.

Wanted—A Few Good Sakha Men: David
On our next trip from Yakutsk, we traveled to two new Sakha villages about four hours across the Lena River. The roads were extremely muddy and dangerous. We went to visit new believers who needed help.

Lyuda, who now lived in Yakutsk, was burdened for her people near Borogontsy. This past summer, God threw open the door when 12 women repented in Mayagas through her testimony. She arranged a meeting at the school. All the students were crowded into one large classroom, plus a dozen teachers. They listened intently during this two-hour break from their classes. They asked many spiritual questions, especially concerning sin in God's sight. Our team presented the Good News of Jesus Christ and distributed literature. After school, we continued for another hour of sharing with the teachers.

Then we drove to a neighboring village of 700 people that was unusually clean. Many brightly painted windows, doors, and fences were seen along the main dirt street. The first order of business was to eat a hearty meal as they rounded up the 12 new believers for our meeting.

When it was my turn to speak, I commented, "Today we saw many men along the road, but where are the men in your meetings?" Most of the women were married, so I encouraged them to live a Christian life before their husbands and pray for their salvation. We read the Bible, but our spouses and others observe how we live. We are the "fifth Gospel" that everybody reads. The women gathered together for weekly fellowship.

Calendars and Christmas Cards
Aita used her artistic ability to design two Christmas cards. One drawing showed Jesus being born in a Sakha barn with Mary and

Joseph dressed in national attire. The front of the card included "Merry Christmas" and "For there is born to you this day in the city of David a Savior, who is Christ the Lord" (Luke 2:11). The backside had "Silent Night" and a picture of a barn. We printed 5,000 copies. Sakha people thought Jesus was a Russian, Jewish, or American Savior. We wanted them to realize Jesus is also a Sakha Savior.

"Sakha Christians Crave Literature"
The David C. Cook Communications Ministry International published an article we wrote about Sakha literature titled "Sakha Christians Crave Literature" in the December 2002 issue of *InterLit*, an international magazine of Christian publishing. Secular books are readily available in Sakha, but only a small handful of Christian books exist in their language. Patrick Johnstone's *Operation World* estimates that over half of the conversions were linked to Christian literature.

2003
A Quieter New Year's Eve
New Year's Eve was spent with a Sakha Christian guest who needed our help and comfort. We welcomed the New Year in prayer as we tried to turn down the volume of noisy fireworks outside. We declined an invite from another Sakha friend. Venturing out in negative 50-degree weather mixed with thick ice fog and party-goers did not appeal to us.

Student Stranded at Home
Andrei had been sleeping on our couch while attending college for two years. He enjoyed the New Year's break in K-X with his family, but his return was delayed due to the extreme cold and ice fog. People everywhere hunkered down as they tried to stay warm. No public transportation was operating.

Youth Conference: Kay
GTE sponsored a youth conference during the end of the New Year's break at their office. They planned for 50 teens, but 90 showed up! A number came from mission churches. I helped in the kitchen by washing dishes, peeling potatoes, cleaning, setting the tables, and filling pop bottles with *mors*, which they prepared in a huge pot. *Mors* is a non-carbonated Russian fruit drink made by boiling cranberries

in water and adding sugar. I quickly washed the limited supply of cups and silverware between the two meal shifts. The dining room had space for only 50.

Traveling During a "Heatwave": David

We left Yakutsk with four Sakha people to go to several villages we had visited last fall with Lyuda. After a three-week cold spell, it finally "warmed up" to minus 30 degrees. We appreciated the bright sunshine, which we had missed for many days with the ice fog.

Our first meeting was in Mayagas, a village of 500 people. Lyuda, a French language teacher, had moved there last fall to work and reach her people. Lyuda was living with a grandmother, her daughter, and two grandchildren.

Upon arriving, we enjoyed a wonderful Sakha dinner of potatoes, beef, chicken, horse meat, beet salad, and a dessert of frozen cream with berries. While we finished eating, 15 people came in for our typical two-hour meeting. Our team of six all shared and sang. The exciting part happened when one man and three women prayed to receive Jesus Christ.

We stayed overnight with Maria and Innokenti, who had prayed to receive the Lord that evening. I asked him when he first heard about God. He replied, "Lyuda told us." More of us need to simply tell others about Jesus Christ.

A Home for Invalids: David

The next morning, we traveled for two hours on a one-lane road with frozen ruts to Us-Kyuel. At the government Home for Invalids with Special Needs, 40 residents listened intently to the gospel for more than an hour. The cooks and workers stood in the hall and doorway. Lyuda translated for me, turning Russian into Sakha. We wanted the message to be understood. We also distributed literature. People's hearts were open to the gospel.

After the meeting, the staff prepared a wonderful meal and commented, "You are the first Christians ever to come here and bring us the Good News. Please, come again when we get our new building next winter."

Lyuda Told Me: David

We arrived at the village of Beidinga after 9:00 p.m., so they dropped us off at Sargylana's home to spend the night. Her house was nice and warm. She was not a believer, but she was seeking and asking many questions. I asked her, "When did you first hear about God?" She responded, "Lyuda told me." Lyuda also told her mother-in-law, one of the 12 women believers.

When people stated, "Lyuda told us about God," we thought about the time Jesus spoke to the woman at the well after she ran home: "And many of the Samaritans of that city believed in Him because of the word of the woman who testified, 'He told me all that I ever did'" (John 4:39).

Definite Flexible Planning

Overnight, the temperature dropped to negative 50 degrees. Our van was parked in an unheated garage and wouldn't start. They had to heat the garage sufficiently before it finally started. This forced us to change our plans, so we visited their village hospital. The staff stopped everything so that those who were mobile could attend our meeting and listen to the gospel. Then we went to the bedridden patients and prayed for them.

We planned to meet earlier with the believers in the community center. Since we had vehicle problems, we were able to include others so they could hear the gospel later in the day. One room was warm where we met with 20 women and a few teenagers. At the end of the service, five ladies stepped forward to repent of their sins! We were thankful to have changed our plans. Then we all went to Elizabeth's home, one of the first group of 12 believers, to enjoy a hearty dinner and fellowship before returning to Yakutsk.

On the Road Again!: David

On the road again, our ministry team headed northwest for 1,000 miles. It was both challenging and exciting to visit growing Sakha churches. Here are excerpts from my diary:

Saturday, Feb. 1st: The group from Oktyomtsi arrived at our apartment to pick up Stas and me. After prayer, our team of three women, five men, and a driver hit the road at 10:00 a.m.

We rode in a Russian *tabletka* (tablet) that looked similar to a pillbox. It was a very plain, basic vehicle. The van was often used for "Fast Medical Aid" services (spelled out in big Russian letters on the sides). I told them, "On our *tabletka,* we should write, 'Fast Spiritual Aid.'"

Our capable driver, Kolya, drove as fast as possible with his eyes glued to the road. He was not a believer, but he observed how we acted 24 hours a day.

Fourteen hours of steady riding were tiring. We passed the time with lots of conversation, singing Sakha Christian songs, listening to Christian music cassettes, and doing Bible quizzes, and I learned how to play the game Peter and Paul while bouncing along. They also sang several Sakha national songs. I recognized the melody of "*Sardaana,*" the name of their beautiful lilies.

We arrived in the Sakha city of Vilyuysk at 1:00 a.m. at the home of Pastor Aleksei and Kidona. The population was 10,000, and the evangelical church was new, with 10 to 15 attending services. After eating a big meal, we caught up on all the news. They scrambled to make nine sleeping spots, and we finally went to bed at 3:00 a.m. In the morning, I noticed that our hosts had had to lay a fur coat on the floor for their sleeping space.

Sunday, Feb. 2nd: We ate breakfast and left at 10:00 a.m. We were sorry to miss their service at 11:00, but we had to make our next destination.

We arrived in Namtsy, a Sakha village near the Vilyuy River. The church had 14 people attending, including three active brothers who led the meetings. We were served a full-course dinner and prepared for their service at 3:00 p.m. Ageet, a vibrant believer, led the meeting.

Sasha and his wife, Vera, were strong believers. He was a songwriter and published a book of original songs three years ago. Sasha used a wheelchair because of a debilitating muscular disease and could no longer play his *bayan*, a Russian button accordion.

We walked to their village clubhouse for a public gospel meeting. About 45 people listened and asked questions from 7:00 to 10:00 p.m. I especially noted their singing of 15 Sakha hymns. I was glad that now everyone had the same hymnal rather than a disposable trial edition as in the past. Then we returned to Sasha's home, ate another meal, and finally dropped into bed at 2:00 a.m.

God gives each of us One Lifetime

Monday, Feb. 3rd: Up at 9:00 a.m. and visited a few small shops. Slava and I watched the video, *Joni*. He asked, "Was she healed?" I replied, "No, but she has brought more glory to God as an invalid rather than as a healed person." He was surprised. I explained that God can and does heal people today. We need to ask God for healing and let Him decide what is best for each of us.

We drove to the village of Tamalakan, 15 minutes away, for another clubhouse meeting. I assumed things would start at 7:00 p.m., even with only a few people. I've learned to wait, trust in the Lord and our Sakha brothers and sisters, and see what happens. Several members of our group talked about the necessity of meeting even if only one person showed up.

Suddenly, 30 people arrived, including several wearing their national outfits. They led our group into a large room. Our team lined up across the front of the hall as we listened to a welcoming speech and one lady throat-singing a national Sakha song. The director of the clubhouse had set up a table with Sakha wooden goblets with three legs, filled with *kymys*, fermented mare's milk, and bread. They gave each of us a goblet from which to drink and a piece of Sakha bread. What an over-the-top welcome!

Stas and I reminisced about the time our gospel team was welcomed to the town of Sinsk by 12 singing and dancing grandmothers with *kymys* and bread 10 years ago.

The clubhouse was packed with 100 for the service from 8:30 to 11:30 p.m. Afterward, we distributed literature and talked with people individually. I noticed one Russian lady. She had moved here from Vladivostok 10 years ago, married a Sakha man, and had four children. They spoke only in Sakha in their home. Yes, the Sakha language is learnable!

After midnight, we drove back to our host's home in Namtsy and ate again at 1:00 a.m. A special meeting with the leaders of the Namtsy church couldn't wait. God used Slava's brother to introduce these people to Jesus Christ, and he was away studying at a Bible school. He was convinced he had the gift of healing. The people in the church commented that he was not humble. They questioned his ministry's healing aspect with several relatives who were not helped since they didn't have "enough faith." Our host, Sasha, was not healed, nor was his son who was born with acute Down syndrome.

That meeting ended at 3:30 a.m. as we formed a circle and prayed for Ageet, Sasha, and Slava, the leaders of this active church. They looked to the Lord for spiritual healing rather than emphasizing physical healing. Stas said, "We're tired, but it's a good tired!"

Tuesday, Feb. 4th: With less enthusiasm, I got up at 7:00 a.m., and we left for Nyurba after breakfast. When we entered the church, several in our group remarked, "They need Bible verses on the wall written in Sakha, not Russian." They've been without a pastor for more than a year. They especially enjoyed singing new Sakha hymns with our group.

They all wanted to know where Kay was and how she was doing. I explained that she was staying in Oktyomtsi with Rosa, one of our Sakha "granddaughters," while her parents were on this trip. Rosa and Kay walked 30 minutes to another village where Rosa taught Sunday school, and Kay gave her testimony. Kay worked with Galya, to create a teaching ministry on prayer. She wanted Kay's input, but Galya did all the teaching.

Yemelyan, their former Sakha pastor who had led many of them to the Lord, asked me to spend the night with him and his wife. Several years ago, he experienced burnout and left the church. He had recently started attending services again.

One issue that confused them was the promises of the "health and wealth" gospel. Certain Christian leaders from another city told him everything in his life would get better after he turned to the Lord. But things got worse. They and their family continued to suffer from serious problems, and a few of their problems were caused by making bad choices.

He explained that life was easier under communism with strict controls in every area of life, where someone else decided every aspect of your life. There were lower crime rates, less pornography, and everyone had jobs. I reminded him of all the freedom we had in Christ, and now we can freely tell others about Him.

I explained that life could get better for us Christians, but sometimes it becomes more complicated. Either way, Jesus is with us to the end of our lives. We discussed pending persecution and dropped into bed at midnight.

Wednesday, Feb. 5th: At 10:00 a.m., we met the staff at the local TV station who had many questions about Christianity.

We walked back to church for lunch. While eating, I heard several ladies in another room singing a translated song from the hymnal using a Sakha melody. We departed at 2:00 p.m. Matryona rode back to Yakutsk with us.

Due to time constraints, we drove past Namtsy as we continued to Lyokyochyon, arriving at 10:30 p.m. I went to bed at midnight, but Matryona talked with our hostess until 2:00 a.m. about spiritual things.

I was awakened at 4:00 a.m. when our host and hostess kept walking back and forth between their bedroom and the kitchen. Then they drove their son to the hospital in Vilyuysk with high blood pressure problems.

Thursday, Feb. 6th: We left at 9:30 a.m. heading back to Yakutsk. We crawled along the "highway" for the first three hours. The road was terrible with deep ruts and under construction. I hit my head only once on the roof, but the toothbrush in my bag broke from all the jostling. This was the main road connecting major areas of the republic.

At a small roadside café, I gave the clerk and several patrons Sakha calendars. I was delighted to find one of our Sakha Christmas cards on their wall among the pictures.

We arrived back in Yakutsk at 7:00 p.m. road-weary but thankful for no vehicle problems. We especially thanked the Lord for the growing church in Namtsy.

Fifth Annual Sakha Christian Conference: David

The fifth annual Sakha conference was held in Oktyomtsi for two days with 150 people. I addressed the group for 15 minutes on their requested theme of helping one's unsaved mate come to the Lord. Many women believers had difficult home situations.

Several Sakha testimonies were made into tracts. This helped meet the need for literature to share with others. Sasha wanted his contact phone number listed on his tract. Many people phoned and asked him questions about his disease, life, and the Lord. Zhenya, our Sakha artist, painted flannelgraph backgrounds and figures in a Sakha setting for the parable of the prodigal son.

Computer Crash: David

First, the good news—the final draft of the Sakha New Testament was checked and completed for the Institute for Bible Translation. The bad news—Sargylana's computer crashed without any backup. She called me, so I tried a few things, but nothing worked. I was meticulous not to erase anything. Finally, I packaged her computer and sent it to Moscow for an expert to work on it. Several months later, they recovered the files.

Tenth Anniversary: Kay

We traveled to K-X to celebrate the 10th anniversary of their church. The snow was melting, but the ice road across the Lena River was still solid.

Sakha Bible verse posters decorated the inside walls. One outside described their purpose: "For My house shall be called a house of prayer for all nations" (Isaiah 56:7b). Welcoming everyone was an oversized poster of Jesus standing with outstretched arms and the caption "Jesus our Savior." Yes, Jesus had come to K-X and the Sakha people.

The Evangelical church had a simple wooden cross on the roof. It was fitting since K-X means "cross on a small hill." Two Russian Orthodox buildings were standing in the village. One had been converted into a museum under communism. The other was beyond repair as a monument to the past.

I participated in a women's meeting as we shared testimonies, songs, and prayed together. Faithful women play a significant role in the body of Christ.

The Celebration: David

The church was packed for the Sunday morning service, which included communion using local cranberry juice and homemade bread. I was asked about substituting juice for wine. The primary purpose was to remember that Jesus died and shed His blood for us. Many recovering alcoholics have a difficult time with even a taste of wine, so cranberry juice is a great substitute.

Then we climbed over the central heating pipes, on our way to the House of Culture. There was a dramatic presentation of their 10 years, with numerous songs by individuals and groups along with testimonies. The two-hour service ended by moving to a bigger room where everyone formed a large circle. Ksenya led the national round dance with call-and-response singing one line at a time while everyone repeated it as we slowly rotated singing the words Valeri had written. Lyuda from Oktyomtsi improvised on the spot and led with the words for another song with more praise to our Creator.

Serving the Lord: David

Later I participated in laying hands on Budimir, dedicating him to assist in the church work with Pastor Valeri. He was a composer and played the *bayan*. Three of his songs were published in the hymnal. Since then, he has composed the music for more hymns. We stayed after the celebration to record additional songs. So we kept growing the collection, one song at a time.

First Believers in K-X: David

The first believer in K-X was Zoya, Valeri's wife, in June 1992. She had visited her aunt, Anna Petrovna, who was once a strong Communist but is now a believer in Khandyga. She went all out for the Lord, witnessing and bringing many into the Kingdom of God. Consequently, the Khandyga church prayed for and evangelized K-X. Two months later, 12 people repented, including Valeri, Zoya's husband.

God tested Valeri and Zoya as young Christians. They were the only believers for a whole year since the others left the church under pressure from their peers and the culture. They stood for the Lord while continuing to be mocked by their family and others in their village. A year later, Natasha repented. Then the pastor in Khandyga came back and baptized the three of them. In general, Sakha people live in some degree of fear of the many spirits in their world. Valeri and Zoya lived in the "fear of the Lord" as they chose to put Him first in their lives.

Happiest and Saddest Moments: David

I asked Pastor Valeri about the happiest moments of his Christian life. He responded, "I'm happiest when a sinner repents, and God changes their life." His most challenging time was when 50 new believers attended their meetings, but no one helped disciple them, and many fell away.

Choose Your Hero

People remembered "Sniper Day" in K-X. They honored their local sniper who had served in the Great Patriotic War (World War II). He killed a record 429 Germans as he was trained to do. Our Savior willingly gave His life and died to save sinners in K-X and Germany. Choose your hero.

Love Is Better

The theme for the Oktyomtsi Bible Camp in mid-June was "Love Is Better." The teachers, leaders, cooks, and helpers demonstrated this by gladly serving the Lord together.

The village was a simple rural community of 1,500 with an active church. No believers lived in this village when we participated in evangelistic meetings here 10 years ago. But now, the church is reaching out to Sakha people in neighboring communities.

Lyuda matched Sakha prose with a lively melody recorded on the audio cassette, which was played for their theme song. Here's a loose translation of the first stanza: "Wake up to a smiling sun, a bird sings. With the bright sun, you wake up." The chorus: "Then you will see the beauty of Jesus and praise His love." Second stanza: "With happy smiles, we meet this day. Oktyomtsi Camp waits for you."

All the campers, cooks, and helpers sang the theme song. In the past, the Young Pioneers (Communist Youth) sang theme songs at their camps.

Nine children and two leaders traveled from Namtsy by the Vilyuy River for two days for their first camping experience. No summer road existed! It was cheaper to travel by commercial boat than to fly. These campers ranged in age from seven to 17 and slept in the church with their counselors.

They called us Grandpa and Grandma as we interacted with the campers. We peeled lots of potatoes, washed dishes, and assisted wherever needed.

At 10:00 a.m., the campers gathered around the flagpole and sang their theme song. Forty campers were divided into small groups for Bible lessons, skits, games, and presentations to the whole group. Five Sakha people led the Bible studies and ran the entire program.

A hearty dinner was served at 1:00 p.m. in a covered area. The afternoons included more Bible teaching, time at the local playground, and snack time. Their day ended at 6:00 p.m. with awards and the theme song. The last day included a trip to the nearby zoo to see bears, moose, reindeer, musk ox, swans, and other animals from the Sakha Republic.

Every day, we heard the cuckoo birds making their familiar calls. It was a bad sign for people who counted the number of "cuckoos" to see how many more years they would live. We were glad to hear a number of the Sakha people say God had removed their fear of this superstition.

Sakha Lovebirds

Nadia, a beautiful Sakha bride, called us at 11:00 p.m., the day before her wedding. She wanted to make sure we were invited to the big event. To their delight, the church in Oktyomtsi was packed with 70 people witnessing their marriage—their first wedding of two Sakha Christians.

The bride was from a small Sakha village three hours south of Yakutsk. Nadia had been a counselor at the Bible Camp earlier this summer.

The groom, Sasha, short for Alexander, was from another small village 12 hours north of Yakutsk. The wedding included a long sermon to tie the knot securely. Individuals presented their gifts, which often included singing a song from the Sakha Hymnal.

Pastor Vasya of the Oktyomtsi church interrupted the wedding ceremony to introduce a local young couple. They were experiencing marital problems and wanted to pray and repent publicly.

After everybody ate, we attacked a massive pile of dishes while the ladies of the church transformed the church's guest room into a honeymoon suite. Then it was decided about 10:00 p.m. for us workers who remained to "drink tea" with the newlyweds. We enjoyed lots of good leftovers with them. This was not our idea of how to start one's wedding night!

The Children of God Testify

The book *The Children of God Testify* reveals how God worked in and through the lives of 27 believers in the Sakha Republic. Lyuda, the editor, remarked, "This book is original, with national believers relating their personal experiences." The illustrated 68-page book was published in July with 3,000 copies. The cover was a photo from Oktyomtsi of horses running freely, a symbol for people who highly value their horses and freedom.

Living by Faith (2002~2004)

Each testimony shared how they met the Creator God and what He was doing in their lives today. All the writers shared one thing in common: they received Christ into their lives since the fall of communism in 1991. "Oh, that men would give thanks to the LORD for His goodness, And for His wonderful works to the children of men!" (Psalm 107:8).

Fast-forward: Their testimonies were published as a book in English in 2015, titled *Native People in Russia Follow Jesus Christ*, with updates on their stories.

"Eternally Frozen"

Yakutsk has more than 200,000 people in this sprawling northern city. Below the surface of beautiful flowers and trees, the subsoil is permanently frozen. In Russia, it is called "eternally frozen," but "permafrost" in Alaska.

Frozen ground thaws unevenly. Consequently, houses and even large apartment buildings can develop cracks and shift so doors don't shut properly, or worse. Newer buildings were built on concrete pilings driven 30 feet into the frozen ground. Hopefully, the pilings will stay frozen and keep the buildings stable. The Sakha and Russian Evangelical churches are built on the solid, unchanging foundation of Jesus Christ and the Bible.

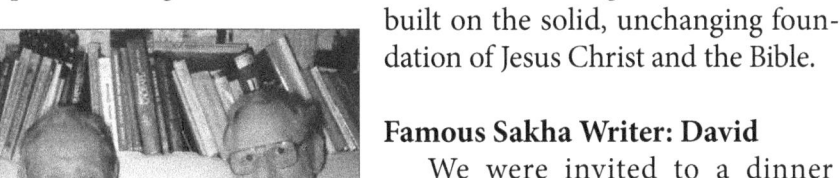

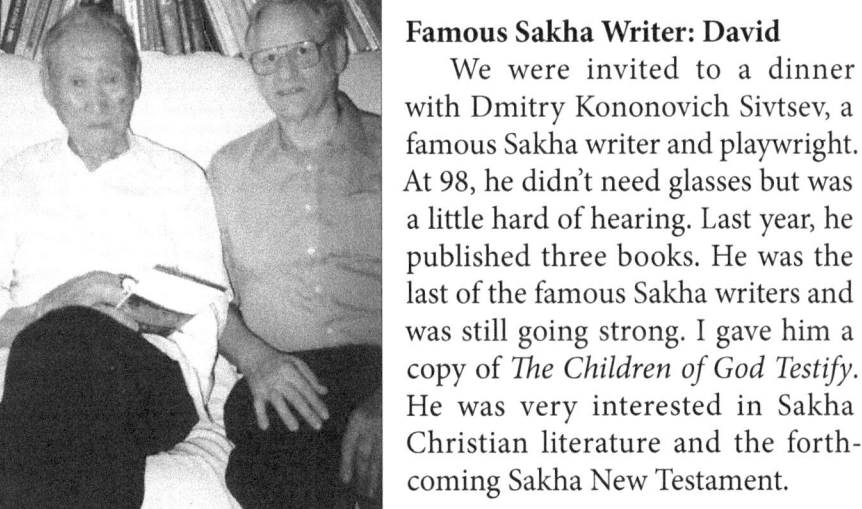

Famous Sakha Writer: David

We were invited to a dinner with Dmitry Kononovich Sivtsev, a famous Sakha writer and playwright. At 98, he didn't need glasses but was a little hard of hearing. Last year, he published three books. He was the last of the famous Sakha writers and was still going strong. I gave him a copy of *The Children of God Testify*. He was very interested in Sakha Christian literature and the forthcoming Sakha New Testament.

Dmitry and myself

Global Consultation on Music and Missions

In September, we flew from Yakutsk to Dallas, Texas, and attended the Global Consultation on Music and Missions for four days with 300 delegates from all over the world. We gained a few insights into reaching the Sakha people with music.

More Calendars

Next year's large calendar included a picture of three swans peacefully swimming in calm water with John 14:27. The pocket-sized one was a picture of a man riding his horse with snow on the ground, along with 1 John 2:1b.

2004
Can You Come to Us?: David

Zoya, the pastor's wife in K-X, excitedly phoned us, "Can you come to us? We found a ride for you on Friday at 7:00 p.m." With equal joy, we responded, "Yes, of course we can!" A half-hour later, we received another phone call from the local dispatcher in Yakutsk saying, "Can you go tonight at 5:00?"

So we sprang into action! I caught a bus to the town center to exchange dollars for rubles and I purchased a few things for this two-week trip. We were ready by 5:00 p.m. Finally, the driver rounded up all his passengers in Yakutsk for our trip, 10 adults and two children. Drivers like a full load since they make more money. It took effort to find a place for all the baggage.

Nine hours later, we arrived at K-X. Pastor Valeri and Zoya were up and dressed to welcome us at 4:00 a.m. We drank tea and caught up on the news before dropping into bed to get a few hours of horizontal sleep. Of course, not all the information was upbeat, but hearing of three men repenting in this church of 25 believers certainly was!

Welcome Back

Friday night was a special meeting for church members. One young man had been excommunicated after several drinking spells. Every member raised their hands as they voted to welcome their brother back into full fellowship. He had stopped drinking and was growing spiritually.

More Togetherness: David

On Sunday, the church minivan from Khandyga brought two loads of believers from nearby villages for a great day of sharing, preaching, and singing. The service began at 11:00 a.m. with 40 people. Two other preachers were more longwinded than I, so the service continued until 1:30.

Cool!

On Monday, Pastor Valeri informed us, "The temperature is negative 60 degrees today." So we put on all the clothes we brought with us before walking to the nearest store to buy groceries for our hosts. Earlier this winter, the central heating system had significant problems. The apartments and homes in one area of town had only 35 degrees inside their homes for several days!

More Small Villages

On Wednesday, the church rented a van, and we drove to Udarnik for an evangelistic service in their Community Center. Then on Thursday, we caught a local taxi to Keskil, about 20 miles away. Later, we dined with eight women who celebrated the beginning of their fellowship.

Five years ago, we stayed overnight in Keskil with Sargylana, a Christian college student who found the Lord in Yakutsk. At that time, she asked us to pray for her village. Now there was a group of believers meeting regularly in answer to prayer and the work of evangelism.

The Regional Center

We hailed another taxi to Khandyga, which had a strong evangelical church. They had discipled and sent out four missionary couples. Their goal was to reach every village in their region. They were in the process of preparing another team to go to a new unreached location.

Afanasy

Afanasy had a burden to reach the people in his hometown of Kempendyay, 700 miles away in the western part of the Sakha Republic. We first met him when he needed work, and the Oktyomtsi church hired him to do odd jobs. He turned to the Lord while working there and he was discipled by their church. Later the church commissioned him as a missionary and supported him. "And the things that you have heard from me among many witnesses, commit these to faithful men who will be able to teach others also" (2 Timothy 2:2).

When Afanasy returned from Kempendyay, he stopped by our apartment to give us a report. He was ecstatic to be able to witness freely in his hometown region.

Granddaughter's Wedding

Our granddaughter Cheryl Marie Cogan (Elizabeth's second daughter) was a beautiful bride who married David Mount in March. David was in the Army at Fort Wainwright, and they lived in Fairbanks. Later he served in Iraq. We were sorry to miss their wedding since we were in Siberia.

The Carrot Farmer: David

Larry DeVilbiss, the "carrot farmer" from Palmer, Alaska, flew to Yakutsk to help the Sakha people. For three weeks, we traveled together to five Sakha towns to record hymns.

Larry recorded Rosa, Lyuda's daughter, singing with Yegor accompanying on the guitar.

The church in Oktyomtsi had Philippians 1:21a written on the wall in Russian, Sakha, and English. Larry produced a 90-minute audiotape of 35 songs titled "Our Father, the Greatest Creator."

Sixth Annual Sakha Christian Conference

The Oktyomtsi church with 30 members was busy. They rented the Community Center to have adequate space with a large stage for the sixth annual Sakha Christian Conference. Christians traveled from all over the republic for the three-day event. The church expected 120 participants, but 180 showed up!

The stage had white, blue, and red curtains, representing the colors of the Russian flag. One Sakha person commented that the white reminds us that our sins are washed white as snow; the blue is our heavenly home; and the red is Christ's blood which He shed for our sins. The curtains gracefully highlighted the conference theme from Matthew 5:14a, "You are the light of the world."

Everybody participated in the Sakha national round dance by forming a large circle in the auditorium and using the words from the hymnal. Joining arms, we gracefully and joyfully swayed in harmony. Three new round dance songs were presented to everyone's delight! These songs praised our Creator and told about the coming of God's Good News to the Sakha people.

Seeing so many Sakha men taking leadership roles was thrilling. Several gave sermons on essential themes such as dealing with occultism, the Christian home, helping disciples grow, and why we believe there is one God. Reports were given from each Sakha church or fellowship.

A New Mother

Since early childhood, Afanasy has remembered his father writing and leading round dance songs to praise nature. Afanasy had two strong desires—first, that his relatives would know Jesus Christ as their Savior, and second, that he could write a Christian song.

On the third day of the conference, seven Sakha people went forward to repent and turn to Jesus Christ. One of those was Afanasy's mother, now a new person in Christ. He wrote the words for a song, which he then led in singing. They lined up on stage with other singers including his mother.

Fast-forward: In 2006, a fuller version of the story "A New Song and a New Mom" appeared in chapter 11 of the book, *All the World is Singing*.

Seven Students

We flew south to Irkutsk and met seven Sakha University students who had turned to the Lord through Campus Crusade for Christ. Christian literature was shared, and we introduced them to Sakha Christians in Yakutsk and Oktyomtsi. They returned to their homes in the Sakha Republic and shared the Good News with family and friends during the summer.

Reaching Relatives

The church in Oktyomtsi reached out to Lyuda's relatives in Mitaakh, five hours away by vehicle. The Sakha people have lots of relatives. Nikolai and his wife, Anya, repented along with a grandmother. Every other week, the people from Oktyomtsi went there to encourage and disciple them. The Christians met regularly in Nikolai's living room and were growing in their faith.

Summer Camps: Kay

Oktyomtsi had an excellent camp for children in July. There were five groups of children with their own leaders and assistants. Thirty-five campers, ages eight to 16 participated. We helped in preparing and serving meals. I even jumped on the trampoline after everyone had gone to bed!

A month later, we received an urgent call to pray. Two female campers rode their bikes into the woods to pick berries and did not return home. Thankfully, they were found after two nights of enduring tons of mosquitoes and no food. They had prayed and depended on the Lord as they had learned at camp.

"Granddaughter's" Big Day

Our Sakha "granddaughter," Marianna, married Riko. Her fiancé had served with the Evenk people as a white missionary from the Republic of South Africa. First, the couple was united in a civil marriage in Pokrovsk. Then everybody traveled to Mokhsogollokh for their church wedding. It was a beautiful and Christ-honoring event. As her honorary grandparents, we sat up front with Marianna's parents. We were also dubbed the groom's "representative" parents since they were in South Africa.

Doing More Time in Siberia: David

We were in Siberia by choice. The Lord had opened the way for us. We were here for the long haul, but only the Lord knew how long that would be. To continue serving, working, and traveling freely, we needed permission from the government.

I was weary from all the paperwork. We prayed for energy, patience, and perseverance. The local police visited our apartment to show us the official Russian law that one must leave on or before the date their passport expires. He reminded us that our Russian passports for foreigners expired on August 28th. And he verified that we were living where it was listed on our passports. We planned to be here another five years.

Last spring, we flew to Vladivostok to apply for new American passports at the eastern regional office of the American Consulate. That was the beginning of the paperwork. Official copies of our marriage certificate from Michigan were required. Our mission wrote a letter stating we had sufficient income to live in Russia. These items had to be translated into Russian by an official translator and notarized.

We had to be tested to prove we did not have AIDS, syphilis, or TB and that we were not on drugs. The results of the AIDS test were on special red paper. We needed a short form to prove the apartment was in our name. For an hour, we stood in line at a bank and paid three rubles (9 cents) for the proper form for each of us.

Of course, there were longer forms to fill out, new Russian passport photos, a passport fee of 40 dollars, plus an official invitation from our Russian partner mission.

We wrote our autobiographies Russian-style. This meant organizing it concisely by dates and listing the significant events in our lives. Our purpose for being here and our activities had to be spelled out.

It was a long process. In the end, we gathered all of the above in a particular folder as instructed. We thanked the Lord for several government officials who were kind and helpful. Such was not always the case.

Ten Years

During our annual Russian Field Conference in Yakutsk, we were honored for serving in Siberia for 10 years. InterAct's general director, Gary Brumbelow, and his wife, Valerie, visited the field. Pastor Vasili and Lyuda from Oktyomtsi were guests, as well as Valentin, the director of our partner Russian mission, and his wife, Lyuba. We sincerely appreciated all the memorable songs, kind words, and gifts.

We celebrated along with other missionaries "doing time" in Siberia. They presented us with a pair of woolly mammoths carved out of their tusks. (These magnificent animals are extinct.) We were the oldest members serving here. With the Lord's help, "They shall still bear fruit in old age" (Psalm 92:14a).

During these 10 years, we've seen the Sakha church grow from a handful of believers to almost 400 Christians out of a population of 432,000 Sakha people. We had a lot of plans and hopes for their future. Our primary vision was to see spiritually healthy Sakha churches.

Shamanistic Conference: David

Lubomir, an indigenous Even (e-VEN) pastor, and I flew to Krasnoyarsk to learn more about the new organization addressing shamanism in Russia. Sixty participants representing different missions and interested parties attended. The goal was to help national believers maintain their ethnic identity without holding onto the shamanistic aspects of their culture.

Our First Great-Grandson

We began to feel our age as we joyfully welcomed our first great-grandson, Caleb Jonathan Mount, into our lives. In November 2004, he was born prematurely to our granddaughter Cheryl and her husband David. We were glad they named him after Caleb in the Bible, a man of God who "wholly followed the Lord" (Joshua 14:8, 9). Caleb and his mother share the same birthdate. Later on, it felt a little strange to hear Caleb calling our daughter Elizabeth, "Grandma."

Another Mug

Jesus said, "I have come that they may have life, and that they may have it more abundantly" (John 10:10b). These words were highlighted on a Scripture mug. Drawings of Sakha lilies, a pair of dancing cranes, and birch trees by a river with an evangelical church on a hill were also on the mug. Sakha people drink a lot of tea, and we made these locally as gifts for the Christians.

Common Sense—Don't Travel: David

Pastor Valeri invited us to spend Christmas and New Year's in Krest-Khaljai with their family. Common sense said, "Don't travel; it's negative 58 degrees!" But we traveled anyway, feeling assured that it was the Lord's will. The church van had comfortable seating for eight. We were dressed for the occasion in case anything unexpected happened.

After spending several hours picking up the other passengers with all their shopping goods to stash inside, we headed off into the frigid cold and ice fog. Zoya welcomed us at 2:00 a.m. with hot tea and pizza. We got to bed at 3:00, in time to catch a few winks before the Sunday service.

In church, one believer told us, "God gave me a national Sakha melody. This is a first for me." I responded, "Wow! The Holy Spirit guides us." Another believer had a similar thought, and the two were working together to match the melody with biblically accurate and culturally relevant words. They sang part of their song the following week at the service.

God Works Freely at Any Temperature

After church, we were invited to drive for an hour to the isolated village of Udarnik for a unique service with two believers or anyone else who would attend. Only one believer was in town. Our group of six was a great encouragement to our sister. Yes, it was worth the rough road and cold for one believer. Both believers were there on our second trip two weeks later.

Cold Weather Law

Zoya was delighted about the new law for the Sakha Republic. First-through fourth-grade students do not go to school at minus 65 degrees or colder. Fifth-through eighth-graders do not go when it's minus 72 degrees or colder, and the school would be closed when it's negative 80 degrees or colder. Church services continued at any temperature.

Teens Stole the Church Van: David

Two teenage boys stole the church van, but it was returned without any damage. The mother of one of the boys sought forgiveness for her son and came to church that night. I found it interesting to meet with four men and a distraught mother at negative 80 degrees. Everything was in Sakha, so I missed the details. Then they switched to Russian, asking for my advice. God used the theft to get this family's attention, showing them that grace and forgiveness were available.

Just Across the River: David

The next day, we rode in the church van across the Aldan River to the village of Megino-Aldan and gathered in Larissa's apartment. Frankly, the low heat in her apartment chilled me. I wore two wool sweaters under my coat for travel at these temperatures. However, we were all warmed by the great fellowship with five believers and one seeker. Their big news was that the husband of one of the ladies recently turned to the Lord.

In the Barn

One notable event was visiting a sizeable barn housing 15 cows and calves. The fragrance inside was self-evident. We thought about Mary and Joseph, who were given a "room" with the animals.

Christmas Traditions

We were asked about our favorite Christmas traditions. They liked the idea of a candlelight service emphasizing Jesus, the Light of the world, with His light spreading through the Sakha believers. They found candles for the service.

To our surprise, guests from Megino-Aldan came for a church dinner to honor our 10 years of service in the Sakha Republic. They specifically expressed gratitude for our part in developing the Sakha Hymnal. We could see and hear it in their singing. They presented us with a pair of carved Sakha mammoths.

To the Regional Center: David

We took the commercial van to Khandyga. Their church was packed with 75 people. The pastor decided I would preach first, followed by 15 Sakha members who sang and shared. After the service, we ate a wonderful dinner.

A Growing Church

In Keskil, we were given a royal welcome in a small cabin with 11 believers and seekers. What an encouragement since our last visit over a year ago when there were only six believers! The place was packed with close fellowship. Jobs were extremely hard to find. One believer from K-X moved here and acted as a missionary to this group of new believers.

"Wait for Me!"

Jacob, a believer in K-X, was 43 and had been abandoned by his father after his parents divorced when he was only two years old. He never knew or saw his father, yet he deeply resented what his father had done to him.

We traveled in the church van for three and a half hours to Chymnaii since Jacob knew his father lived in this village. A group of five Sakha women believers met there. One woman at the service knew where Jacob's father lived, so we drove by his home after the service. She went to the door with Jacob, introduced him to his father, and returned to the van. The rest of us prayed and waited in the van as he met his father after 41 years! The reaction could be good, bad, or indifferent, depending on his shocked and unbelieving father.

A joyful Jacob returned to the van 15 minutes later. He had told his father, "God loves you. Now I'm a believer in Jesus Christ. I've quit drinking, and I have forgiven you." His father welcomed him with open arms! He told his lost son, "I want to have a long visit with you the next time you come to Chymnaii." God brought the two of them together, in His time and way.

Wait for Me was the title of a well-known Russian TV show. The goal was to reunite family members who had been separated but were still waiting and looking for one another. They showed deep emotions and tears of joy.

New Year's Celebration

Valeri's whole family worked to prepare a New Year's Eve feast. Two of their children, a son-in-law, plus a grandson arrived from Khandyga to celebrate.

We finally sat down to feast after preparing an abundant supply of food all day. It was a wonderful family time that also included us. The parents reminisced about periods in their lives when there was very little food under the communist system. Their children needed to hear and appreciate what they have now.

At the stroke of midnight, we all went outside to enjoy their fireworks and everybody else's. The view was great from their hill overlooking the village.

Kempendyay Evangelism: David

Afanasy, four others, and I went on an evangelistic trip for a week to Kempendyay and stayed with his brother, a Sakha artist. We met in their House of Culture, and three people repented. This fired up Afanasy, who moved back to his hometown as their missionary.

God enabled him to live in his mother's house, which was large enough to hold weekly meetings with the new believers. God also provided him with an excellent job with a company harvesting salt from a nearby pond.

Two on Crutches (2005)

Ice-Skating without Permission

A Sakha children's book, *Ice-Skating without Permission*, was about a girl who went skating when her mother told her to stay home and take care of her sick little brother. She disobeyed, and there were consequences. Her brother grew worse, so they called for "Urgent Medical Help" to help him recover.

Zhenya, a young Sakha Christian artist, drew 14 color pictures realistically reflecting their culture. After reading the book, one young girl told her mother, "Now I'll be kinder to my little brother."

A Broken Leg: Kay

At a major stop in the center of Yakutsk, while changing buses on March 16th, I slipped on the icy road and fell. I couldn't get up! I had broken my leg! I called on God to help me. He responded by sending a Sakha couple, who carried me to their car and drove directly to emergency care. I don't know their names, but I thanked God for sending them to help me.

David quickly caught a bus to the large regional hospital. Someone had phoned him from the hospital to let him know where I was. He kept praying he could find me as he walked through the first floor of the hospital maze. He recognized my fur boot sitting on a table through the half-open door in the X-ray room and walked in. My leg was hanging at a strange angle.

The doctor said I had broken both bones in my left leg just above the ankle. I had minimal pain and a swollen leg. They applied traction to my foot and elevated it. The Sakha doctor had been trained in Austria and planned to operate the next day.

The Lord knows and determines what will happen to us. God's plan is best, and that was what we ultimately wanted. We thought about my blood not clotting properly 16 years ago in Alaska when they removed my spleen. Complications in a foreign country are much more challenging. This also presented an opportunity to witness to a different group of people than we usually see.

How Long?: Kay

The hospital staff informed us I would be there for a month! We both groaned loudly! They said that was normal. We wondered about the cost! A staff member told us that since we were Americans without Russian insurance, we had to pay for everything—in advance with cash—before they would do anything. The hospital would not accept credit cards, either.

David went to a bank the next morning to exchange money and returned with the rubles. It was 1,000 dollars for the operation, cast, leg care, and extended stay in the hospital. He immediately took my paid-in-full receipt to the doctor, who was waiting for him. I was already in the operating room.

Surgery: Kay

My surgery lasted for two hours. The X-ray showed a metal strip seven inches long on my tibia leg bone, with eight long screws to hold everything securely. The smiling doctor emerged from the operating room, gave David a thumbs-up, and said, "OK" in English. But the full explanation was in Russian.

My Hospital Room: Kay

My hospital room was 20 feet by 20 feet with six beds. I had both Sakha and Russian female roommates. The thin mattress had distinct signs of having served others as well. David brought sheets from home.

A Different Room: David

All of the roughly 100 patients in this hospital wing had broken bones. Kay was moved into a different room with two Sakha roommates. One roommate was the wife of the leader in a region with several Sakha churches. The other was a Sakha *babushka*, age 75, from an area with very few believers. Both of them appreciated Sakha literature. She took the opportunity to witness to her roommates, who kept changing, and to the staff.

Kay was the only American in this huge state-run hospital complex. I could visit her at any reasonable time, so every day, I took the

10-minute bus ride. Then I dropped off my coat and hat in the coatroom, put on a pair of thin, blue plastic "slippers" over my shoes, and went up to the third floor to her room.

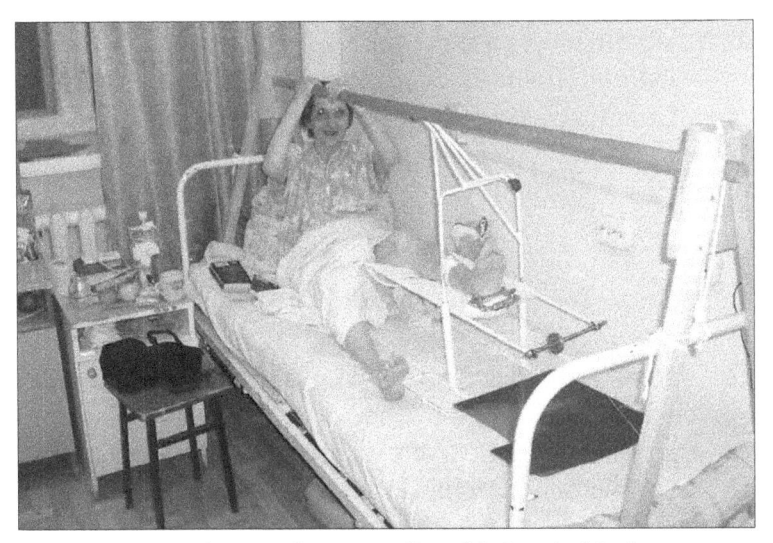

Kay is lying in her non-adjustable hospital bed with her broken leg elevated.

Hospital Food: David

Frankly, hospital food was bland and the same every day. Cooked cereal was served for breakfast; soup and cooked cereal and a scoop of pale, ground-up fish with their bones in it for lunch; and more cereal for supper. All meals were served with two slices of bread and tea plus compote (juice from dried, cooked fruit) and cranberry juice. Kay didn't know so many varieties of cereal existed, but she enjoyed most of them.

I brought Kay a plate, cup, and silverware. Many things in Russia require self-service for added comfort. I collected a variety of items to supplement her hospital "rations." She had many visitors and gifts of food. Sakha friends brought her cooked and also thinly sliced raw horsemeat, which she shared with the *Babushka*. Kay offered her cooked horsemeat, but she declined, saying, "No teeth."

Kholodets for Kay: David

Several people insisted that I needed to give Kay *kholodets* (meat aspic or jelly). It jiggled and looked like jelly. This would strengthen her bones and help in the healing process.

At our outdoor market, I found the lower section of a cow's leg with part of the hoof and hair trimmed off. One Sakha lady sold me a complete leg that was already cut into four convenient pieces. It weighed five pounds and cost five dollars.

I trimmed the pieces a little more at home and soaked them in a big pan overnight. The next day I boiled them for eight hours. Salt, pepper, and whatever seasoning I could find were added. After eight hours, I removed small pieces of bones and anything unwanted before pouring the thick broth into jars. I let them set and solidify, then took a container to Kay.

Sent on Errands: David

The doctor sent me on several errands. First, I needed to find crutches. After several hours of looking, I finally found a pair. They were too long, so I hacksawed them to size.

Next, the doctor told me to buy a long sock to go under Kay's cast. I don't visit women's shops, but the clerks were accommodating, and I explained why I needed a pair of women's socks.

Another assignment was to purchase medicine to keep her leg swelling down. This one was easy because her roommate wrote the name on a piece of paper. The pharmacist understood.

Once, the nurses asked me for assistance to help offload a good-sized lady from a hospital stretcher. They needed my help with this one!

Prognosis

Kay's prognosis looked good, but it took time. They waited for her leg swelling to go down before putting on her cast. She learned to use her crutches, but we lived on the fifth floor with no elevator.

Our "Five-Year" Plan: David

There were only two more hoops to jump through. First, we needed to obtain a residency registration with a stamp on our Russian internal passport. So I went to the regional registration office. They sternly warned me, "You have 10 days or else," but they would not give me the registration. So I went back to the head office. They also reminded me, "You only have 10 days or else!" Then they sent me to a different office. This office stamped our documents with a smile.

The second hoop was to present our documents to the head government office for their approval. This was the final step to receiving our five-year Russian passports for foreigners. We thanked the Lord that our passports were valid until 2010.

Kay Returned Home: David

After a month in the hospital, Kay returned home. There was no elevator, and I couldn't carry her up our five flights of steps. Ramil, a young man from the church, offered to drive us home and carried Kay up to our apartment.

Sakha New Testament Presentation: David

The official presentation of the Sakha New Testament was on May 11th at the National Academy of Science in Yakutsk. Sargylana had worked with the Institute for Bible Translation on this project for 12 years. The audience displayed awe and approval when they heard that Dmitry Sivtsev was the literary editor.

Now the New Testament is available to 450,000 people. The translation complimented the Hymnal published in 2002, strengthening the growing church.

I received an advance copy of the New Testament. I gave it to Lyuda, from Oktyomtsi, since she was so eager to read it. A shipment of 10,000 copies arrived during the summer.

I was the only evangelical Christian at the presentation. The local Russian Orthodox Bishop, Zosima, gladly welcomed the new translation. He read a congratulatory message from Patriarch Alexie II in Moscow.

Babushka Pasha was among the first to eagerly read the Sakha New Testament. She had earnestly prayed for Sargylana during the long printing process and was eager to share the written Word with her family and friends.

Traveling on Crutches: Kay

We flew back to Alaska on June 2nd. I was concerned that my leg would swell during the flight, so I took several strolls down the narrow aisle with minimal problems. Our children and grandchildren welcomed us back. An American doctor checked out my broken leg and answered all my concerns. He stated, "The Sakha surgeon has done a good job on your leg."

Our Firstborn Grandchild Married

Our first granddaughter, Melissa Marie Cogan, married James Cameron Hendrickson in June at the Friends Church in Fairbanks. Melissa's father, Andy Cogan, performed the ceremony. After the wedding, we all watched as James helped Melissa gracefully climb a tall kitchen stepstool to get into their getaway high-rise Alaskan truck while wearing her gown.

Cheryl Mount, James, Melissa, Milo Cogan, and Andrea Cogan

Needing Knee Work: David

Climbing five flights of steps every day to our apartment in Yakutsk was tiring, and my right knee kept bothering me more and more. My doctor in Fairbanks replaced it in June. He removed all the areas of my leg bone that had arthritis, which had bothered me since our days in Koyukuk. He said I should be good for another 20 years, provided I stay off trampolines. All I wanted was to walk and work freely until the Lord returned.

On Crutches: David

Now we both used crutches. I couldn't drive, since my right leg needed healing. Debbie chauffeured us to doctor appointments for the first week. After that, Kay drove. It was her left leg that had broken. I went to rehab and tried my best to return to normal. Both of our bodies were being "fine-tuned" to enable us to continue working with the Sakha Church in Siberia.

The first Sunday after surgery, I stayed home. The second Sunday, we both went to church on crutches. But the third Sunday, I walked into the church without any help, although I was still weak-kneed—with a literal weak knee.

Return Trip

We returned to Yakutsk and the work God had called us to do via Frankfurt, Germany, and Moscow in August. We were welcomed with excellent reports about the summer camping ministry in Oktyomtsi and that the Sakha New Testaments were being distributed.

Wearing a New "Hat": David

All the members serving in Siberia, along with Gary Brumbelow and his wife, Valeri, rode the ferry across the Lena River to the Druzhba Museum in Soltinski. At our annual Russian field conference, Gary asked me to serve as the acting Russian field director for InterAct's work in Siberia. This included the work in the Sakha Republic and the new outreach in the Tuvan Republic on the Mongolian border. My goal was to help every member fulfill the will of God in their lives.

We stayed at the rustic guest facilities which included all the necessities of life. On the first day, we were served Sakha soup. I called it "Gut soup," made from the stomach and innards of the cows. The color was a shade of army green. It did not look appealing but it tasted good. I even took seconds.

An Active Church

We were part of the first evangelization team in Oktyomtsi 12 years ago. Christians from Mokhsogollokh and elsewhere continued to pray for people to be saved during these outreaches to various villages and traveled there as witnesses for our Lord.

Now there is an active Sakha church with 27 members. They were growing spiritually. They also sponsored the Sakha Christian Conferences, summer camping programs, and continued outreach to other towns in the Sakha Republic. Vasya, their pastor, was a Russian from Mokhsogollokh who used to bicycle there to participate in the services. Then he married Lyuda, the Sakha Hymnal editor, and they worked together to serve the Lord.

Calendars Are Ready

The Sakha Christian calendars for 2006 were printed locally. The full-size one had a picture of a large open field with contented cows and horses grazing. The accompanying Bible verse quoted Jesus saying, "Come to Me, all you who labor and are heavy laden, and I will give you rest" (Matthew 11:28). The pocket-sized calendar had a picture of beautiful Sakha flowers overlooking a serene lake, along with the Bible verse John 10:10b.

Golden Wedding Anniversary (2006)

2006
Meeting in Oregon and California: David

I flew to Oregon and participated in the general director's meetings and candidate interviews in mid-January. This was a new and great experience for me to be with the directors for the work in Alaska and western Canada. Kay flew out with me, but she went on to Michigan.

After the business was over, I flew to California to see my brother Martin, Cuca, and their three daughters. Martin was still laying carpet and doing well. I celebrated my 70th birthday with them. I spent three days with my sister Miriam and family near Sacramento.

Quality Time: Kay

In Michigan, I had quality time with my father and other family members. I helped celebrate Dad's 95th birthday along with all my siblings (Shirley, Nancy, Jim, and Sheryl), a few grandchildren, plus Uncle Ed and cousins Diane Pedersen, Barbara Johnson, LuAnn Mahan, and other close friends. My father loved having his family with him. I was glad two of our children, Debbie and Dan, with granddaughter Suzanna, came.

Reconnecting: David

We reconnected at the Seattle Airport. I flew from California and picked up my luggage. Ten minutes later, Kay arrived from Detroit and met me in the baggage claim area. We continued together back to Yakutsk.

Ministry Trip to the West: David

I joined a ministry trip to the west of Yakutsk with five Sakha believers for a week. We stopped in Kempendyay to support the new house church with Afanasy serving as their missionary. Two more ladies repented while we were there.

Next, we stopped in Nyurba, with the biggest Sakha church, followed by Vilyuysk, a Sakha town of 10,000. Pasha and Anna, two young, single Sakha ladies, moved here as missionaries. Pasha's parents were both believers and moved into the large house to assist in the church ministry.

Eighth Sakha Christian Conference

"The Harvest Is Great, But the Laborers Are Few," was the theme for the eighth annual Sakha Christian Conference in Oktyomtsi. This offered good reasons for praise as well as tremendous challenges ahead to reach more Sakha people. Two men and five women repented at this time. We enjoyed the wonderful fellowship with over 100 Sakha believers.

Golden Anniversary

We flew from Yakutsk to Michigan on May 29th, rented a car, and drove to Vassar to see my father and family. Then our golden wedding anniversary was celebrated on idyllic Mackinac Island in northern Michigan on June 8th. We parked our car in Mackinaw City and took the passenger ferry across Lake Huron. Vehicles were banned in 1898, so our taxi to the Murray Hotel was a horse-drawn wagon. We heard the clip-clops of many horse hooves while sightseeing and exploring this beautiful island of old-world hospitality and charm. Fragrant lilacs of many colors were in full bloom. It was a relaxing weekend.

More Celebrating

Our 50th wedding anniversary was celebrated with all of our children, grandchildren, and great-grandson at Dan's home in Fairbanks. Mark and Darla Grisez, David's niece, came from California for a week. We thanked the Lord for His faithfulness through many years of being together. Every day in our lives is precious, so we need to make the most of our time with them for the Lord's glory. We'd had 49 years of parenting four children and 24 years of grandparenting a dozen outstanding grandchildren. Then we returned to Yakutsk.

Dorie, the Girl Nobody Loved

The book *Dorie, the Girl Nobody Loved* deals with the evils of sexual abuse and not being loved as a child. It was the life story of Dorie Van Stone, through her hard years of growing up in an orphanage and foster homes. Once she turned to Jesus Christ, God gave her victory over the effects of sexual abuse and enabled her to live a Christian life.

The text was sent to Austria for the layout with Operation Mobilization, then to Saint Petersburg, Russia, for printing. When Dorie was in her 70s, she spoke at a church in Fairbanks, and she was delighted to hear of the Sakha printing. Many people relate to her early childhood.

Sakha Bible Study: David

We attended the Sakha Bible study on Saturdays. Sixteen Sakha people listened attentively while Saeed taught from the Bible. Then I preached a series on "What Nature Reveals about God." We were glad to see the success of this group which developed into a church.

Developing Christian Literature

September was wet and muddy. We traveled by taxi to K-X to assist Zoya in a translation project. The mud ruts were so deep that we bottomed out several times. Then the driver hollered, "Everybody out. You have to walk so I can get through this section of road." The empty vehicle rode higher, and he waited for his 10 passengers two miles down the road.

Kyzyl

The Russian Field Conference was held in Kyzyl, the capital of the Tuvan Republic on the Mongolian border, in October. We flew 1,500 miles from Yakutsk to attend. Jim and Kari Capaldo served there. We were booked in a quaint, Soviet-era hotel across the street from a former KGB building.

We learned about the Tuvan people. Our time included a Tuvan throat-singing concert. It was interesting, but we could not imitate that kind of singing.

Our special speaker, Bob Crane, served with SEND in Khabarovsk, with his wife, Robin. He shared from 2 Timothy about encouraging Timothy to live the Christian life, which also applies to us today.

Our First Great-Granddaughter: Kay

Kristen Nickole Mount, our first great-granddaughter, was born to Cheryl and David on my birthday in October. We get to celebrate our birthdays together. We were anxious to meet her in person and hold her on our next trip to Alaska.

Set of Children's Activity Books: David

A set of children's books from Ukraine looked promising for use with the Sakha people, especially for those teaching Sunday school or for the children to complete independently. These 32-page books were well illustrated with various activities and cutout stickers teaching biblical truths. The author graciously permitted us to translate the books into Sakha.

Zhenya did an excellent job adapting the illustrations to the Sakha culture. The titles were: *God Created the World,* telling of creation and God's purpose for us; *God Is with Me All Day,* that God is present everywhere watching over us; and *God Is with Me All Year,* building on the concept of living in God's everlasting love rather than in fear of spirits or anything else.

House Guests

Saeed, the director of the Sakha work, and his wife, Natasha, had a baby boy. Her pregnancy was difficult, and the child was born by C-section with multiple congenital disorders, living only for a day and a half. For two months they lived with us while Natasha healed from the tragic loss of their first child.

Balancing Rubles and Dollars: Kay

Win and Gracia Stiefel moved back to Alaska. We appreciated all their help with the two InterAct interns for whom we now assumed responsibility. This left gaps in the workload. I took over Win's job as the bookkeeper, juggling rubles and dollars and making them balance.

Another Director's Meeting: David

I flew to Portland, Oregon, and participated in InterAct's General Director's Council in December. Our meetings were held in Lincoln City overlooking the Pacific Ocean.

Dan and Steve surprised me by showing up at the end of our meetings. I gladly jumped in their car as we saw several sights along the coast and visited old Fairbanks friends. We drove through the Wildlife Safari in Winston, Oregon. It was great seeing animals from around the world in a more natural setting, but the best part was spending time with my sons.

Let's Sing Sakha (2007~2008)

2007
Goodbye, Interns

We accompanied the two InterAct interns, Stephanie and Heather, on their flight to Vladivostok to help them catch their flight to the US. They had stayed a year in Yakutsk, observing, learning, and working with the young people.

Three Trees

The book *Three Trees* was translated into Sakha and included 18 original drawings for the children to color. The story was based on a traditional folktale about three trees and their aspirations in life. The first tree wanted to be a great treasure chest and became the manger. The second tree wanted to be a great ship carrying kings and queens while it became the boat carrying Jesus and the 12 disciples. The third tree wanted to be a great and mighty tree to show the glory of God and became the lowly cross pointing the whole world to Jesus Christ.

Audio Recording of the Sakha New Testament: David

The Yakutsk Center for the Blind had a presentation of the audio recording of the Sakha New Testament. Sargylana, the main translator, and I spoke at the public event. A professional Sakha reader took 23 hours to record the entire New Testament. The center distributed them to 2,000 blind and visually impaired people in the Republic.

Sakha Songbook for Youth and Children

Lyuda compiled songs for a special songbook for young people. A growing number of summer campers and other youth gatherings needed them. We printed a trial edition in time for the summer camps in Oktyomtsi, Namtsy, Nizhny Bestyakh, Kempendyay, and several day camps.

Let's Sing Sakha (2007~2008)

Sin Has Consequences: David

I traveled with Vasya and Lyuda to visit Edic, the Sakha pastor from Ulakhan-An, in the hospital. He had been saved from a life of drinking, like many other Christians. He faithfully followed the Lord with Vasya and Lyuda serving as his mentors on his spiritual journey. We had a wonderful visit with Edic but were saddened when he died from liver cirrhosis. God forgave his sins, but he suffered the consequences of his sinful lifestyle choice to drink, dying early in life.

Super Special Guests

Our oldest daughter and husband, Debbie and John, with their children, Danielle and Joshua, visited us in Yakutsk at the end of May. Their teens had entered the First LEGO League International Robotics competition in Norway. After the weeklong competition, they flew to Moscow.

We purchased tickets for them from Moscow to Yakutsk and paid extra to have them waiting at the counter. Thankfully, the Russian who brought their family to the airport knew the system and scrambled to search for the tickets. They were found just before their scheduled departure.

The local people were glad to meet our family. We included them in our daily activities. All of us were interested in seeing the Mammoth Museum, and taking a tour of the underground tunnel at the Permafrost Institute with a replica of a baby mammoth carcass found fully intact. We visited the church in Oktyomtsi. At their picnic, the Rathbun family sang and signed for the song "He Keeps Me Singing." Each one shared their testimony.

In front of the cathedral: Kay, Danielle, John, Debbie, Joshua, and David

In Moscow Together

We flew together to Moscow on our way back to America. We walked through GUM (the main department store) next to Red Square. It was a magnificent piece of architecture with a wide variety of quality stores.

One of the home school books the family read together was *God's Smuggler* by Brother Andrew—who secretly smuggled Bibles across closed borders—recounting the miraculous ways in which God provided for him. GUM was one location where Brother Andrew parked his van during the communist era. Red Square includes Saint Basil's Russian Orthodox Cathedral, completed in 1561.

Saint Petersburg, Russia

At the end of August, we returned via Moscow, then rode an overnight train north to Saint Petersburg, Russia. We visited Tom and Cristy Slawson, new InterAct members, on our way to Yakutsk.

Church Multiplication

The Evangelical Church in Yakutsk enjoyed great growth. Two churches met on the original mission property, while two more churches met on the other side of town. The fifth church continued to meet on Saturdays. So now, five Evangelical churches work together in this thriving city of almost 300,000 people. We worshipped with the Sakha church on Saturdays and the Fourth Church on Sunday afternoons with Russian believers.

Pastor Valeri moved to Yakutsk to become pastor of the Sakha church. His wife, Zoya, suffered a life-altering stroke. In October, they broke the ice on White Lake to baptize two Sakha men. They kept adding more people.

Russia Field Conference: David

This year, our Russia Field Conference was held in Krasnoyarsk in early October. Southern Siberia was closer to InterAct's expanding work with the Tuvan people. Our special speaker, Ralph Alexander with SEND International, shared from the Psalms about giving both praise and thanksgiving to God. I passed the baton of Russian field leadership to Jim Capaldo.

To the West: David

Seven Sakha believers and I ministered in five churches to the west of Yakutsk. I was exhausted from traveling every day for a week with little time for sleep. The temperature hovered between minus 40 degrees and minus 47 degrees, so we were thankful for only one flat tire. The heater quit working for part of the trip. That was good for the ice cream one pastor bought for our village celebration.

Bus Stop Meetings

We were invited to meet a college student at the bus stop called "Old Folks Home." We thought they were having a special service at the nearby Elder's Home. But it was a group of 12 college-age adults meeting at Nadia's apartment for a special holiday meal. They asked us what we did for Christmas when we were children growing up in America. Then they told us about their childhood and all the changes since they and a few of their parents had become believers. It was a wonderful fellowship.

2008
Christian Youth Conference: David

Forty young people attended the Christian Youth Conference at the mission office in Yakutsk the first week of January. They were excited to be with other Christian youth studying biblical creation. It is important to understand that God created everything versus the theory of evolution.

At one session, I shared 10 logical observations from the natural world around us pointing to evidence of our Creator and His work. For example, we know a moose walked by a certain place because we see the evidence of his tracks. We could tell if he was walking or running, had stopped to eat willow branches, or had lain down, as well as which direction he was going. We could determine if it was a cow or calf, and how fresh the tracks were. While we can't see our Creator, we know He is here. We can also realize many things about His character.

Tenth Annual Sakha Christian Conference: Kay

In March, 200 participants came to the 10th annual Sakha Christian Conference in Oktyomtsi. We participated in all 10 conferences. This year's theme supported Sakha Christian families. We wrote a paper on

"Raising our Children for the Lord" with practical suggestions. I taught a session with the women about preparing our children for marriage, while David taught a similar one with the men. The love and unity displayed were a great demonstration of God's grace and goodness. We were thankful that two regional Sakha gatherings followed a similar pattern in Nyurba and Khandyga.

Granddaughter Kara Married

Dan's oldest daughter, Kara, married Aaron Matthew Fields in June in Fair Haven, New York, on the southern shore of Lake Ontario. Their first home was nearby in Red Creek, New York. We missed their wedding, but we got to see more of them in due time.

A Christian Celebration: David

Twenty miles from Yakutsk, the Sakha Christians planned a two-day summer solstice festival as an alternative to the secular one. Special events included the stick pull, carrying a heavy rock for a distance, and several jumping games. A large banner with Psalm 118:14a, "The Lord Is My Strength and Song," was displayed at a special concert. Individuals and groups shared their songs and testimonies. Pastor Valeri baptized nine people in a small stream of water. Much of the time was spent fellowshipping and eating while enjoying God's great outdoors. We praised our Creator while other celebrations praised His creation.

Kay in her unique Sakha outfit, standing between Galina and Lyuda from Oktyomtsi

Two pastors and I laid hands on Nikolai and prayed as we set him apart to be the new Sakha pastor in Tabaga. Later that day, he baptized three people. Sakha Christians were discipling others.

Slow Internet: David

Our apartment landline phone had slow internet, but it worked for short messages. The publisher for the expanded hymnal in Saint Petersburg sent us a mockup of the book in a huge file. I attempted downloading it at home, but the line kept dropping after several tries and I lost everything. So I took my flash drive to an internet café and in 40 minutes their "high speed" internet downloaded it. I paid 70 dollars for the service, but I never told Lyuda or others how much it cost. We were willing to do "whatever it took—within reason" to speed up the process and get the hymnal in print. We thanked the Lord that He had supplied the money for such expenses.

Back Home

We flew back to Alaska at the end of July. Our route took us through Moscow, Amsterdam, Minneapolis, and Fairbanks after 28 hours of flight time. Significant benefits of international flights include more legroom in economy class and good meals. While in Fairbanks, we saw our family and many friends, reported to churches and supporters, and consulted with doctors for minor issues.

Grandson Tim Married

Timothy Daniel Henry and Bambi were married in August at the Bible Baptist Church in Fairbanks. Tim is Dan's middle child. We were grateful to witness this happy occasion in a Christ-honoring ceremony uniting their lives. Both mothers sang to them. Tim's mother wrote the words and melody for her song. Two people in the audience made decisions for the Lord through the testimony of their wedding!

Another Great-Granddaughter: Kay

God blessed us with another great-granddaughter, Angelica Marie Head, born at Bassett Army Community Hospital on Fort Wainwright in September. She looked "angelic" as her name implied. Andrea's husband, Joseph, was deployed two weeks after her birth for his second tour with the Army in Iraq. I knitted her a hooded sweater with a prominent giraffe figure on the front.

God Is Faithful

Bethel Baptist Church hosted a big celebration remembering our 50 years of missionary service on September 28th. Dwayne King, a long-time friend with SEND International in Alaska and Siberia, was the master of ceremonies. It was a fantastic evening!

We sincerely appreciate everyone who shared or wrote a kind note about God's faithfulness and the work over 50 years. All four of our children shared their reflections as missionary kids. Steve prepared a timeline with pictures. Dan reprinted a daily devotional for Natives that we had written in 1988.

Daughter-in-law Cheryl sang "Find Us Faithful" with the challenging refrain, "Oh may all who come behind us find us faithful." The celebration ended with Paul Sauer playing "My Tribute (To God Be the Glory)" by Andraé Crouch on his saxophone. We felt overwhelmed by the display of gratitude for us and our service to the Lord. "Now to our God and Father be glory forever and ever. Amen" (Philippians 4:20).

Our family: Steve and Cheryl Henry, Andy and Elizabeth Cogan, Kay and David, John and Debbie Rathbun, and Dan and Linda Henry

We are ordinary Christians with our ups and downs. We are not worthy, but our Lord gave us strength for each day. We praised God for His faithfulness to help us do the work He called us to do these past 50 years.

Connecting with Cousins: Kay

My father in Michigan was happy to see us. Then we drove north to visit a church near Traverse City, Michigan, the cherry capital of the world. We spoke at New Hope Community Church and appreciated their support.

Cousin Janet Grundas and her husband, Mike, in Prudenville, were close by. She made beautiful quilts, and he carved wooden birds.

We visited another cousin, Bob Temple, and his wife, Karen, who sold double-wide house trailers. We inquired about his sister, Joanne Temple, who was at a friend's house nearby, so we had to find her.

Joanne is four years younger than I and still calls me Marilyn Kay like she did when we were kids. She never married because her mother, Alma Temple, was a widow and needed help. She had a boyfriend and was seeking God's will for her to marry or not after being single all her life.

Cousin Ruth: David

While in Michigan, I received an urgent message from Lois (Meyer) Clark that her mother, Cousin Ruth Lydia (Bachanz) Meyer, was critically ill in Saint Louis. I immediately talked with her on the phone. Ruth stated, "I am trusting in Jesus Christ. I am ready and anxious to go to heaven." Those were encouraging words to me as well. She was 84 and ready. She went to heaven that week.

The Long Way to Yakutsk: Kay

Thanksgiving was celebrated early with my father and sisters, Shirley and Nancy, and their husbands. Then we left Detroit that afternoon, flying via Amsterdam, Moscow, Novosibirsk, and finally home.

The baggage-retrieval area in Yakutsk was bedlam! There were no lines—just a mad, mass rush! Everybody searched for and dragged baggage to the small exit door as two security agents checked the tags.

We arrived at our apartment at 7:00 a.m. and woke up the young couple staying at our place. We quickly caught up on all the news and crashed into bed. Then Vasya and Lyuda from Oktyomtsi phoned and said that they were coming soon. They brought a huge meal of leftovers from Galya's 50th jubilee party. What a fantastic way to be welcomed back! Galya's celebration was an evangelistic outreach to many of her friends and relatives.

Sorting through Other People's Things: David

One of my projects was to sell the InterAct apartment. This included sorting through boxes of stored items that missionaries had left. It took many trips to consult with a Christian friend, a real estate agent, and government offices dealing with paperwork to sell the apartment.

More Than a Carpenter

Josh McDowell's book *More Than a Carpenter* which includes the Four Spiritual Laws, was printed in Sakha. Yes, Jesus was much more than a carpenter! Josh paid for printing this unique book, which he wanted to be available throughout the former Soviet Union. We received 5,000 copies.

What Is Most Important?

Completing the final edition of the Sakha Hymnal dragged on and on. The project was on our minds much of the time. It is most important to keep Jesus Christ in the first place in our thoughts and lives. Second, we followed what God wanted us to do. The Lord gave us the Apostle Paul's admonition to Archippus: "Take heed to the ministry which you have received in the Lord, that you may fulfill it" (Colossians 4:17). So, we completed our part in facilitating this project.

Singing in Sakha: David

The hymnal, *Praise to God,* was a tremendous Christmas gift for the growing Sakha church. It was a hardcover compilation of 200 Sakha hymns that was used in all the Sakha churches and fellowships.

Dense fog and negative 50 degrees did not stop the shipment from Saint Petersburg, Russia, via train and truck. Vasya, Lyuda, and I picked up the books. We wanted to see the result of our prayers and hard work.

Lyuda and Pastor Vasya sang a duet out of the new hymnal.

She Stuck with It

Lyuda had accepted the challenge to serve as the head editor—and stuck with it. She was totally committed to using her gifts for the Lord. There were 71 original songs. The Sakha people wanted a wide variety of original and translated songs. All of the songs required editing to be biblical and flow freely in the Sakha language. Lyuda was very capable and up to the long-running project.

The hymnal was the most essential piece of literature for Christians, other than the Sakha Bible. Songs influence what people believe. It is easier to remember a song than to memorize a Bible verse. Songs were a powerful teaching tool to reinforce biblical concepts memorably. The hymnal was also used by all the other denominations in the Sakha Republic.

Following the Lord's Leading: David

God placed us with the Sakha people at the right time, according to His wisdom and plan. We had the training, with 30 years of experience producing literature with Indians in Alaska. I felt like Abraham's servant who was sent on a special mission to find a wife for Isaac in Genesis 24:27 simply following the Lord's leading.

We lived in Siberia working with Lyuda and many Sakha Christians to facilitate the publication of the hymnal. Some people believe that only a gifted musician must work on such a project. The Sakha people realized we were there to promote and use their ways for the Lord's glory among them.

Visitors from Alaska (2009)

2009
Conference in Krasnoyarsk

Our group in the Sakha Republic had shrunk to three members, consisting of Sue Hauge and ourselves. However, the group in the south was growing, so the annual conference was held in Krasnoyarsk in February. Our colleagues were all studying Russian in Novosibirsk and making significant progress.

Native Visitors from Alaska: David

Native believers from Alaska came to visit us! Gilbert and Margaret Huntington from Galena on the Yukon River had talked about making this trip for a long time. It was a giant leap of faith to fly to Yakutsk and go through customs in different countries on their own. Their itinerary took them through Fairbanks, Los Angeles, Seoul, Korea, and Khabarovsk, Russia.

They did not want any financial help. Gilbert stated, "From the moment we decided to go on this journey, the Lord started blessing us. I received a call to work on a construction project which lasted long enough to cover any financial burden incurred during this trip."

We had prayed for Natives from Alaska to come and be a witness for our Lord to the Sakha people. Other Koyukon Natives have traveled here in a tour group or with government delegations.

Eleventh Annual Sakha Christian Conference

"Walk as Children of Light" from Ephesians 5:8 was the theme of the 11th annual Sakha Christian Conference held in Oktyomtsi in March. Church members worked to provide for 150 Sakha participants for three days.

Gilbert spoke using Yegor, a Russian translator, relating about his former life as Sakha people closely identified with him. His spiritual journey went from "I can do it to only the Lord can do it." Margaret sang two songs in typical Alaskan Native "country and western" style, which everyone appreciated. The Sakha people were interested in seeing their photos and asked many questions about Indian life.

Gilbert shared with Yegor translating.
Margaret was seated on the left, waiting for her turn to speak.

Alaskan "Cowboy": Kay

Walking to the Lena River with Gilbert and Margaret, we spotted a man with his horse pulling a small wagon. Gilbert suggested, "Maybe we could pay him for a ride." There was only room for two in the wagon. Another man leading a horse suggested Gilbert ride his horse. Gilbert was delighted to ride the horse but later remarked that he was sore from riding bareback.

The Road to Salvation

The Sakha Christian newspaper, *The Road to Salvation*, started in June 2006 as an eight-page paper. One issue included an article with pictures about us and our work with the Sakha people and their hymnal.

Translation Help: Kay

Galina Woods, a translator for the Altai New Testament in southern Siberia, spent several days at our apartment. She came specifically to work with Lyuda on key biblical concepts and the children's book *The Lamb*, which was in the final editing phase.

The day she left, we heard that my father was not doing well. We hurriedly packed and caught the plane to Michigan. My father, Robert David Temple, wanted to live to be 100. He made it to 98. He would say, "I'm ready to go, but not anxious!"

The Final Goodbye: Kay

We said our final goodbyes at my father's funeral service in Vassar. We were sad to see the family patriarch go on May 14th, but we knew he had loved and served the Lord, so we did not grieve like those who have no hope. I was saved when I was eight, and at the same time, my father committed his life to the Lord. Everybody saw the change in his life. When my time comes, we will meet again in heaven. He pointed to 2 Timothy 4:7 which was underlined in his open Bible inside the casket.

His funeral was more like a family reunion with many relatives we had not seen for a long time. All of my siblings, their spouses, and our four children attended. We stayed an extra week to help sort through his possessions.

Some Sad News

After returning to Fairbanks, the Lord arranged for us to connect with Paul and Mary Starr from Tanana. The first thing they told us was that Paul had stomach cancer. It was a shock for all of us! It left them with a lot of questions for their immediate future. This prompted Paul to say, "I know that I will die soon, but I know where I am going, and I'm ready."

A New Knee: David

I was having problems with my right knee while in Siberia. It was replaced for the second time on July 13th in Fairbanks. The healing process required a lot of physical therapy and time before I was fully functional.

Praying with Old Friends: Kay

Before returning to Siberia, we visited Johnson and Bertha Moses at Denali Center in Fairbanks. Bertha and I used to pray together. Johnson, who was resting in the adjoining room, got up and asked, "What are you doing?" I told him and his immediate response was, "Well, I want to pray, too." So, the four of us prayed together.

Great-Granddaughters

Great-granddaughter Abagail Joy Mount, "Abby," was born to granddaughter Cheryl and David in August. She was named after a woman of beauty and brains in the Bible (1 Samuel chapter 25).

Another great-granddaughter, Audrina Faye Fields, was born to our granddaughter Kara and her husband, Aaron in September. She was a healthy girl born in upstate New York.

A Henry Wedding

Grandson Wesley Reuben Henry married Lara Burrows at her home church in Lebanon, Oregon, in September. The new couple then moved to Fairbanks, where Wes enrolled in pre-medical courses at the University of Alaska.

Flying Back to Russia: Kay

We missed Wes and Lara's wedding since we flew back to Russia via Seattle, Chicago, Moscow, and then to Yakutsk. While in the airport, we noticed a Sakha girl looking at us. She introduced herself as Nadia. I had given her a Christian book on a flight from Vladivostok four years ago! She was 17 years old and from Churapcha, a Sakha town of 10,000 with no Christian witness.

Church Building Dedication

We rode the bus to Pokrovsk for the dedication of their "new" church building. The church had outgrown the living room of their founding pastor, Valeri, and his wife Valentina. A well-used building became available, so they dismantled it and moved it, log by log, to the pastor's backyard. They renovated it to meet the immediate needs of 40 Russian and Sakha believers.

Our Time Has Come (2010)

Visas Denied

All the paperwork to extend our five-year Russian visas for foreigners was submitted. But the visas were denied and no reason was given—as was customary.

This made major changes to our plans. We had to leave before March 22nd, 2010 when our current visas expired. We were grateful for the past 16 years of serving in Siberia with the growing Sakha and Russian churches.

Last of the Mohicans

We felt like "the last of the Mohicans." We were the first missionaries with InterAct Ministries to move to Siberia in March 1994. And now we're the last to leave in 2010. Other members who had served in the Sakha Republic had returned to America. The newer members were in different regions of Russia. Even though we kept aging with every heartbeat, our goal was to continue supporting and encouraging spiritual growth and outreach among the Sakha and Koyukon people.

Last Christmas in Siberia

This would be our last Christmas in Siberia. We celebrated with the Russian church in Yakutsk on December 25th. Then on January 7th, we traveled to Oktyomtsi for their Christmas program. They performed the traditional biblical account in Sakha with actors in full costumes.

2010
A 16-Year Collection

We started the New Year by sorting through our 16-year collection of ministry and personal belongings in earnest. Only the bare minimum went back to Alaska. One man's trash is another man's treasure!

Our problem was determining which to keep and finding the right person for everything left behind.

Five Sakha Churches

We took a six-day trip to visit five Sakha churches in western Yakutia. The weather was cold, but our hearts were warmed by these churches' love, interest, and growth. Misha, the former director, made a return visit to Yakutia, so we traveled together with others in a minivan.

The Right Choices Bible

Josh and Dottie McDowell wrote *The Right Choices Bible* to help their grandchildren make the right choices in life. This 368-page book gives 63 examples from the Bible of people experiencing the blessings of God or reaping the consequences of their choices in life. Two thousand hardcover copies were printed in full color in Saint Petersburg, Russia.

Trivia question: How many e-mails did it take to prepare and get *The Right Choices Bible* off to the printer? This includes all the e-mails concerning the layout, financing, and shipping. Answer: Ninety-one!

"A Drop of Ink"

"A drop of ink may make a million think," as famously stated by Lord Byron, and we wholeheartedly agreed. Before we left Siberia, the Sakha people wanted a complete list of everything we helped them print. They published the list in their Sakha Christian newspaper. We saw the big picture of the tremendous need for Christian literature in their language and what God was doing as we traveled extensively throughout the Sakha Republic. Everywhere we went, people asked many spiritual questions that needed biblical answers. Russia experienced the dissolution of communism in 1991, and now people are openly searching for spiritual answers.

We assisted and facilitated the publication of Christian literature with eight books geared for children and youth and four for adults, including two Sakha hymnals. There were gospel tracts, Scripture mugs, Scripture calendars for 11 years, and assorted smaller items.

Looking back, we can see how God provided the funding for various projects through vitally interested Christian individuals, churches, Operation Mobilization, the Josh McDowell Ministry, and the Tyndale Foundation.

Growing in Our Christian Lives: **David**

One day before the annual conference, the booklet *Growing in Our Christian Lives* came back from the local printer. Sakha colts wanting to run and grow were pictured on the cover. We felt a heavy burden to give the Sakha people practical advice in writing on this vital subject. God wants each of us believers to keep growing spiritually throughout our lives. The final editing, layout, and printing had been sandwiched between sorting and packing.

Goodbye, Fellow Sakha Believers

More than 200 Sakha believers met on the first day of their 12th annual Sakha Christian conference, which was moved from Oktyomtsi to Yakutsk. We don't like being in the spotlight, but they insisted we stand upfront while every church and group of believers came forward to sing a Sakha hymn and say something to us. We were completely overwhelmed by their love and appreciation for all the Sakha Christian literature, hymnals, and years of service with them. Saying goodbye was not easy after our 16-year relationship with these outstanding believers.

"Not unto us, O LORD, not unto us, But to Your name give glory, Because of Your mercy, Because of Your truth" (Psalm 115:1).

Home in Alaska
Fairbanks (2010~2012)

Goodbye Siberia

We left Siberia on March 19th with two suitcases apiece. Excess baggage was too expensive. Vasya and Lyuda made us an offer we couldn't refuse—to do the final cleaning and disposal of whatever was left. That took a huge load off of us. We sold our apartment to a wonderful young Sakha Christian couple. Our flight was through Moscow, Chicago, and finally to Fairbanks, just in time for Easter.

On the Way Home: Kay

We saw Uncle Ed Erb, who was 97 with a sharp mind when we stopped in Chicago. We stayed with my cousin Diane and her husband, Gerry, who lived an hour from O'Hare International Airport.

Adjusting to Alaska

Our ranch-style home on the outskirts of Fairbanks was more spacious compared to our small apartment on the top floor in Siberia. The stores had so many choices! It was easier to buy groceries and put them in our car rather than catching a bus and carrying them up five flights of stairs.

Now, to working with Alaskan Natives with no language limitations. We fit back in at the Fairbanks Native Bible Church.

Another benefit to Fairbanks was living in the same city as most of our children, grandchildren, and great-grandchildren. We cherished our God-given family.

On the Yukon River

In early June, we flew to Tanana on the Yukon River for one last visit with Paul and Mary. We were thankful for their commitment to the Lord in spite of Paul's physical problems with cancer. We had previously attended his 70th birthday celebration in Fairbanks with his family and friends.

Our trip was scheduled to attend the annual Denakkanaaga Conference of Native Elders and Youth. The youth listened and learned from their elders as they discussed various Native issues.

End of the Trail: David

A week later, we returned to Tanana to conduct the funeral for Paul Starr, who had come to the end of the trail. We met him in 1958 at Arctic Camp in Nenana, when he was a new teenage Christian. 1984 was a momentous year for Paul—when he made Jesus Christ the Chief of his life, quit the occasional drinking sprees, and got married. He left this life as a victorious Indian Christian warrior.

Another Great-Granddaughter

Great-granddaughter Nyomi Gwendolyn Hendrickson was born in June 2010 to our oldest granddaughter, Melissa, and James. She was named in honor of the biblical Naomi, the great-grandmother of King David, with a variant spelling to make her unique. We were thankful to hold and enjoy her as a beautiful newborn at Fairbanks Memorial Hospital.

Churchgoing

We attended Bethel Baptist Church on Sunday mornings and grabbed a quick bite to eat. We then picked up people who needed a ride to the Fairbanks Native Bible Church for the service from 1:30 p.m. to 4:00 p.m. Sundays were busy.

Visiting

We visited Native friends at the hospital and Denali Center. We kept seeing them at funerals, fundraisers, or in the grocery store. God gave us many opportunities to reconnect with "old" Indian friends and meet new ones.

Another Wedding: Kay

Elizabeth and I flew to Murfreesboro, Tennessee, to witness my niece Tiffanie Dawn Temple being united in marriage to "Rudge"

Shandcler Kee Rudgley in July. I also spent time with my brother Jim, Tiff's father.

Alaskan Conference: David

InterAct's Alaska Conference was held in Palmer. The emphasis was to involve more Natives to reach their people. Our director appointed me to an advisory board called ATM (Alaska Team Ministries).

Bible Study: Kay

I taught an inductive Bible study on the Gospel of Luke with two Indian ladies. One was a believer, while the other was interested in the Bible. It started once a week, but then one requested we meet twice a week, so I obliged.

Packing Again: David

Packing again, we returned to Siberia on October 23rd with one-month tourist visas, the easiest kind to get. We felt like the Apostle Paul, who said to Barnabas: "Let us now go back and visit our brethren in every city where we have preached the word of the Lord, and see how they are doing" (Acts 15:36).

Our goal was to strengthen the growing Sakha church. One of the first reports we heard was that the police had illegally raided the home of two Sakha women and confiscated their Bibles and Sunday school materials. They also lost their jobs. This was illegal. Yet it was better to suffer for being a Christian than for doing bad things.

I preached at the Second Church in Yakutsk and emphasized the importance of keeping Jesus and His Word first in our lives. I gave the same message the following week in Oktyomtsi. We saw everybody we could during our brief visit.

The Lamb

We were glad to see printed copies of *The Lamb* by John R. Cross, which was written to explain the gospel to children five to 10 years old and show their need for the Savior. A clear link was made between Jesus and the Old Testament sacrificial lambs with the use of many illustrations.

A Celebration for Galina

In Oktyomtsi, we helped celebrate Galina's 50th jubilee. The event was a great testimony to her unsaved relatives. She had chosen to follow the Lord 15 years earlier. Life was a struggle, raising a disabled son, having a teenage daughter who stopped going to church, and having an alcoholic husband who died three years ago. Yet, she was faithfully living for the Lord. The retirement age in the Sakha Republic was 50 for women and 55 for men. Her meager pension helped her make ends meet.

Traveling Farther: David

We traveled farther to the village of Ulakhan-An. This Sakha village now has a growing house church with 17 regular attendees. The number of men and women was almost equal. I preached about a growing relationship with the Lord, using examples from Scripture. We first visited this village in 1992, when there was total darkness—no believers. Lyuba, an early believer, opened up her home for services and taught Sunday school.

As we were ready to leave, a man ran out to us, asking, "One of the women wants to repent; should she now or later?" That was a no-brainer, and our group immediately went back inside.

Psychological Invasion: David

A psychological approach to life's problems called *The Murray Method,* had invaded the Sakha Republic. The founder was an American who had left her husband to pursue her dream—an obvious red flag that she was not following the Lord. At first, I did not take a hard stand against its influence, but later I did.

Her approach emphasized oneself as the center of one's life. It blamed other people or bad events for one's problems rather than promoting personal accountability like in Scripture. It caused much division and many hard feelings throughout the churches in the republic as two churches left the association and a larger church split. This continued to be a nagging problem. God wants us to follow Jesus Christ and biblical principles.

More Paperwork: David

Our apartment was sold when we left Yakutsk in March, but we needed to remove our names from the official registration. The new owner contacted us while we were in Alaska. Our outdated passports for foreigners proved we no longer lived there.

We spent one whole day running around to different offices scattered throughout the city. The top boss had no idea what to do, but the Lord sent a Sakha lady, whom I recognized from a previous trip, to calm him down and redirect his thinking. Then he signed the form. Next, we went to a different office, arriving at 5:00 p.m., their official closing time. We were the only customers, yet the head lady took our documents to another room to make a phone call. Meanwhile, the workers continued playing solitaire on their computers. The head lady returned and had one of the workers complete our "checkout" after closing time. Such was the document work and answers to our prayers.

Home for Thanksgiving

We flew back to Alaska in time for a wonderful Thanksgiving celebration at our house with our children, grandchildren, and great-grandchildren. This trip took a lot of energy, but we were delighted to have a part in helping the Sakha church in any way we could.

2011
Another Great-Grandson

We were blessed with another great-grandson named in honor of two outstanding people in the Bible. Elijah Joseph Head was born in February to our granddaughter Andrea and her husband, Joseph.

Telling Our Story

In Michigan, we participated in the annual mission conference at First Baptist Church in Vassar, our original sending church. We were happy to share what God was doing in Alaska and Siberia. We visited four other churches and individual supporters.

Seeing Relatives: Kay

Our next stop was in Evart, Michigan, to see my cousin Joanne, who had married Richard Greeley in June 2010. She had been single for 73 years. They lived in the country. Then we drove through upper

Michigan to Blind River, Ontario, where we had a great visit with my sister Nancy and Louie, plus their boys.

Old Timers: Kay

Shortly after his 99th birthday, we visited Uncle Edgar Francis Erb in Illinois. While there, we stayed with Cousin Diane and Gerry. Our family was very close to Uncle Ed and Auntie Leola. We grew up together in Vassar, Michigan, and attended the same church most of our lives. Uncle Ed almost made it to 100 but passed away in December.

A California Wedding: David

Then, we flew to California to attend the August wedding of my niece Helen Ann Avila and Nathan White at Mount Baldy Village Church. It was a beautiful, Christ-honoring wedding on a scenic mountain. My sister Miriam and brother Martin were there with other family members for this happy event.

We stayed nearby in a motel. While eating breakfast, someone stole Kay's purse. We searched and asked the staff but to no avail. The thief immediately went on a shopping spree with her credit cards, so I spent several hours on the phone to report the theft, cancel our cards, and order new ones.

Family History: David

Returning to Chicago, we drove to Sheboygan, Wisconsin. We were anxious to meet my second cousin Annemarie (Bachanz), her husband, Lothar Burchardt, and two of their five children from Drehnow in East Germany. This was their first trip to America. Lothar knew the Bachanz family's history. Cousin Lois (Meyer) Clark and her sister-in-law, Beth Meyer, joined us.

My grandfather Martin Bachanz Jr. had immigrated to Sheboygan in 1887 when he was 16, with his half-brother, Friederich Schimlick. The rich farming country of Wisconsin was similar to what they had left in East Germany. Other relatives also lived in this region.

A local German Christian, Marlin Miller, who enjoyed digging into genealogies, escorted us to visit family graves, farmlands, and several locations with regional records. It took time to assimilate all the family information.

Cousin Charles: David

In Sheboygan Falls, we saw Cousin Charles Bachanz, who is five years younger than I am. He lived in a home for disabled people and used a wheelchair. He instantly remembered me, even though it had been 15 years since our last visit. He showed us an Indian plaque in his bedroom, which I had given him.

Uncle Carl and Naomi (Collins) Bachanz from Plymouth, Wisconsin, had only one child, Charles, who had cerebral palsy. His umbilical cord had been wrapped around his neck at birth, limiting the flow of oxygen to his brain. I asked Charles if he believed in Jesus Christ. He shot back, "Yes, I don't want to go to hell."

Visiting Israel

Then we flew to Newark, New Jersey, in October to join our 10-day tour group with The Friends of Israel. The most outstanding sites were the Garden Tomb, the Great Isaiah Scroll, and the Masada Fortress, as we mentioned in our account of our first visit to Israel in 1984. We learned more the second time. We even saw Josh Czech, a friend from Fairbanks, who worked in Israel.

It was a unique experience to "float" in the Dead Sea. But if you rub your eyes, they will burn. The salinity is around 34 percent, which is 10 times as salty as the Pacific or Atlantic Oceans.

Holocaust Museum

This time we toured "Yad Vashem," the Hebrew name for the memorial in Jerusalem, remembering the six million Jewish victims of the terrible Holocaust during World War II. They honored the non-Jewish people who helped them. There is no Jewish blood in our family, but we respect and support the Hebrew people.

One haunting quote by Kurt Tucholsky, hit us while visiting the memorial: "A country is not just what it does—it is also what it tolerates." It initiated thoughts about the tolerance of anti-Semitism in America and around the world.

Back in America

We flew back to Newark, New Jersey, and continued mixing business and pleasure. We reported to churches in New Jersey and Tennessee. This gave us more opportunities to visit relatives between Sundays.

Cousin Bill Fox and his wife, Joyce, lived in the rural community of Narrows, Virginia. There wasn't much vehicle traffic, but we heard every freight train running close to their home. We also attended their midweek church service.

At our alma mater, Bryan College, in Dayton, Tennessee, our granddaughter Danielle was studying hard and playing soccer. We rode with her team to Athens, Tennessee, where they played against Tennessee Wesleyan University. Then we traveled to Virginia, where we visited another granddaughter, Kara, her husband Aaron, and our great-granddaughter Audrina. It was fun to walk to the ocean and spend time with them.

Another Big Wedding

Next, we drove to Gaithersburg, Maryland, to witness our granddaughter Suzanna Grace Henry's marriage to Andrew DeCarlo in November at the Church of the Redeemer. It was a great Christian celebration with a radiant bride. Many relatives from both sides of the family were present.

2012
Honoring Great-Grandma Kay

In February, we flew to Newark, New Jersey, and drove to Virginia to see Kara and Aaron. We held and played with great-granddaughter Alana Kay Fields, born in December, who was named in honor of Great-Grandma Kay.

Sarah Kay Henry was born to grandson Tim and Bambi at Fairbanks Memorial Hospital in March. She was named in honor of a famous woman in the Bible. Her middle name is "Kay" like her Great-Grandma Marilyn Kay.

Fairbanks Native Bible Church (2012~2016)

Easter Day: David

Easter at the Fairbanks Native Bible Church started with a brunch. I worked with the men to cook, while adding fried moose heart to the menu. We were thankful for a full church as Pastor Harry explained and emphasized the importance of the resurrection of Jesus Christ.

Saying Goodbye

We visited Diane Williams from Allakaket who was critically ill with cancer in Fairbanks and nearing the end of her young life. She made a brief comeback, even walking again. Her sister, Priscilla, said, "It's a miracle, an answer to prayer!" God gave Diane time to tell her many visitors about the Lord and say goodbye.

Citation Award

The highest award granted by AWANA was the Citation Award. We were delighted to watch our grandson Joshua receive the trophy. To earn it, he completed 10 AWANA books. Joshua memorized 648 Bible verses, summarized the books of the Bible, and did four service projects.

His sister, Danielle, received this award in 2010 and completed the same requirements. They both received a 1,000-dollar scholarship each year they were students at Bryan College for this achievement.

Connecting: David

We drove to Victory Bible Camp to participate in a networking summit at the end of April. The purpose was to bring together Christ-centered ministries from across Alaska to develop a network to further God's Kingdom in Alaska. Alaska Freedom Journey (AFJ), a Native-led organization, had asked me to present a paper on building trust between Native and non-Native people in Alaska.

InterAct's Annual Conference

Our next event in the same region was InterAct's annual conference in Palmer. Partnering and networking were the major topics. Each group had different skills and ways of solving problems as we collaborated for the common goal of bringing honor and glory to Jesus Christ.

ICU

Kitty David from Allakaket was flown by medevac to Fairbanks with serious heart problems. We visited and prayed for her with others around her bed. We saw her family and many friends while waiting in the hospital. Kitty survived as God gave her more time.

Cousin Diane: Kay

Cousin Diane was like a sister to me since our families have done things together all of our lives. She and her husband Gerry came to visit us in Alaska for a week. Gerry enjoyed seeing the Alaska Railroad since he was a train enthusiast.

Up North

We flew north to Allakaket in the middle of July. It was wonderful to be with old friends. Our Native hostess asked if we could have a church service while there. We were encouraged by excellent spiritual conversations with several people.

Monument for World War II

A large monument in Fairbanks features a Russian and an American pilot standing in front of an aircraft propeller. They commemorated the spirit of friendship and cooperation between allies during World War II. Soviet pilots received 8,000 US aircraft loaded with supplies in Fairbanks and flew them via Yakutsk to the war front.

Ordained to Serve: David

Pastor Harry Hafford Jr. was ordained as the pastor of the Fairbanks Native Bible Church on November 18th. I participated with several others in praying over him. Since then he has been using and developing his spiritual gifts of preaching and teaching with a shepherd's heart. Harry and his wife, Heather, returned to Alaska in 2010 and adopted three Native children from Venetie. They are all musical and have played in the church band.

2013
Another Wedding: Kay

In January, we flew to Flint, Michigan, and loaded our baggage into a rental car to drive to the Rosseau Muskoka Resort and Spa in Minett, Ontario, a picturesque location. Our nephew Robin and Susan Horwath's daughter, Melissa, married Freddy Jenson with a lovely wedding in the hotel facilities. It was terrific to spend time with my sister Nancy and her family.

Cousin Carol: Kay

Touring the frozen version of Niagara Falls, it was easy to see why this was a favorite site for honeymooners. It was a little cool in January, but we enjoyed seeing all the ice. Then we drove to Fort Wayne, Indiana, to see Cousin Carol Bower. She demonstrated her large weaving loom and spun wool yarn.

Carol and Kay

Back to Siberia

From Detroit, we caught our flight overseas via Moscow. We had Russian visas, good for three months, as we headed back to Siberia on February 22nd. Our first stop was Krasnoyarsk, a bustling city of one million people on the Trans-Siberian Railway. We stopped to visit fellow InterAct members Lucas and Jamie Orner with their three young children. They had an apartment on the eighth floor with two working elevators.

How many doors did we have to go through to enter their apartment? First, we went through a steel door with a unique lock and then two wooden doors to help keep the cold from entering the building. Next, we rode the elevator to the eighth floor, with another locked steel door to enter their complex of apartments. Finally, we encountered a steel door to their apartment with three locks, which they lock each night or when they are gone, and another wooden door. That was a total of six doors, typical for big-city dwellers. But no matter how many security doors and locks there are, our absolute safety is in the Lord.

Sakha Students

Two Sakha students were at a weekly class for learning English in the Orners' living room. We spoke one week about Alaska. One of the girls, Olga, was from a regional town in the Republic of Sakha.

Southern Detour

The overnight Trans-Siberian train took us west for 12 hours to Abakan. We appreciated the time to sleep with a comfortable ride, arriving before 7:00 a.m. We found a minibus taxi for the six-hour ride through the beautiful Sayan Mountains to Kyzyl. It was a pleasure to be with the newlyweds and fellow members Scott and Vanessa Gilbert, working with a local Tuvan church.

The Train to Kansk

After returning to Krasnoyarsk, we took the Trans-Siberian train east for four hours to Kansk to visit Misha and Olga and their family of five children. We had traveled and ministered all over the Sakha Republic with Misha when he was single. Misha showed us the new church building. We met many of the 40 church members.

Making Disciples: David

We returned to Krasnoyarsk and then flew northeast to Yakutsk. We were thrilled to attend two Sunday services at the Sakha Church with 100 people at each service. At their Monday night discipleship meetings, I met seven young men Pastor Valeri discipled each week. He encouraged them to report on their outreach activities and preach a short sermon on Sundays.

When one of these young men preached on Sunday, he publicly told everyone about meeting me seven years ago in his hometown. At that time, this young man was intoxicated and staggered into the Sakha church service with two buddies. I gently escorted them outside. The meeting came to a halt as the ones inside prayed for them. Once outside, he asked me for money, but none was given. Shortly thereafter, he turned to the Lord

"Path of Life" Church

One Sakha couple owned a construction company and was in the process of building a nine-story concrete apartment complex near a bus stop. The main floor became the new meeting place for the Sakha church called "Path of Life." It was dedicated on December 29th, 2013.

In Russia, space in a new building means a rough, bare concrete floor, walls, and ceiling. The church added lights, flooring, wallpaper, and bathroom fixtures to make it usable.

Another Sunday in Yakutsk

On another Sunday morning in Yakutsk, we shared the service with Dwayne King and several other men from Alaska supporting the flight training program. It was good to see the interest in going out to more isolated people off the road system in the Sakha Republic.

Other Churches

We were privileged to visit the Sakha church in Ulakhan-An, two hours south of Yakutsk over rutted, dirt roads. Our host told us this road would be impassible the following week due to the spring thaw. Now, 20 believers meet each Sunday in the pastor's living room, including many men.

Easter Celebration

We shared in all the excitement of another Easter in Yakutsk. At the Russian Evangelical Church, special singing groups of adults and children led the worship. It was uplifting to see so many Christians together and excited about the resurrection of Jesus Christ.

Picnic Time

The Oktyomtsi Church had a picnic on the edge of town where many yellow crocus flowers bloomed. This gave us another opportunity to interact with their members. Our prayer is for more Sakha people to respond to Jesus Christ and share our joy in this life.

Going-Away Parties: David

We were the honored guests at a going-away party in the main Russian church. Pastor Valentin, the mission director, and his wife, Lyuba, especially thanked us for hand-carrying an accordion for her from America on one of our early trips to Yakutsk. A few of the people were those we met on our first visit to Yakutsk in 1992, still being faithful witnesses for our Lord, along with many newer ones.

The Sakha church hosted a going-away party. It was wonderful to see the stability and growth in their church. Members shared many interesting memories and highlights. We wished that our three-month visas were longer.

We were invited to the homes of several people. After enjoying a great Sakha meal together, Pastor Valeri and their family gave us gifts. Kay had to model the Sakha dress along with a large silver pendant and headband. I received a handmade Sakha hunting knife.

Hello, Germany: David

Flying from Yakutsk to Moscow was great, but our connecting flight on Air Berlin to Germany had a mechanical problem, so we returned shortly after takeoff. Russian customs officials came through the plane, canceled everybody's exit stamp in their passports, and then two hours later, they stamped new exit stamps.

Tegel Airport in Berlin was a welcome sight, even in the dark. We caught the next available taxi with a driver from Iran, arriving at our hotel at midnight, much later than planned. I gave him a gospel tract and we had a friendly conversation. We got a little sleep before meeting

Lothar Burchardt, my cousin's husband, in the morning after a big German hotel breakfast.

A major attraction was the 12-foot-high Berlin Wall which was in place from 1961-1989. It became an iconic symbol of the Cold War. A line of cobblestones in the sidewalk and road marked its former location. Only one section of the wall was left standing and is now covered with graffiti. We walked freely through the Brandenburg Gate.

Cousin Annemarie: David
Lothar's Ford Fiesta moved right along on the Autobahn (Expressway) with no speed limits. He generally cruised at 80 mph or more on the smooth highway. Several cars flew past us like we were parked! It reminded me of when I was a teenager riding with my brother Martin in his 1939 Ford pickup going over 100 mph!

The Bachanz family's hometown of Drehnow (DREH-no) was southeast of Berlin. The highway sign was written in both Wendish and German in this traditional location of the Wendish/Sorbian people. The town's stork nest, 30 feet high, stood out as did a maypole over 100 feet tall. Drehnow was a clean, simple town of 600 people with well-built homes in a rural farming community.

My second cousin Annemarie Burchardt knew some English, but I only knew a few words in German. Lothar served as our translator because he also spoke English, Russian, and Wendish. We met all five of their children: Esther, Maria, Johannes, Ulrike, and Tabea.

The Church in Drehnow
Their church sponsored an annual youth conference and evangelistic gathering attended by several hundred people. Their menu featured German bratwurst. The church was the only one in Drehnow, built in 1930. When my grandfather Martin Bachanz Jr. immigrated to Wisconsin in 1887, everybody in Drehnow walked three miles to Peitz for church services conducted in the Wendish language. Now all the regular services are in German.

The Church in Pietz: David
We drove to Pietz to see the town and the Holy Cross Lutheran Church. We were shown old church records, including those of my grandparents, that had been hidden during the Russian occupation

of Eastern Germany. Communists destroyed all church and historical records in Drehnow, including the Bachanz heritage.

In neighboring Cottbus, we saw several houses that my great-uncle Christian Bachanz had built. The Communists reassigned them to someone else since residents could only own one. He stayed in Eastern Germany during the occupation. If he had fled to free Western Germany, he would have taken only the clothes he wore but no family possessions, money, or anything of value and would forfeit his home. We met more distant relatives there.

Dresden

Allied forces in a massive air raid firebombed and leveled the historic center of Dresden in 1945. The bombing was controversial since the city was a cultural center, not an industrial complex. We attended a wonderful service in the large reconstructed Lutheran Church with a statue of Martin Luther out front.

Boating Through the Canals

Lothar drove 27 miles to the Spreewald region, where we rode in an open gondola boat. The gondolier used a pole to propel us through the canals, forests, and wetlands. Everything about the town and trip reflected the homeland of the Wendish/Sorbian culture.

Moving On: David

Our time in Germany came to an end. Lothar drove us to Berlin, where we flew to Frankfurt, New York, Boston, Seattle, and Fairbanks. We reflected on all the blessings the Lord had given us with a Christian heritage.

Annemarie, Kay, David, and Lothar in front of their home.

Father's Day

Father's Day was an excellent time with Dan and Linda and their family at their cabin on Harding Lake. We had a cookout, enjoyed the lake, played in the water, and spent valuable time with family.

Early Birthday Party

A combination of early birthday parties for Kay and great-granddaughter Nyomi, plus a family get-together, took place at the end of June at our house. We picked a time when most of our family could be there. Nyomi turned three while Kay turned 80 in October. We enjoyed celebrating with our family.

Salmon Are Running: David

I rode to Chitina with Dan and Tim to dip for sockeye salmon in the Copper River. This was a yearly event with top-quality red salmon coming up the river from the ocean to spawn. The harvest was limited to Alaska residents for personal use. Dan and Tim were tied off with a rope to prevent them from falling into the swift current. They used dip nets, three feet in diameter with 10-foot-long poles, to catch the fish. After returning to Fairbanks, we worked together to cut and package the fish to freeze for the long winter ahead.

Fast-forward: As we grew older, I gave Dan my proxy permit to get a winter supply of red salmon for us. We appreciated all his hard work and generosity.

My Friend, Olga: Kay

Olga Solomon lived in Kaltag on the Yukon River. Another Native friend, who had attended Bible studies in Fairbanks, passed away. In August, I flew there to pay my respects at her funeral and spend time with Olga.

The Fairbanks Equinox Marathon

Two of our children ran in the 51st running of the Equinox Marathon in Fairbanks.

We gave them credit for running and crossing the finish line along with 960 other racers. Elizabeth had been training and participating in several races earlier in the year. Dan ran track in college and also in the 1980 Boston Marathon but he had not competed recently.

More Family Events

We enjoyed a variety of family events in Fairbanks. One was a special Ethiopian dinner at Steve and Cheryl's house with their boys. The masters of authentic Ethiopian food were Tadesse and Frey with their three young children. They were family friends from their days of serving as missionaries in Ethiopia. The food was eaten with your fingers—like Adam and Eve did.

Elizabeth and Dan—just before the grueling race of 26.2 miles with climbs and uneven trails of Alaska adventure

Great-granddaughter Abagail Mount celebrated with a special tea party at her home. It was a fun excuse to do something special on her fourth birthday.

There was a back-to-school party at our house for Danielle and Joshua. She was enrolled in Bryan College. Would Danielle find a husband there as Kay and Debbie had?

We picked blueberries with Kara and four-year-old Audrina. She was learning the valuable Alaskan fall art of picking berries. Kara invited us to her house for a Thanksgiving dinner. It was so much fun spending time together.

2014
Alana's Birthday Party

The year started with a big birthday party for Alana Kay at Aaron and Kara's home in North Pole, Alaska. They waited until after the holiday rush to celebrate her second birthday since she was born in December.

Native Musicale: Kay

The annual *Native Musicale* was always a happy time to reconnect with old Native friends and meet new ones. I made sure we saw Virginia Maillelle. Friends who lived faithfully and consistently for our Lord meant so much to us. The Fairbanks church brought a group of singers, but all the singers needed people like us in the audience to cheer them on, pray for their witness, and thank the Lord for them.

On Saturday, we laughed while watching the "Running of the Reindeer" on Fourth Avenue in Anchorage. Running among the reindeer, people wore wild costumes as they dodged and ducked the flying hoofs. This was a tamer version of Running of the Bulls. We even donned foam antlers.

Praying in Sakha: David

I was asked to pray in Sakha at a gala dinner celebration at Raven Landing, a senior apartment complex. I contacted Olga in Yakutsk for help. I gave her my prayer points; then I worked on saying them smoothly enough to be understood by the Sakha guests. Others must have thought I was praying in an unknown language.

Leslye Korvola sponsored the event to honor the sister city organization's 25th anniversary and the visiting dignitaries from Yakutsk. I gave Alexander Savvinov, chairman of the City Council of Yakutsk, our last coffee mug with Psalm 121:2 printed in Sakha and Russian. It was more meaningful for him to have a verse in his own language than for me to keep a nice souvenir. He showed me a book about Sakha horses, saying, "They are bold and strong!" to which I responded, "And tasty too." The meat from colts is a delicacy for the Sakha people.

Preaching: David

I preached at the Fairbanks Native Bible Church about once every two months. When needed, I led the Wednesday night Bible study. We always transported people to church.

Going to the Dogs

The Alaska Dog Mushers Association runs many races on their property, which borders ours. Watching the three-day Junior Dog Sled Races, we connected with many Native friends. Then we joined the crowd for the Open North American Championship race, which started and finished on Second Avenue in Fairbanks. The city hauled in snow for this event, which drew a larger audience in town. Dog mushing is the official Alaskan sport.

Dinosaur Era Audio Cassettes

Our ancient cassettes were recorded in the 1980s for two Koyukon dialects of the Gospel of Mark. With assistance from personnel at UAF, they converted these to DVDs and thus made them accessible to the Koyukon people. Both narrators were highly respected elders among their people.

Our Creator Bet'o hudeełt'aaye

The 32-page booklet, *Our Creator Bet'o hudeełt'aaye*, with the Koyukon word meaning "the one on whom things depend" was ready. The booklet is based on Romans 1:19-20 and how Indians knew about their Creator through His creation before the coming of the gospel to Alaska. It was designed to help them turn to God with 10 reasons in narrative form with historic photos.

Birthday Party for Sarah Kay

Sarah Kay turned two in March, and we were invited to help her celebrate at Tim and Bambi's house. One of her fun things was jumping with friends in their bounce house set up in the garage.

Cancer Is a Deadly Disease: Kay

My sister Nancy Lou Horwath was dying of cancer. We moved up our departure date in hopes of seeing her alive, but she died before our arrival. Our daughter Elizabeth and Andy had visited her in Canada. We made it to her funeral in Vassar, where she was buried near my parents. She was a wonderful, godly sister and was missed by all.

Meeting in Murfreesboro: Kay

Next, we traveled to Murfreesboro, Tennessee, to visit Jim, his daughter, Tiffanie, and her family. Tiffanie and Rudge had an addition built on the backside of their home with a private bedroom and bathroom for Jim.

On Sunday, we visited their large church. It was unique being in a junior church with a large stage, lots of kids, and colored-coded chairs to identify where each age group sat. Great-niece Soma was a talented singer who performed solos or with a group on stage. And we ate at Demos' Restaurant, one of their favorites.

Graduation Day: Kay

We drove to Dayton, Tennessee, to watch our granddaughter Danielle graduate from Bryan College with a BA in Education. We also had the benefit of meeting older classmates from our generation.

In front of the Chapel: David, Kay, Joshua, Danielle, Debbie and John

More Relatives: Kay

Then we continued to Thomasville, North Carolina, to visit my cousin Robert Kauffman and his daughter, Carolyn (Kauffman) Winslow. I am grateful to have a large Christian family.

While in Greensboro, Georgia, we had lunch with my niece Tamie (Temple) Ruark, who was in a terrible head-on collision several years ago. We praised the Lord that she was now able to drive her car and has recuperated! Her healing took years of physical therapy and prayer.

Reconnecting

We stopped at the Wycliffe Discovery Center in Orlando. As we approached the welcome area, a lady our age stepped forward, exclaiming: "I know you!" It was Corine Hatch from 56 years ago! We trained together at SIL in Grand Forks, North Dakota. Then she and her late husband planned to go to Vietnam while we headed north to Alaska.

While driving from Orlando toward Bradenton, Steve and Cheryl called! They were temporarily in Saint Petersburg, Florida, for a job interview. We were surprised they were nearby! We enjoyed a few days with them at the beach.

More Relatives: Kay

We saw my cousin Wilaine (Kauffman) Northway in Grand Blanc, Michigan. It seemed that everywhere we went, we could find family.

After a missionary conference in Vassar, Michigan, we drove through the Adirondack Mountains to Specular, New York, to spend time with my sister Shirley and Tom. They have a beachfront property with sand, sailing, and boating activities on Lake Pleasant. All the fall colors of the oak and maple trees were brilliant.

Three More Great-Grandchildren

Adelaide (Addy) Evertt Henry was born to Wes and Lara in June. Addy was Steve and Cheryl's first grandchild. Our family was expanding!

Great-grandson, Emerson Matthew Fields, was born to Kara and Aaron in October in Fairbanks. His middle name was the same as one of the four gospels. We saw him at their home in North Pole, Alaska.

And another great-granddaughter, Clara Belle Henry, was born to Tim and Bambi in December in Fairbanks. We held her the next day in her hospital room.

Rocket Man

Grandson Joshua was taking courses at UAF toward earning his engineering degree. One course was building a rocket and then launching it on campus, going several hundred feet in the air.

Christmas Celebrations: Kay

Our first big celebration with a full house was for Christmas. Years ago, I started the family tradition of having the youngest child light a candle on a birthday cake for Jesus and have all the grandchildren take their turn. Then we all sang "Happy Birthday" to Jesus and enjoyed the cake. We continued this tradition each Christmas and enjoyed every minute of celebrating!

Then we went to Andrew and Suzanna's new house in North Pole for more family time. We compared the latest great-grandchildren side by side. We all agreed that the boys were bigger.

2015
A Test of Endurance

The Iditarod Sled Dog Race had a ceremonial start in downtown Anchorage. But the official "Restart" happened in Fairbanks this year due to a limited amount of snow in the south. We watched the 78 racers start their 1,000-mile trip. Nine grueling days later, the winner crossed the finish line in Nome. All the racers needed lots of endurance.

California Travels: David

Two weeks in California was barely enough time to see all my relatives. In San Diego, Kay and I picked up a rental car. It was always fun to visit my brother Martin, his wife, Cuca, and daughter, Angela, with her family and busy household in Ramona. We saw my nieces Julie and Beto in Ramona and Darla and Mark in El Cajon. These times included reminiscing about our history and sharing good memories.

Then we drove to Porter Ranch to see my niece Abby (Llewellyn) and Jeff Bailes with their four children: Lilly, Dylan, Matthew, and Hudson.

Santa Barbara was on our way north. Touring the mission site helped us understand more of the plight of the local Indians as they were thrust into the American way of life.

Next, we stayed overnight with my niece Liz (Llewellyn) and Jeff Krulick and their four children: Kara, Stasia, Josie, and Evan. Then we drove to Fair Oaks to spend time with my sister Miriam and her husband David. They had my parents' china cabinet with a beautiful set of mother's hand-painted dishes.

AWANA Award Night

We flew back to Alaska to see all the excited kids at the AWANA Award Night at Bethel Baptist Church in Fairbanks, including great-grandchildren Caleb, Kristen, and Abagail. Debbie and John were senior helpers with 20 years of experience. Both of their children,

Danielle and Josh, were assistant leaders. Other grandchildren and great-grandchildren have participated in the program.

Two Great-Grandsons Are Here

Great-grandson Vincent (Vinny) Nicholas DeCarlo was born in Fairbanks in June to granddaughter Suzanna and Andrew. Vinny was their first child, and they lived in North Pole.

Azryl Sebastian Hendrickson was born in November to granddaughter Melissa and James. He was named in honor of an Old Testament man whose name meant "Helped by God," using a variant spelling for Azriel or Azreel. After years of waiting, they now have a girl and a boy.

Thanksgiving at Andrea's House

Everybody enjoyed our family Thanksgiving dinner at Andrea's new two-story home in the country. Everyone took their turn holding Azryl as we welcomed him into our family.

God's Frozen Chosen (2016)

2016
Sakha Silver Anniversary

The Association of Evangelical Christian Churches (ACEC), formerly Gospel to the East, in the Sakha Republic, invited us to participate in their 25th anniversary in January. We were honored to be invited to "Praise God" with them.

We highlighted the national believers by printing a book of their testimonies in English for English-speaking supporters and friends. The book *Native People in Russia Follow Jesus Christ* included 26 people telling their stories of how they came to know Jesus Christ, their struggles along the way, and how God forever changed their lives. They were God's work in progress, serving their people for His glory in Siberia. These testimonies had been printed previously in Sakha.

Winging Our Way to Yakutsk: David

Buying our tickets and choosing the route was challenging. The Russian government required us to have medical insurance during the trip. It was much more expensive for Kay since she had already turned 80. We flew through Moscow, departing Fairbanks on December 30th.

Meeting in Moscow

In the middle of the night, while we were sleeping at the Domodedovo Airhotel in Moscow, Larry DeVilbiss knocked on our door. Even though he woke us up, we were happy to see him. We ate breakfast together which included all the fresh carrot juice anyone could drink.

The 25th Anniversary

The celebration in Yakutsk began on January 4th in the enlarged church building near the airport with 400 registered participants. Seeing old friends and meeting the guests was great. On the first day,

Sergei, the former mission leader spoke and presented Larry and us with an honorary red rose for our long-term assistance. We've seen a lot of mission history and changes.

Dedicated to the Lord's Service: David

Valentin, the director of ACEC, called for us to come forward, and we joined them in dedicating Stas to be the next leader of the organization. I talked about knowing him as a young teenager serving the Lord in 1993. Then I laid my hands on him and prayed over him. Valentin and two other pastors did the same.

In the back row are Misha, Valentin, Valeri, David, and Kay; in the front row are Nadia and Stas with their three daughters.

Four days of feasting, hearing field reports, singing, sharing their talents, and visiting were part of the joyous 25th anniversary occasion. Everyone enjoyed celebrating God's blessings in the Republic of Sakha.

South Korea

Once a week, a direct flight flew from Yakutsk to Seoul, South Korea. This gave us an easy connection to see our granddaughter Danielle near Seoul. She lived in Uijeongbu (WE-jong-boo), only 15 miles from North Korea.

The airport was huge and intimidating but well-marked and loaded with modern technology. Danielle and two young men guided us through the maze. We stayed at her one-room apartment for foreign school teachers on the fifth floor. Debbie and John with their son, Joshua, were there waiting for us. We rolled out air mattresses and slept wall-to-wall as one happy family.

We visited the Seoul Central Evangelical Church planted by TEAM, where Pastor Kang and his wife ministered. They were family friends when John's parents, John and Joyce Rathbun, served here with TEAM in administration, Christian education, and radio for 20 years. We toured sites around Seoul, including the traditional Korean Palace. We enjoyed authentic Korean cuisine in several restaurants and each other's company.

Back in the USA

After another long flight from Seoul to Seattle, we met Steve and Cheryl. He was the airport engineer at Paine Field in Everett, Washington. They were assembling Boeing aircraft on the other side of the field. A new Boeing Dreamliner was parked just outside his office window with no lettering or painted design on the plane since it was waiting for a mega-million-dollar buyer.

Solomon Is Here

Another great-grandson, Solomon Thomas Henry, was born in February to our grandson Wes and his wife, Lara in Fairbanks. He has an older sister, Addy. We liked both of his names from the Bible, which causes us to think about wisdom.

Solomon has Duchenne muscular dystrophy causing progressive weakness and loss of muscle mass, so he is unable to run fast. But he lives for the moment and enjoys life!

Fast-forward: A later photo shows Solomon, his sister Addy, and Grandfather Steve enjoying a snack.

Outreach Bible Study: David

The Native Church received permission to hold weekly Bible studies at the Bertha Moses Patient Hostel across the street from the entrance to the Fairbanks Memorial Hospital. The audience varied depending on who was at the hostel. Natives from outlying villages came to Fairbanks for medical attention and stayed there. I led the study if Pastor Harry was not available.

Special Graduates at UAF

Our daughter Elizabeth Cogan earned an associate degree in applied management. She was already a grandma but she put in the effort to earn her degree. The graduation ceremony was at the Carlson Center, with adequate room for all the guests.

Photo time with Elizabeth, Kay, Kristen, Abagail, Angelica, Nyomi, Caleb, Elijah, Andy, and David

Finding a Lifetime Mate: David

Choosing our lifetime mate is the second most important decision in life. We do that while we're still a little wet behind the ears, but with the Lord's help and direction, it can be done for His glory.

At our 60th celebration, I shared three principles to follow: 1. Choose a growing Christian who wants to walk closer to the Lord; 2. Choose one who has similar goals and basic values in life. Dogs on a team all need to pull in the same direction; and 3. Choose the one you want to live with for the rest of your life, even after the honeymoon period, which reveals a few faults and weaknesses. The formula is one man + one woman = one lifetime together.

It Was a Blast!

We celebrated our 60th wedding anniversary with our family at our place. We had a big cookout with lots of food and a special cake.

God gives each of us One Lifetime

Our family, standing: Bambi and Tim Henry, Kara Fields, Andrew and Suzanna DeCarlo, Joshua Rathbun, Danielle Rathbun, Wes and Lara Henry, Jon Henry, Curtis Henry, David and Cheryl Mount, Milo Cogan, Melissa and James Hendrickson, Andrea Head; seated: Dan and Linda Henry, holding Clara Henry, Debbie and John Rathbun, David Henry holding Vinny DeCarlo, Kay Henry holding Solomon Henry, Steve Henry holding Addy Henry, Cheryl Henry, Elizabeth and Andy Cogan. Sitting on the ground: Audrina Fields, Emerson Fields, Alana Fields, Angelica Head, Kristen Mount, Sarah Henry, Nyomi Hendrickson holding Azryl Hendrickson, Abagail Mount, and Caleb Mount. That's 41 of us! One great-grandson and two grandsons-in-law were not there.

Remembering Jesus Christ

The year ended with more celebrations as we remembered the coming of Jesus Christ to earth for all of us. We watched as our great-grandchildren played their parts in the birth of Jesus at our house. Finally, we went to Dan's house in North Pole and listened to Dan read the remarkable Christmas account of the birth of Jesus from Luke chapter two.w

Finishing Well (2017~2020)

2017
Anthropological Presentation: David

I was asked to give an informal presentation based on two articles we wrote about Koyukon locationals/directionals and Koyukon classificatory verbs. The meeting was held at the Westmark Hotel in Fairbanks. It was good to be there with the Koyukon expert and longtime helper, Eliza Jones from Koyukuk, as well as Jim Kari, who contributed significantly to the overall Northern Athabascan work.

Family Birthdays

All the family birthdays for May were celebrated together at our house, which included Steve, Linda, and Bambi. We tried to keep up with our growing family in the Fairbanks area.

More University Graduations

The graduation exercises for UAF were held at the Carlson Center on May 6th. Out of 1,400 graduates, we focused on two exceptional young men. Wes Henry received a BS in biological sciences: physiology as he advanced and became a registered nurse. Joshua Rathbun received a BS in civil engineering. Joshua presented his "Water Revitalization" engineering recommendations for Snowed Inn RV Park in Delta Junction, Alaska.

Exciting Family Trip

We flew from Fairbanks to Nashville, Tennessee. Two cars were needed to accommodate our group of eight—our daughters Debbie and Elizabeth with four great-grandchildren: Caleb, Kristen, and Abagail Mount, plus Nyomi Hendrickson. We brought a wheelchair for Kay to conserve her energy with all the walking we had planned.

Seeing More Family: Kay

Murfreesboro, Tennessee, was our first stop to see my brother Jim, his daughter Tiffanie and Rudge, and their daughter Soma. We reminisced about our past.

Kay Henry, Jim Temple, David Henry, Debbie Rathbun, Tiffany Rudgley, Elizabeth Cogan, Soma's friend, and Soma Rudgley

Another Great-Granddaughter

Evelyn Marie DeCarlo was born in June to Suzanna and Andrew in Knoxville, Tennessee. We were excited because God allowed us to see and hold this precious little life soon after she was born.

Noah's Ark: David

I planned the weeklong family trip to see the life-sized Noah's Ark replica and the Creation Museum in northern Kentucky. The ark was 510 feet long and three stories high, with much to see and comprehend. Why visit? Because the information in Genesis chapters 1 through 11 is foundational for understanding the rest of the Bible.

Elizabeth and I tested how much a camel could carry, with the Ark in our background. Frankly, the camel had an odd gait and was not smooth to ride.

We spent a full first day at the Ark and its zoo. The second and third days were spent at the Creation Museum. It was unique, fun, and educational. We highly recommend visiting these sites.

With Cousin Denise

It was an easy drive to see Denise and Rodney Phipps. Denise invited us to spend a night at their house in New Carlisle, Ohio. They even had a backyard pool which all the kids enjoyed. Nyomi turned seven, so we celebrated with a pool party.

Another Great-Grandson: David

Ethan Cameron Fields was born to Kara and Aaron in June at their home in North Pole, Alaska. They gave him the same middle name as his grandfather, Daniel C. Henry, his great-grandfather David C. Henry (me), and his great-great-grandfather, Luther C. Henry. Cameron was my grandmother Henry's maiden name. He was born while we were exploring the Ark in Kentucky.

Three cousins by the pool: Elizabeth, Denise, and Debbie

Golden Days Parade

Each July, we enjoyed watching the annual Golden Days Parade, celebrating Fairbanks's rich mining history and the local gold rush of 1902. Many of our family members gathered together to watch.

A Major Change: David

On October 31st, we officially became retired missionaries. It's been a tremendous 59 years serving our Lord in Alaska with Wycliffe for 29 years and in both Alaska and Siberia with InterAct for 30 years. We would gladly do it again—if God gave us another lifetime. The retirement parties came later.

Mounting Health Problems: David

Kay's physical problems had increased to the point where I became her full-time caregiver. She suffered several mini-strokes that affected her vision. She also had hearing loss and limited mobility. Last year, Kay had two MRIs that showed "white matter" in her brain (chronic small vessel ischemic disease and leukoencephalopathy). The buildup of plaque limited proper blood flow to all areas of her brain.

"Therefore we do not lose heart. Even though our outward man is perishing, yet the inward man is being renewed day by day" (2 Corinthians 4:16).

Our Caregivers

God has given us many wonderful caregivers, including churches, individuals, families, and special friends over the years. An amazing group of prayer supporters, helpers, encouragers, and financial givers have made our lifetime ministry in Alaska and Siberia a reality and very rewarding. We couldn't have done it without each of these dear people. Of course, our ultimate Caregiver is God Himself, who cares for us and orchestrates the events of our lifetime.

Thanksgiving Dinner at Our House

We were thankful to have most of our family at our house for Thanksgiving to enjoy one another and all the excellent food. Family times together mean so much to us.

North Pole Activities

North Pole has much to offer. The Christmas on Ice sculptures were beautiful. We viewed them with Dan and the three Field children: Audrina, Alana, and Emerson. Then we celebrated with Clara who turned three. We ended this year with Christmas dinner at Dan and Linda's house, which included Tim and Bambi's family plus Aaron and Kara with their four children.

2018
Bright Blue Fingernails: Kay

Granddaughter Danielle accompanied me to Top Nails in Fairbanks to trim my nails. The girl at the salon asked, "What color do you want your nails?" I chose a bright blue to match my jacket. This was a first for me!

Two Birthdays: David

Curtis Henry and I celebrated our January birthdays together at our house. I'm 55 years ahead of Curtis. Little helpers blew out our candles.

Family Helpers: David

I drove to Anchorage while Debbie, Elizabeth, and Danielle stayed with Kay. Each day adds variety to life. "Old age is like a plane flying through a storm. Once you're aboard, there's nothing you can do." So said Golda Meir, the former prime minister of Israel. We can't reverse the aging process, but we can keep on trusting our faithful Lord in our golden years.

Marianna and Family

On the way to Anchorage, I stayed overnight in Palmer with Marianna, Riko, and their family at the InterAct Center on Lazy Mountain. She had lived with us in Yakutsk. Her husband, Riko, studied at Alaska Bible College and completed a flight training program. It was fun to meet their five children who spoke both Russian and English while being homeschooled in Russian.

Native Musicale: **David**

Many Natives enjoyed the weeklong *musicale* in Anchorage. The annual event brought people together to see and hear what the Lord was doing around the state with their people. I was one of 2,000 guests at Changepoint Church.

A Native-Led Church

The Fairbanks Native Bible Church has become Native-led rather than a mission-led church. It has been years in the making, having begun back in the '80s in our log house. Mission involvement laid a strong foundation. Pastor Harry Hafford Jr. preached and led full-time while the Church Council gave guidance and direction behind the scenes.

Retirement Parties: Kay

Our first retirement party was at the Native Church, where we actively participated every Sunday and Wednesday. All four of our children attended, plus our grandson Tim with his family and our granddaughter Melissa. It felt strange to have the spotlight turned on us, but we credit the Lord for our lives and ministry.

Our second party was at Bethel Baptist Church in Fairbanks. The mission committee organized a fun retirement party at the church.

At a display table showing highlights of our 59 years of ministry, I wore my Indian kuspuk.

    The church has been a generous supporter for many years, allowing us to continue ministering in Alaska and Siberia.

    Our three older children, Debbie, Dan, and Elizabeth, who live in or near Fairbanks, shared their unique personal memories of village life and ministry. It was like the Gospels telling about Jesus, all accurate and based on their own experience with Him. Steve was employed in Washington State and unable to attend.

    My home church, First Baptist in Vassar, Michigan, put together a unique memory book and gave us a large basket of tasty Michigan goodies. They were our sending church as well as being great prayer and financial supporters since 1959.

    Our third retirement party was at the annual InterAct Conference in Palmer in May. No gold watches, but we received a large-print study Bible with gold letters on the cover with "InterAct Ministries 30 years." They also gave us money for a woodshed to store firewood for the winter. Now we're retired, but there's still much to do. As long as the Lord gives us the strength and health to do it, we plan to continue working for His glory.

    Our first 29 years were spent working with Wycliffe and the Koyukon Indians in Alaska, translating portions of the New Testament and other forms of literature into their language. The next 30 years were spent with InterAct Ministries, continuing with the Koyukon Indians in discipling and church planting. This also included a 16-year stint in

Siberia, serving with the Sakha people in discipleship and developing different forms of literature as well as the Sakha hymnal with 200 songs.

Summer Fun

A big July party at our house included most of our family. Even Isaac Reynolds, Danielle's fiancé, was here from Alabama. We rented animals from a petting farm which they set up in our backyard with a donkey, two sheep, two goats, two ducks, two rabbits, and guinea pigs. Everybody had fun petting and feeding them.

We tossed water balloons over a volleyball net. All joined in while a few got wet. Two people working as a team held a large towel to launch or catch the balloons. Even the younger children enjoyed filling the balloons and loading them into a towel. It was a beautiful day for eating and playing outdoors.

Fast-forward: The Field and DeCarlo cousins enjoyed being together.

Vincent DeCarlo, Emerson Fields, Audrina Fields,
Evelyn DeCarlo, Ethan Fields, Alana Fields, and Noelle DeCarlo

A Baptism in the River

Danielle was baptized in the Chatanika River, 30 miles from Fairbanks, by her father, John Rathbun, and Mark Holmes, the youth pastor at Bethel Baptist Church. We were glad to witness the event as she continued to move forward in following Jesus Christ.

The Watermelon Seeds

Danielle led a CEF presentation at our house using six large flashcards to tell the story of the stolen seeds that grew. She concluded by saying, "This was a true story about my Grandmother Kay when she was a child." That experience helped Kay recognize she was a sinner needing God's forgiveness.

Big Coal Mine

Elizabeth took us with Angelica and Elijah to tour the massive Usibelli Coal Mine 120 miles south of Fairbanks in Healy, Alaska. The 75th anniversary was celebrated with a huge picnic and a display of their enormous equipment. It was fun to spend an active day with two great-grandchildren.

Ready for a Cold Winter

Since our primary heat source is a wood stove, we appreciated our mission's woodshed retirement gift. I ordered three cords of split birch to further cure and store in it for the long winter. Strong, young, and energetic great-grandchildren—Angelica, Elijah, Caleb, Kristen, and Abagail—stacked and filled it in a few hours.

Bob's Funeral

We drove to Nenana for Bob Mitchell's funeral and to support his wife, Elaine. We knew them in Tanana before and after Bob gave his life to Jesus Christ in 1974.

End-of-the-Year Celebrations

Thanksgiving was celebrated at Elizabeth and Andy's home on Haystack Mountain, 25 miles from our house. Their children and grandchildren came, plus Steve and Cheryl with our grandsons Jon and Curtis, Debbie and John, and grandson Josh. There was also a cake for Azryl's third birthday.

Early Christmas was enjoyed on December 16th at our house with family guests. We feasted on good food, read the story of Jesus's birth from the Gospel of Luke, and had a birthday cake for Jesus.

God gives each of us One Lifetime

Kay is surrounded by some of her favorite ladies dressed in red: Danielle, Debbie, Elizabeth, Linda, and Bambi.

A Major Change

That night, Kay experienced a stroke which dramatically changed her life. She could no longer feed herself, talk above a whisper, or walk.

My Advice: David

It is a difficult experience to watch the person you love decline mentally and physically. After several years of living with Kay's dementia, I recommend the following principles:

1. <u>Keep on trusting in the Lord and looking to Him.</u> He is dependable and will see you through to the end (Hebrews 12:1-2).
2. <u>Pray for guidance and assistance.</u> Accept those whom the Lord sends your way with help and advice.
3. <u>Share Christ with those you love while you can.</u> Praise the Lord that Kay accepted Jesus Christ as her Savior when she was eight and lived for Him. At this stage of dementia, it would be too late to explain Jesus to her.
4. <u>Live for the present and enjoy it to the fullest.</u> We celebrated Christmas early this year. We had included her on our trip to see her brother Jim and other family members in Tennessee when

we toured the Ark and Creation Museum. Her wheelchair and slowness took time and effort, but it was her last family trip!
5. Don't keep thinking about death. The end will come in God's time. "My flesh and my heart fail; But God is the strength of my heart and my portion forever" (Psalm 73:26).

Hospice Care: David
Debbie and I took Kay to the Fairbanks Memorial Hospital to deal with a urinary tract infection and her sudden overall decline. The doctor assigned to her was no help. He merely said, "She has end-stage dementia. Put her in hospice care. She will only live a week or two!"
Hospice delivered an adjustable hospital bed to our home on Christmas Eve and set it up in the center of our living room. Then, a hospice worker came for an hour each day.

Christmas Dinner
Kay was exhausted from the stroke. Debbie and her family brought Christmas dinner on December 25th to our house. Tim, Bambi, and their children joined us.

2019
Caregiving: David
I was Kay's full-time caregiver. The work was not difficult, but the emotional stress was draining. She occupied all of my attention. I also needed to feed her. I was available to help her anytime, even though she couldn't speak above a whisper. Often, we could get only her first word or two. I couldn't go anywhere unless I had someone to watch her, in case she needed anything. The Fairbanks Resource Agency provided a helper for 20 hours per week.

Separation: David
We moved Kay into Hope Haven, an assisted living home in Fairbanks at 1511 21st Avenue. The facility had home-cooked meals and good care. Anybody could go there to visit. It was difficult emotionally to live separately after 62 years of marriage and always doing life and ministry together.

Wedding Bells Were Ringing: David

Our youngest granddaughter, Danielle, married Isaac Solomon Reynolds in March at Bethel Baptist Church in Fairbanks. They invited Pastor David Kang from Seoul, South Korea, to fly here and conduct the ceremony. I started giving Danielle's vows to Isaac to repeat, but it didn't sound quite right. I stopped and corrected the situation. They both had written out beautiful, personalized vows. Their parents and grandparents gathered around and prayed for them.

After their honeymoon, they lived in Prattville, Alabama. They met at Bryan College, where we had met 65 years earlier.

California Bound: David

Dan and I flew to San Diego and rented a car to visit our relatives scattered across California. The first stop was in Ramona, where my brother Martin and Cuca lived. They were slowing down like the rest of us. Unfortunately, Martin had fallen and shattered his shoulder. My niece Angela lived there with her daughter, Amanda and Pedro. Mexican food was always on their menus, which we enjoyed. We absorbed the sun and warmer weather. We stayed overnight with my niece Julie, her husband Beto, and their son Neal.

Meeting on the Sidewalk

It took two Alaskans to navigate the busy freeways in the big cities as we drove toward Porter Ranch, north of Los Angeles. En route, we got a phone call from Steve, saying, "I'll meet you in front of the Los Angeles International Airport (LAX)." We pulled up to the main entrance and saw him on the sidewalk. We couldn't believe it was that easy at a busy airport.

My niece Abby and Jeff welcomed us, neighbors, and friends to a big Mexican party in their backyard with their four children: Lilly, Dylan, Matthew, and Hudson. My sister Miriam and David were there from Fair Oaks, California. We got to see many relatives in one short week!

Remembering President Reagan

In Simi Valley, we toured the Ronald Reagan Presidential Library and Museum. The Air Force One aircraft he used was retired there, so we walked through it. We were thankful Reagan was our 40th president as we viewed memorabilia, touched a piece of the Berlin Wall, and heard parts of his speeches. His favorite book was the Bible.

End of Hospice Care: David

By law, Hospice reevaluated Kay at the end of six months. They determined there was no change, so she was removed from their service.

Another Great-Grandson

George Watson Henry was born in July to Tim and Bambi in Fairbanks. He has two helpful older sisters to play with him.

The Henry four-generation photo with his grandfather, Dan, father Tim holding George, and great-grandfather David.

A Friend Moves On: David

Paul Milanowski Sr., a longtime friend and colleague with the Upper Tanana Athabascan Indians, passed away in August. I rode to Tok, Alaska, with Dan and Linda for the graveside service. Debbie and Elizabeth were present. Grandson Joshua lived and worked in Tok, so we all stayed with him.

Visiting Kay: David

I preferred to see Kay in her room rather than in the living room with the other patients and their big TV. It was challenging to live with the consequences of a debilitating stroke combined with dementia as her brain continued to shut down. When I told her, "Jesus loves you," her response could be a smile and a whispered, "I know He does," or a blank stare. Yet, she still reflected being made in "the image of God." She was well cared for and not taking any medications.

Happy B-day, Kay!

Kay's 86th birthday celebration was at Hope Haven with an Alaskan blueberry cake. Our family let Kay know she was loved. She sang "Jesus Loves Me" and "Happy Birthday" along with us. It was a fun party.

Standing in the back row: Dan Henry, Milo Cogan holding Azryl Hendrickson, Tim and Bambi Henry with George, Steve and Cheryl Henry, John and Debbie Rathbun, and Melissa Hendrickson; front row: David and Kay Henry, Sarah Henry, Clara Henry, and Nyomi Hendrickson

Another Christmas: David

I enjoyed hosting a Christmas get-together at our house. We were not permitted to bring Kay home for the celebration. We retold the Christmas story from the Gospel of Luke. The younger kids stood by our manger as we remembered Jesus' birth. We had a birthday cake for Jesus.

2020
In a Holding Pattern

Our lives were like flying on an airplane. We knew where we were as our captain announced, "We are continuing to circle in a holding pattern." There had been some decline for Kay since her stroke over a year ago. We were still waiting for the Lord's time. I don't know if she knew who I was, but I knew who she was.

I continued helping at the Fairbanks Native Bible Church by transporting people to Sunday services and weekly Bible studies. I preached when Pastor Harry was out of town.

Young Henry Birthdays: David

I had fun watching all the kids with boundless energy at Sarah Kay's eighth birthday bash in March. Her father, Tim, took all of us in his big green bus to Go-Karts of Alaska. Everybody drove the indoor electric go-karts. Good thing they had rubber bumpers! Bambi decorated a beautiful cake in the shape of an eight, which doubled as a racetrack.

In July, I attended her brother George's first birthday party. He was a little too young to appreciate all the extra activities, but the rest of us did.

Then in December, I helped their sister, Clara, celebrate her sixth birthday at their home. Being older, this party included a trip to ride the go-karts. So much fun! These birthday parties caused me to think about our past celebrations.

Another Great-Granddaughter

We were blessed with another great-granddaughter, Abigail Grace Reynolds, born in July to our granddaughter Danielle and her husband, Isaac, in Alabama. She was named after a woman with beauty and brains in the Bible (1 Samuel chapter 25). This made our oldest daughter, Debbie, a new grandmother.

Picking Raspberries

Two long rows of raspberries in our backyard kept reproducing every year with minimal care. Kay has liked picking and eating raspberries since childhood.

Debbie and Priscilla have been close friends since we lived in Allakaket in the 1960s when they were classmates.

Happy 87th, Kay!

Hope Haven allowed us to bring Kay outdoors to celebrate her 87th birthday. This birthday was less enthusiastic than her last one but enjoyed by all. Kay stayed in her wheelchair. A brightly colored pink and red quilt kept her warm, crocheted by Brenda Taylor, remembering their long Christian friendship.

Kay's COVID-19 Experience: David

Debbie Rathbun and Priscilla Williams

Kay tested positive for COVID at Hope Haven but quickly recovered from the Chinese virus. During one lockdown period, I had to don a medical gown, gloves, and a face shield to visit Kay. I'm sure she didn't recognize me! For five months, I could not even enter the facility to see or encourage Kay. That was very disappointing!

Kay's Heavenly Home (2021)

2021
Nieces from California: David

My nieces and their families, less one, came from California to visit Alaska under the midnight sun. Abby and Jeff came with their children: Lilly, Dylan, Matthew, and Hudson. Lilly had just graduated from high school. Abby was working with Child Evangelism Fellowship, and her family participated.

Also, Liz and their children: Kara, Stasia, Josie, and Evan, visited. Her husband, Jeff, was serving in the Air Force and adjusting to his new position as Lieutenant Colonial at Travis Air Base.

Four cousins standing together: Liz Krulick, Elizabeth Cogan, Abby Bailes, and Debbie Rathbun

We had a big day at our house with food, games, fun, and getting to know each other better. Steve flew in from Wisconsin for Father's Day to celebrate with his cousins. Our California family saw a lot of Alaska in a week. I accompanied them on the Riverboat Discovery trip to the Native village near Fairbanks.

More Great-Grandchildren

Noelle Rose DeCarlo was born at home to our granddaughter Suzanna and her husband, Andrew, in Knoxville, Tennessee, in May. Noelle makes us think of Christmas and a happy time. Her middle name, Rose, is after her Italian great-grandmother, Rose DeCarlo.

Two years later, Cyrus Daniel DeCarlo was born at home to our granddaughter Suzanna and her husband Andrew in Knoxville, Tennessee, in June. He was given his grandfather's name, Daniel, as his middle name, keeping it in the family and after a great man in the Bible.

Lydia Kay Reynolds was born to our granddaughter Danielle and her husband Isaac in Montgomery, Alabama, in October 2021. Lydia was named in honor of a godly woman in the book of Acts. Her middle name was just like her grandmother's, Debbie Kay, and great-grandmother's, Marilyn Kay.

Their family: Isaac, Abigail, Lydia, and Danielle

In March 2024, our granddaughter, Danielle, and her husband Isaac, welcomed Timothy Joseph Reynolds to their home in Millbrook,

Alabama. They gave him two biblical names. Timothy means one who honors God.

Double Eights!
Kay turned 88. Eight was great! When Kay was eight years old in the third grade, she made the most crucial decision affecting and changing her entire life for the next 80 years. She turned to Jesus Christ.

The End Is Near: David
Two days after her birthday, I received a phone call informing me that Kay had difficulty swallowing, eating, and drinking. She was ready to meet Jesus Christ face-to-face, and her departure was imminent.

At Kay's Bedside
We, as a family, were with Kay for her last hours. Steve and Cheryl had flown into Fairbanks. Debbie was visiting her newest granddaughter, Lydia Kay, in Alabama and returned to Fairbanks. Dan and Linda live nearby, as do Elizabeth and Andy.

Our children and spouses spent several days at Hope Haven. Kay was not in any pain. We gathered around her bed, prayed for her, sang hymns, and recalled memories. This brought us more comfort than it did for Kay, who was in a semi-conscious state. We don't know how much she heard. We were all at peace, but it was sad to see her go as we shed a few tears.

"But as for me, I trust in You, O LORD; I say, 'You are my God.' My times are in Your hand" (Psalm 31:14-15a).

Her Time Had Come: David
I spent nine hours on Sunday by Kay's bedside. She was breathing calmly throughout the day. I drove home that evening but was soon awakened by a phone call stating, "Kay is gone!" The news still hit me as a shock as I quickly drove to Hope Haven. Our four children were already there. I prayed at her bedside, committing her body to the Lord. We realized she had breathed her last breath around 11:00 p.m.

Once the family was at Kay's side, Gladys, the owner of Hope Haven, phoned the police to officially confirm her death. At 12:30 a.m., several people from the Fire Department and a policeman came and pronounced her dead. Consequently, her records indicate that she departed on Monday, October 18th.

Celebration of Kay's Life

Our family worked closely together for Kay's celebration of life at Community Baptist Church in North Pole, Alaska. I felt overwhelmed by the generous outpouring of love and respect for Kay from so many friends. All four children and I spoke during the service. See "Family Sharing" in Appendix D for what the five of us shared. Pastor Harry Hafford, Jr. with the Fairbanks Native Bible Church gave the message.

In Kay's coffin, her finger was pointing to this verse which was underlined in an open Bible "I have fought the good fight, I have finished the race, I have kept the faith" (2 Timothy 4:7). The beautiful pink and red quilt Brenda made was draped over her.

Kay loved children! During the service, granddaughter Kara led some of our great-grandchildren in singing "Jesus Loves Me." The three verses they sang are given in Appendix C.

Kay's Final Wish

Kay would love to see all of you again in heaven, but you must first go through Jesus Christ. The way to God and how to know Him personally as Kay did when she was a child is written in Appendix A. How to Know God.

Her body was interred at Birch Hill Cemetery near Fairbanks. She is waiting for the imminent return of Christ in the air at the Rapture of God's church. Kay's soul and spirit are already with Jesus in heaven. The tombstone includes Philippians 1:21 "For to me to live is Christ."

Kay's Name Lives On

We liked Marilyn Kay's middle name and passed it on to our daughter Debbie. Three of our granddaughters also honored her name with three of their children.

Alana Kay Fields, Sarah Kay Henry, and Lydia Kay Reynolds held by Deborah Kay Rathbun. They are standing in front of a long horsetail fly chaser and Kay's Sakha national dress from Siberia.

The End

Our one lifetime together has crossed the finish line. May we all live our lives for Jesus Christ and point others to Him.

Appendices

Appendix A. How to Know God

Bad News

First, the bad news: All of us have done things God does not like. We have had bad thoughts, spoken words we would like to take back, and committed things we are ashamed of before God.

The Bible informs us, "For all have sinned and fall short of the glory of God" (Romans 3:23). Which means all people "fall short" of being perfect (or have missed the bullseye as when shooting an arrow). God knows and sees everything we think, say, or do.

Good News

Jesus was tempted in every way like we are, but He never sinned. God the Father took all of our sins and placed them on Jesus, who willingly died in our place on the cross. "For He (God the Father) made Him (Jesus) who knew no sin to be sin for us, that we might become the righteousness of God in Him (Jesus)" (2 Corinthians 5:21).

After three days, Jesus arose from the dead proving that God is stronger than sin, death, and the devil. Jesus is alive today.

What Should We Do?

A **Admit we are sinners.**

"For all have sinned and fall short of the glory of God" (Romans 3:23).

B **Believe that Jesus Christ died on the cross for your sins.**

"For God so loved the world that He gave His only begotten Son (Jesus), that whoever believes in Him should not perish but have everlasting life" (John 3:16).

C Call upon Jesus Christ in prayer.

For "whoever calls on the name of the Lord shall be saved" (Romans 10:13).

Prayer is simply talking to God like this:

> **"Dear God, I know I am a sinner. I am ashamed of what I have done. Please forgive me. I am thankful Jesus took my punishment when He died on the cross. Help me to obey You. I pray this in the name of Jesus, Amen."**

What Does God Give Us?

In place of fear, God gives us His peace. Jesus said, "Peace I leave with you, My peace I give to you; not as the world gives do I give to you. Let not your heart be troubled, neither let it be afraid" (John 14:27).

We become one of God's children. "But as many as received Him, to them He gave the right to become children of God, to those who believe in His name" (John 1:12).

What Should We Do Next?

Now you are a child of God. Reading and obeying the Bible helps you get to know Him better. Begin with John's Gospel. Understand what you can. Pass over things that you don't understand yet.

Throughout the day or night, talk with God. He wants to connect with you. This is called prayer.

Go to a local church and Bible study that honors Jesus Christ and the Bible.

Tell your family and friends about your decision. Some will like it, and others will not. Ask Jesus for wisdom.

When you sin, tell God about it. "If we confess our sins, He is faithful and just to forgive us our sins and to cleanse us from all unrighteousness" (1 John 1:9). New believers are like little children who trip, fall, and get a few bruises. They have so much to learn but they will get up and grow stronger.

B. Letters from Grandma Kay

First Letter to My Grandchildren—1995

I want to write a few letters to you, my precious grandchildren, once in a while. I've been thinking about it for a long time, but this is the first one. I hope you enjoy it and learn a little more about me when I was young.

I am also writing this in Russian, but you'll get the English version! A lady wants Grandpa to preach without an interpreter, so she is helping him. She is also helping me, but mine would be for children. So what better children to tell first than my grandchildren.

My name is Katya, but Kay is my English name. My husband, David, and I are from Alaska, and we have four children and 12 grandchildren.

When I was a little girl, my mother (your Great-Grandma Vera Temple) planted a garden in the spring. I wanted to plant a garden, too.

I went out to our sandbox and made rows just like my mother did. Then I went inside our house and looked at the seeds on the shelf. I chose the package of watermelon seeds, took them out to the sandbox, and planted them. Then I covered them.

One day my mother wanted to plant watermelon seeds in her garden, but she couldn't find them. She asked me if I knew where they were.

I answered, "No, I don't know where they are."

Two weeks later, my mother said, "Marilyn Kay, come with me." She took me by the hand, and we went out to our sandbox. She asked me, "What is that?" The seeds had sprouted, and tiny green leaves came up in rows all over the sandbox. My mother gave me a spanking because I took the seeds without asking.

Every night my mother read a Bible story. I liked to hear about real people who lived so long ago. One day just before Easter, she explained that Jesus is God and lived in heaven. He loved us so much that He came down to earth to live like we do. He was perfect and never sinned. He died for my sins. He didn't stay dead but came back to life. He is alive today in heaven.

I knew I was a sinner. I had taken the seeds without asking. I knew, too, I lied and said, "No, I didn't know where they were."

She asked my sister, Shirley, and me if we wanted to pray, tell Jesus that we were sorry for our sins, and ask Him to forgive us. We wanted to be forgiven for our sins! So we kneeled beside our mother and prayed.

God forgave me for my sins. I'm so glad I started following Him and living for Him while I was young. I pray each one of you confess your sins to Him and ask for His forgiveness. Most of you have. I would be pleased if you would write and tell us when you became a Christian.

With my deep love for each one of you,
Grandma Kay

Second Letter to My Grandchildren—1999

It's been a long time since I wrote a "Letter from Grandma." I miss each one of you. Both Grandpa and I think of you often and pray for you. Most of you are now teenagers or soon will be. I think about my teen years and how fun it was to get together with our Young People's Group, as it was called back then.

In 1949 Great-Grandpa Bob Temple, who owned Vassar Electric, bought the second television in Vassar. We were so pleased to have a TV though it was just a tiny 13 inches and only black and white. We had excellent programs you may have seen in the 1950s reruns: "Leave It to Beaver," "Lassie," and "The Lone Ranger."

The young people gathered at our home on Sunday, just after the evening church service. Two Christian programs were watched. Percy Crawford, who founded Kings College, sponsored one. The other was Jack Wyrtzen with Word of Life, who supported a camp for young people in New York State. I never went there, but your Uncle Steve studied at Word of Life Bible Institute.

These two programs were geared toward teens and those in their early 20s. They included many singing groups, male and female quartets, along with choirs.

Certain issues discussed were boy-girl relationships. They emphasized keeping ourselves pure without having sex until the day we were married. God tells us, "Flee also youthful lusts; but pursue righteousness, faith, love, peace with those who call on the Lord out of a pure heart" (2 Timothy 2:22).

I'm so glad for their teaching and that Grandpa and I were both virgins when we married. Sex wasn't as openly discussed back then as it is now, but some slept around, and everyone laughed or joked about them. They didn't have much respect from the other kids, though they were popular with the fellows who wanted to sleep with them. I remember one friend getting pregnant and feeling ashamed.

It's like buying second-hand clothes—they're not new. Probably all of you prefer going to the store and buying a new blouse, shirt, blue jeans, or skirt that no one has worn. I have read that this is one reason why the divorce rate is so high.

I wish I could be with you and talk to each of you face-to-face—that's the best—but you now have my letter in writing to read and keep.

With my deep love for each one of you,
Grandma Kay

C. "Jesus Loves Me"

Sung by our great-grandchildren during Kay's funeral service. (Anna Bartlett Warner, "Jesus Loves Me," 1859, public domain.)

1. Jesus loves me this I know
For the Bible tells me so.
Little ones to Him belong
They are weak, but He is strong.
Yes, Jesus loves me!
Yes, Jesus loves me!
Yes, Jesus loves me!
The Bible tells me so.

2. Jesus loves me when I'm good
When I do the things I should.
Jesus loves me when I'm bad
'Tho it makes Him very sad.
Yes, Jesus loves me!
Yes, Jesus loves me!
Yes, Jesus loves me!
The Bible tells me so.

3. Jesus loves the Eskimo
Mushing dog teams through the snow.
Indians and Aleuts
Wearing mukluks for their boots
Yes, Jesus loves me!
Yes, Jesus loves me!
Yes, Jesus loves me!
The Bible tells me so.

D. Family Sharing

1. (Debbie Rathbun)

Good Afternoon. I'm Deborah Kay Rathbun, one of Kay's daughters.

My mother was one of my best friends. Several years ago, I asked Mom how many times she had read through her whole Bible. She replied, "I lost track; I've read it so many times over the years."

Mom probably would not win a beauty contest. She was not perfect, and sometimes she would ask us to forgive her. BUT she was a wise woman!!

Prov.31:26 "She opens her mouth with wisdom, and on her tongue is the law of kindness."

Mom had much wisdom and gave good advice to many people. Some of the best advice came when I was in high school. One evening Mom came into my bedroom, and we talked about several things. She recommended I start praying for my future husband. I thought about my parent's marriage and what she said and I decided to begin praying for my future lifemate. God blessed me with a godly husband, John!

Because Mom had recommended this to me, I counseled my daughter, Danielle, to pray for her future husband while she was a teen. To remember each day, she made her computer password, "Pray for a godly man." God blessed her with Isaac, a man of God. Now she has two daughters and can pass down Grandma Kay's words of wisdom to them.

My mother loved her grandchildren but often lived far away. She always sent birthday cards with personal notes to each one. She decided to write about her life when she was a young girl. She titled them "Letters to My Grandchildren." One contained the "watermelon story," which tells of her being forgiven of her sins.

Thank you for your consistent Christian life all seven days of the week!

Thank you for your great advice and wisdom. I love and miss you, Mom.

2. (Dan Henry)

During the last few years, Mom's physical body was in decline. Mom's years of service to people have stopped. Today, she knows God and can enjoy Him for a long time. Jesus was the firstborn from the dead. Her hope. Our hope.

A few years ago, Dad and Mom were in Russia telling people about Jesus. They could have disappeared. Alone. However, God allowed Dad and all four of us kids to be with Mom on her final day.

In Fairbanks, I've met people who knew Mom. Rarely do people ask about Kay Henry. Usually, people will say, "Yes, I know Dave and Kay." The two of them had a sense of oneness. A bond. Team spirit.

When I was 20 years old, my mom drove me to the airport to attend college. We were running late. Our vehicle would quit every mile and Mom was frustrated. I was the last man to board the plane. As I reflect back, Mom did not cuss or scream, and she never criticized Dad.

Mom practiced hospitality. Often a simple meal together. Laughter. Contentment. Friendships started with a simple bowl of soup. Mom loved dessert. Sometimes just a spoonful of molasses or jam.

Mom was blessed with a great mind. She loved Scrabble. She tried to help ordinary people in a quiet way.

Once I attended a statewide political convention and met a lady named Susan Lockwood. She had been a teacher in Koyukuk. She told me mom had helped her come to Jesus. I appreciate Mom's life, legacy, and love for a spoonful of jam.

3. (Elizabeth Cogan)

It was difficult to decide what to say about Mom as there were many things she did. I'd start on a subject and was down a rabbit trail before I knew it and off-topic. Finally, I decided on Mom's ability to provide clothes for us.

Mom was good at crafting clothes for us, especially while living in the village with limited supplies. She couldn't just go to the store to purchase our clothing or supplies. We lived 350 miles from the nearest department store. Our items arrived either by mail plane or on the yearly barge. She sewed on her treadle sewing machine and then later an electric machine, which was purchased with a huge pile of Betty Crocker coupons, but only on the days when Dad ran the generator.

My favorite parka was a red velvety fabric lined with dark gray fake fur. Mom had been given a large fake fur coat, and she used the fabric to line the parka she made. She even used the leftover fur to line the outer pockets to keep my hands toasty. A fur ruff was attached to the hood and some pretty trim around the bottom, sleeves, and pockets. Another nice touch was a knitted cuff on the inner sleeves. I'm not sure if these were store-bought cuffs or knitted by Mom, as she did both ways depending on which ones she had on hand. I was toasty warm and very proud of my parka.

Besides sewing parkas, Mom knitted us socks. Later when her grandchildren arrived, she hand-knit socks as part of their Christmas present. She took the time to hand-knit a special sweater for each grandchild's birth. These baby sweaters usually were hooded with a zipper down the back, making it easier to dress the little ones.

When I was young, Mom attempted to teach me to knit. Alas, as I watched her, I reversed everything and needed three hands which didn't work so well. I'd hold the yarn in my left hand, a knitting needle in my right, and tuck the other knitting needle into my belly. Needless to say, it wasn't an enjoyable pastime for me as it was for Mom. And forget about trying to visit while knitting, as it would turn out completely wonky. Thankfully sewing lessons went much better, and I sewed parkas for all my children as well as other clothing.

I'm very thankful my mom taught me the life skill of sewing and attempted to teach me knitting. Maybe now would be a great time to get some knitting needles and reminisce about Mom knitting for us. Thanks, Mom, for taking the time to sew and knit for us and passing on those life skills.

4. (Steve Henry)

A couple of things stand out about Mom's life. She was grounded in the Word and was always reading her Bible. She wanted to tell everyone about Jesus. She encouraged others to read their Bible and to share the Good News about Jesus Christ.

My observations about Mom started when I was young. God sent a flood to Koyukuk 10 days after I was born. The village was airlifted to Fairbanks and lived in the Barnette Elementary School gym for a while. I mostly yelled and screamed, I couldn't communicate, but I observed Mom's actions before I knew anything.

D. Family Sharing

As a child, our family always picked people up for Sunday School, church, AWANA, and Bible study. Mom shared Christ with others and she wanted us to take part in the sharing.

Our home was open to anyone to stay or eat. When people were passing through Fairbanks, you were invited. If you had a hangover, were drunk, or had other issues, you were welcome, too. People traveling for ministry came. Matthew 5:14 "You are the light of the world. A city set on a hill cannot be hidden." **She was a beacon of light to the world.**

During junior high, I remember the military personnel coming to our house for several pizza parties. I was surprised since Mom was ministering to Native Indians. Why would she keep inviting the military to our house? Her focus was on the Native population, but if God opened the door to sharing with the military, she shared Jesus with them.

As a teen, my friend who had been to our home many times over the years commented one day, "Doesn't your mom ever finish reading that book?" That book was the Bible and she read it all her life. **Reading the Bible was the foundation of Mom's faith in Jesus.**

In my college years, a friend and I came home. When we turned off Farmers Loop onto our driveway a couple was arguing on the road and one of them had a gun. We tried not to make eye contact while driving by. When we told my mom, she jumped into action and insisted we needed to go help the couple. She wasn't afraid to act. Her size didn't slow her down. We drove to the couple, and then Mom talked to them and helped them.

When I was an adult, we were headed to Ethiopia as missionaries and Mom asked me, "Why are you going?" It was important to her that we went for the right reason—to share Christ. It was not about us, making a name for ourselves or our comfort.

She wanted us to involve as many people as possible in the ministry. It was hard, but we involved many people in the process and God blessed it. She wanted everyone to know Jesus and pass the gospel on from friend to friend and from generation to generation.

Thank you, Mom, for not being caught up in the world but dedicating your life to Jesus Christ.

5 (David Henry)

I want to thank everyone for coming today and for the outpouring of love and generosity for Kay and our family. Thank you from the bottom of our hearts.

A lot of water has gone down the Yukon River during Kay's lifetime of 88 years. I knew Kay better than anyone. We were married for 65 years. Together, we have four wonderful children. Then there are grandchildren and great-grandchildren. I always wondered how our genes would look in future generations.

Our life verse was: "For this God is our God forever and ever: He will be our guide even unto death" (Psalm 48:14 KJV). This awesome God, who created the world, has become our personal God. This is true today and forever in the future. Throughout this lifetime, He has been our guide, and Kay has reached the end.

We are not perfect. We have always tried to work together as a team in our ministry with the Native people in Alaska and the Sakha people in northern Siberia.

Here is one example from our time living in the small Indian village of Kokrines on the Yukon River, 30 miles upriver from Ruby. In the spring of 1960, we decided to move to Koyukuk. I bought a fixer-upper cabin from Edward Pitka Sr. for 30 dollars down and a handshake.

I contacted Yutana Barge Lines in Nenana. But the Taku Chief only had four inches of freeboard and couldn't land at our beach. So I told Kay about my plan B, to build a raft. Immediately she asked, "Have you prayed about it?" She was hesitant! She asked the right question, and I responded, "Yes, I prayed about it."

I made a raft out of 10 drift logs and attached our 16-foot open boat. Kay, six months pregnant with Beth Ann, climbed into the boat with our children, Debbie and Danny.

It took us three and a half days to go 110 miles down the Yukon River. We only discovered one unmarked sandbar on the way. Kay and I did not always see eye to eye on everything, but the Lord took care of us.

We have already grieved for Kay. Two and a half years ago, she had a major stroke on top of end-stage dementia. The doctor at Fairbanks Memorial Hospital told us, "She will only live about two weeks." But God knows and determines the exact number of our days.

Now Kay has a new body and mind in heaven. She's living face-to-face with Jesus. We know where she is. I am very thankful to God for having known, lived with, and worked with her for all these years.

www.ingramcontent.com/pod-product-compliance
Ingram Content Group UK Ltd.
Pitfield, Milton Keynes, MK11 3LW, UK
UKHW032217171224
452513UK00011B/733